Access Scaffolding

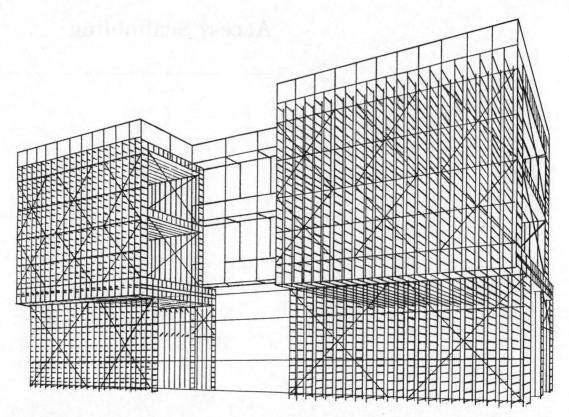

Frontispiece A pavement gantry with a lorry tunnel and a pedestrian tunnel providing support for two levels of contractor's huts, an access scaffold and two loading platforms, each capable of supporting 70 tonnes of precast concrete cladding units. The gantry incorporates 414 ladder beams. By courtesy of Deborah Scaffolding Services.

Access Scaffolding

STEWART CHAMPION
M.Sc., Ph.D., C.Eng., F.I.C.E.,
F.I.Struct.E., M.Soc.C.E. France.

Joint editors
Winifred P. Landsell, B.A.
Harold L. Landsell, F.C.I.O.B., F.C.I.A.b.

Longman

The CHARTERED
INSTITUTE OF
BUILDING

Longman Group Limited
Longman House, Burnt Mill, Harlow
Essex CM20 2JE, England
and Associated Companies throughout the world

© Longman Group Limited 1996

First published 1996

British Library Cataloguing in Publication Data
A catalogue entry for this title is available from the
British Library

ISBN 0-582-25436-1

Library of Congress Cataloging-in-Publication data
A catalog entry for this title is available from the
Library of Congress

Set by 4 in 9/11pt Times/Melior
Produced by Longman Singapore Publishers (Pte) Ltd
Printed in Singapore

Contents

Preface

This book is intended to assist the scaffold designer to prepare designs for unusual access scaffolds and the practical scaffolder to build safe scaffolds in cases that do not warrant the attention of a designer.

It is hoped that the book brings together the best features of the British Standard and international codes of practice and specifications, the various government advisory notes, the instruction sheets of the various training establishments and the manuals of the manufacturers and contractors specialising in scaffolding.

A scaffold is a temporary structure built of previously used components and intended to be altered from time to time and eventually dismantled. It has to service a building that is changing in shape and height day by day. Accordingly, it is not possible for any code of practice, guidance note or book to cover every circumstance that will be encountered. The technical information, however, does not change from job to job and so is collected together in Part IV.

Unlike a permanent structure, which is designed by one team and then constructed by a second team to the designs of the first, a scaffold frequently has to be designed by the craftsmen as they build it. The consequence of this is that different gangs will tackle a job in different ways, and a different structure may result. It will perform the same purpose in the end. The fact that building façades are so different also results in variations in the way scaffolders tackle the work.

Only the most elementary knowledge of the theory of structures is assumed. The properties of struts and beams are tabulated so that all building personnel can apply the data either by way of designing a scaffold or checking a design by others. This does not preclude a designer carrying out sophisticated calculations, but as the assembled scaffold will contain in-line joints irregularly placed all over the structure, and connections with a wide range of unpredictable stiffness and lines of force not passing through set points, he will be obliged to make some allowance by way of a correction factor for these indeterminates.

Part I deals with normal access scaffolds, which do not need the attention of a designer.

Part II deals with special scaffolds, which need more skilled attention, site supervision and possibly detailed calculations.

Part III deals with some of the administrative requirements of the scaffolding industry.

Part IV is a collection of the technical data that designers will need from time to time. Conversion tables are given. Weights of materials and imposed loads are tabulated. A chapter is devoted to wind loads.

Throughout the book, paragraphs have been inserted under the sub-heading 'Hazard', in which attention is drawn to common errors. This is an unusual feature in a textbook, but it is hoped that designers, inspectors and scaffolding contractors will be alerted to the problems in each type of scaffold and so avoid the pitfalls.

Acknowledgements

The writer is indebted to the following, who have assisted in the text and provided some of the diagrams and photographs from which the diagrams have been made.

Deborah Grayston Ltd for the frontispiece.

The British Standards Institution for all the extracts from the relevant standards and codes.

The Building Research Establishment for the technical information assisting the code drafters, and particularly Messrs Newberry and Eaton for their wind loading handbook.

The National Association of Scaffolding Contractors and their members for their material's properties, particularly Messrs Scaffolding Great Britain, GKN Kwickform Mills, Palmers, Deborah Grayston, Bolton, Presco, Raylor, Barton and London Midland.

PART I
General construction
of ordinary access
scaffolds

1 Background, scope, definitions, terminology and units

1.1 Background

From the time building began, working platforms have been necessary to provide access for the workmen and to support plant and materials. Ramps of earth are in evidence in buildings of antiquity. At some point in time, platforms made of timber members must have been introduced. Lashings are quite adequate for connecting the members. This system has survived for many thousands of years.

In the mainland of Europe, 'pole and lashing' scaffolds were superseded first by sawn timber poles and bolts, dogs or cleats. In Britain, the pole and lashing scaffold gave way directly to 'tube and fittings' from about 1920, because of the shortage of timber. Poles and lashings are still used in Britain for roof rigs for suspended scaffolds and in places where steel or aluminium is undesirable.

With the advent of the steel tube, a whole variety of means of coupling tubes together came into being, and subsequently many supplementary fittings have been added to the range to perform special duties.

Shortly after the introduction of tube-and-coupler scaffolds, prefabricated systems using frames or straight elements were developed. Prefabricated beams and other prefabricated components are now extensively used.

1.2 Scope of the book

The three materials commonly used in scaffolding in Europe are timber, steel and aluminium. Information is included on all of these materials. In Britain, timber is widely used for boards, stagings, platforms, gangways and stairs. Some notes are included on the selection of the timber for scaffolding purposes. No reference to timber scaffolds is made in the British Standard BS 5973 1990. BS 2482 1981 deals with timber scaffold boards.

Steel and aluminium scaffolds can both be subdivided into two classes, namely tube-and-coupler and prefabricated systems. BS 5973 1990 is essentially concerned with tube-

and-coupler scaffolds, but has Section 4 on prefabricated elements. BS 1139 1982 and 1983 deals in several parts with the metal components.

Aluminium scaffold tubes have the same external diameter as steel tubes and are used with the same scaffold couplers. The strength of the aluminium tube is sometimes less than that of steel, its weight is less and its deflection under load is greater, so that some special rules are needed for it. Nevertheless, the basic rules of scaffolding, and the statutory regulations governing the way working platforms are constructed, relate to aluminium as well as to steel. Some information on aluminium tubes is given in Part IV.

Prefabricated systems and towers are not dealt with in detail as each type has its own rules for assembly and use. Where there are special rules, the adequacy of the structure may be assessed using the same principles as set down in this book.

1.3 Definitions and terminology

Access This term includes egress.

Access scaffolding A scaffold providing both working places and access. It may provide a place for the storage of materials. It may provide support for part of the permanent works.

Adjustable base plate A base plate with an adjustable screw jack to raise or lower the level of the bottom of a standard.

Adjustable leg A screw jack without a base plate that gives an adjustable distance between two coaxial tubes or provides adjustment of the height of the bottom or top of a standard.

Anchor A ground anchor or wall bolt for stabilising a scaffold by ropes, tubes, tie assemblies, etc.

Base plate A steel plate to distribute the load from a standard on to a sole plate, or the ground if it is hard enough.

Base frame or base lift   The ledgers and transoms forming the lowest approximately horizontal framework close to the ground or, in the case of a frame scaffold, the bottom frame resting on the ground.

Bay and bay length   The space between the centre lines of two adjacent standards measured along the face of the scaffold.

Bay, storage   A bay used exclusively for materials storage.

Bay, loading; loading platform   A bay used for temporary storage and deposit of materials from cranes and hoists before they are repositioned inside the building. The loading bay may be an extension of the width of the working platform.

Beam   An assembly of tubes and couplers, or of tubes welded into a prefabricated unit.

Beam, battens and braces   A beam batten is a cross-member in the plane of the beam at right angles to the chord. A beam brace is a member in the plane of the beam placed diagonally between the chords and the battens.

Beam, spine   A longitudinal main beam spanning a roof, usually at the ridge or the eaves.

Beam chord   The top or bottom member of a beam, which carries the thrust or tension to resist the bending moment.

Beam chord brace   A tube that prevents a beam from rolling over sideways.

Beam chord plan brace   A tube that prevents sideways displacement of a beam.

Beam chord stiffener   A tube that is coupled to a chord to prevent lateral displacement.

Birdcage   A scaffold covering a wide area as opposed to one serving a wall, e.g. a scaffold to give access to a theatre ceiling. A birdcage scaffold may be single-lift or multi-lift. It may contain side access scaffolds.

Board   A board or plank of a specified width and thickness for use on a scaffold.

Board clip and toe-board clip   Clips to hold platform boards and toe boards in place.

Board bearing transom and board end transom   Transoms to support boards at their mid-sections and at their ends.

Crawling board   A board adapted for access on sloping roofs.

Board on edge   A scaffold board used on its edge, sometimes as a structural member, rather than on the flat for platform decking.

Fleet of boards   Several boards side by side making up the width of a platform.

Inside board   A board placed on the extended end of a transom between a scaffold and a building façade or on a hop-up bracket designed to support it.

Board retaining bar or clip   A bar or strip over the top of a fleet of boards to hold them down when necessary.

Clips serve the same purpose.

Short board   A board that has been cut down from the standard length of 3.96 metres.

Sole board   A board used on the ground as a load spreader.

Toe board and end toe board   Boards on edge along the sides and ends of a platform.

Trap-ended board   A board that projects dangerously far beyond its end support.

Box tie   A tie assembly completely surrounding a column in a building.

Brace   A tube coupled diagonally across a rectangular framework to give it stability. It may be a diagonal tube joining a scaffold to a building.

Face brace or façade brace   A brace in a vertical plane parallel to a building. In an access scaffold, it is usually fixed to the outside line of standards. In a birdcage, it may be fixed to intermediate lines of standards.

Internal brace and external brace   An internal brace is one within the perimeter of the scaffold structure. An external brace is one outside the perimeter but performing the same function. (See Raker.)

Knee brace   A tube diagonally across the corner of an opening in a scaffold to stiffen the end connections of a beam, a long transom or a ledger.

Ledger brace, cross brace or transverse brace   Braces in the vertical plane at right angles to a building. In an access scaffold, they are usually fixed to alternate pairs of standards. In a birdcage, they may be fixed to intermediate lines of standards.

Braced pair of standards, unbraced pair of standards   Two standards in an independent tied scaffold to which or near which the ledger bracing is attached. Unbraced pairs of standards are the two intermediate standards between the braced pair.

Plan brace   A horizontal brace that transfers the rigidity of an outside façade braced standard to the inside standards and that forms horizontal lattice beams to give the scaffold stability at right angles to the building.

Bracket   A prefabricated unit that forms the base for a scaffold on a wall surface.

Bracket scaffold   A scaffold based on wall brackets.

Hop-up bracket   A unit that extends the width of a scaffold either at the level of a lift or at an adjustable intermediate level.

Brickguard   A panel, usually of wire mesh, on the outside end of a platform, usually between the guard rail and the toe board, to prevent stacked material falling from the platform.

Bridle   A tube, usually horizontal, across an opening in a building façade to support the inner end of a putlog or transom. Bridles may be placed horizontally or vertically inside the building for attaching tie tubes.

Building   The main structure that the scaffold serves or is built against.

Butt   A short tube.

Butting tube and butting transom   Tubes that are fixed end-on to and in contact with the building surface.

Cantilever tube or beam   A tube or beam extending outwards from a scaffold with no further support at its end.

Propped cantilever   A cantilever with the free end supported but not coupled into a structure.

Cantilever loading platform or storage bay   A bay extending from a scaffold at the level of a lift for storing plant or materials or for receiving materials from a crane. The platform is bracketed from the scaffold without support from the ground, and it may be bracketed from the building façade.

Cantilever scaffold   A scaffold built on beams or a framework cantilevered out from the floors of a building in places where it is impracticable to found it on the ground.

Counterweight, Kentledge   Weight added to a cantilevered beam supporting a scaffold or a slung scaffold to achieve a balance.

Coupler   A component for fixing scaffold tubes together.

Brace coupler   A coupler for fixing bracing tubes into a scaffold.

Check coupler   An extra coupler fixed to support another coupler to increase the safety factor and to provide emergency support.

Finial coupler, fixed and swivel   A coupler to attach a tube across the end of another tube, e.g. a guard rail to the top of a guard rail post. A fixed finial gives a right-angle connection, while a swivel finial enables a cross tube to be at an angle, as in a staircase.

Girder coupler and roller girder coupler   A coupler to attach a scaffold tube to the edge of a rolled steel structural member. A fixed girder coupler gives a rigid connection, while a roller girder coupler enables a scaffold tube to be moved along a structural steel member as part of a trolley.

Joint pin, expanding joint pin, spigot pin, internal coupler   Internal pins used to join one tube to another coaxially. They may be in two parts, with an expanding mechanism offering some grip on the bores of the two tubes.

Parallel coupler   A coupler used to join two tubes side by side or in parallel with a small space between them.

Putlog coupler   A coupler for locating putlogs or transoms on to ledgers. This is not a load-bearing coupler for joining heavily loaded tubes together at right angles. Colloquially termed a 'single'.

Right-angle coupler   A coupler used to join tubes at right angles. Colloquially termed a 'double'.

Safety coupler   An extra or check coupler added as an emergency measure to support another coupler.

Sleeve coupler   An external coupler used to join one tube to another coaxially. It can be adjusted to grip both tubes.

Spigot coupler, fixed or loose   A non-expanding pin used to join tubes coaxially, particularly in a prefabricated scaffold frame or system.

Spigot pin and wedge   A means of fixing two tubes joined by a spigot together.

Supplementary coupler   An additional coupler added to a tube to back up the main coupler at a joint when the load in the joint exceeds the safe working load of the main coupler. Several supplementary couplers may be used in series.

Swivel coupler   A coupler used to join tubes at an angle other than a right angle. It retains its ability to swivel after being clamped to both tubes, and it has the same load-bearing ability as a right-angle coupler.

Cross tube and cross butt   Short tubes fixed across scaffold tubes to form anchor points, bearing points, footsteps or tie restraints, or to enable a scaffold to be butted up against a building.

Drop scaffold   A scaffold built downwards from an elevated frame projecting from a high level on a building.

Eccentric loading   This term generally implies an eccentricity of loading of 50 mm resulting from one tube being coupled to the side of another.

End guard rail   A rail fixed across the end of a working platform.

End toe board   A toe board fixed across the end of a working platform.

Engineering scaffold   A scaffold for support and similar load-bearing applications.

Falsework and shoring   These are support scaffolds and are not dealt with in this volume.

Façade   The face of a building up against which a scaffold is built.

Fitting   An item, other than a coupler, used in the construction of a scaffold. This includes base plates, tie components, fork heads, stair tread attachments, sheeting hooks, gin wheels, board clips and retaining bars, ladder clips, ropes, lashings and anchors.

Fan   A sheeted protection structure usually projecting from a scaffold to prevent damage and injury from falling material.

Fan wire   A rope used to support and stabilise a fan.

Façade, normal   A façade that permits the fixing of through ties or non-removable ties in the normal pattern at the normal spacing.

Façade, abnormal   A façade that does not permit the usual tie fixing.

Foot lift and foot tie   The lowest frame of horizontal tubes in a scaffold.

Foundations   This term refers not only to foundations on the ground but also to alternative places of support for a scaffold, e.g. roofs, balconies and podiums. Bracketed scaffolds have their foundations on brackets or on truss-outs.

Free-standing scaffold A scaffold that is not attached to any other structure and is stable against overturning, if necessary being assisted by anchors, guys, rakers and buttresses.

Front face The main face of a scaffold, from which return scaffolds may be built around the ends of a building.

Grid Numbering and lettering locating the standards on a scaffold drawing.

Ground-based scaffold A scaffold founded on the ground, as opposed to a cantilever, drop, slung, suspended or truss-out type.

Guard rail, hand rail, guard rail post Tubes fixed to the side or end of a working platform to afford protection to persons on the platform. Guard rail posts are the uprights and intermediate uprights supporting a guard rail.

Hanger A prefabricated link for a slung scaffold.

Hoist tower A frame containing a builder's hoist or lift and, where necessary, supporting a hoist mast or runners and lifting gear, and access gates.

Hop-up bracket A prefabricated bracket for attachment to a scaffold or a building to permit a platform to be supported at various levels outside the scaffold or the building.

Horn The portion of a standard between the lowest lift and the ground or between the top lift and the top end of the standard.

Inside The inside of a scaffold is that side nearest to the building.

Joint, staggered A joint in a standard or ledger that has to be staggered into a lift or bay.

Lateral forces Horizontal forces along or across a scaffold from wind effects, plant surges, known applied forces and stability restraints.

Lacing Tubes that join assemblies of scaffolding together. The ledgers and transoms in an independent scaffold are laces, as are tubes joining towers.

Lap tube A short length of tube used to lap across an in-line joint in a scaffold tube.

Lapping tube Tubes that overlap and are joined side by side.

Ledger A horizontal tube, usually parallel to the face of a building and attached to standards. Transoms are normally fixed to and above the ledgers.

Length The total length of a scaffold.

Bay length The length of one bay of a scaffold, measured centre to centre.

Strut length The length in bending of a compression member, measured either between its ends or between points of reverse curvature.

Lift An assembly of horizontal tubes, ledgers and transoms forming the lacing of standards and providing support for platform boards.

Foot lift and head lift (or top lift) The lifts nearest the ground and uppermost on a scaffold, respectively.

Lift height The vertical distance between two lifts, measured centre to centre.

Lift head room The clear distance between the top of a platform and the underside of the lift above.

Mast scaffold A scaffold consisting of one or two masts up which a platform is moved by manual or power operation.

Mobile scaffold and mobile tower Scaffolds on wheels, which may be moved manually or towed by a vehicle, or scaffolds mounted on towed trailers or lorries.

Motorised scaffold A scaffold with a motorised means of raising and lowering the platforms.

Net, safety A rope net to protect the area below it or to catch falling persons or debris.

Node Structural nodes are intersections where vertical, horizontal and inclined tubes meet. In scaffolding, nodes are not specific points where the centre lines of the members intersect. They are usually dispersed in space.

Non-removable tie A tie that is fixed to the surface of a building to stabilise a scaffold. It is never taken away for any building operation.

Outside The outside of a scaffold is that side farthest from the building.

Prefabrication Members welded into units prior to delivery to a site.

Protection scaffold A scaffold that provides protection against falling objects or prevents persons falling.

Puncheon A vertical tube not founded on the ground but starting at a height and coupled on to the framework.

Purlin A sheeting rail on a roof.

Rafter A tube or beam fixed up the slope of a temporary roof.

Raker An inclined tube, usually outside a scaffold, to widen its base or to form a buttress. An inclined tube at a higher level to increase the width of a scaffold. An inclined tube resting on part of a building to carry some of the scaffold loading to that point.

Removable tie A tie that may be removed and replaced during the construction of a building.

Reveals The opposing faces of an opening or recess in a building wall.

Reveal tie A tube with an adjustment device to enable it to be expanded between the reveals of a window, doorway or recess to form a firm tube for the attachment of a tie tube.

Roof edge protection A scaffold around the perimeter of a roof.

Right-angle coupler See couplers.

Rope termination A means of preventing the end of a rope becoming frayed or unlaid.

Roof rig The part of a suspended scaffold on and fixing it to the roof.

Scaffold A scaffold is defined in the Construction (Working Places) Regulations 1966 as 'any temporarily provided structure on or from which persons perform work or use as access'. The term includes working platforms, gangways, runs, ladders and stepladders.

Scaffold couplers See couplers.

Scaffold types See Tables 2.1 to 2.5.

Sheeting rail A tube to which protection sheets are fixed in roofs and on walls of scaffold structures.

Slung scaffold A scaffold hanging down from overhead supports but not capable of height adjustment.

Sole plate, sole board A board on the foundations to support the standards of a scaffold.

Standard (upright) A vertical or near-vertical tube forming an upright in a scaffold framework.

Storage bay A bay of a scaffold or an extension of the width of a platform that is used as an area for the storage of material.

Suspended scaffold A scaffold hanging on ropes from an overhead support and capable of being raised and lowered.

Tie The means of attaching a scaffold to a building.

Tie assembly All the tubes and couplers making up a tie.

Box tie A tie assembly completely surrounding a column in a building.

Tie cross tube A butt tube at right angles to a tie tube that prevents the tie tube moving towards or away from a building.

Lip tie A tie consisting of a tie tube and a cross butt that fixes the tie tube to the front and back of a wall.

Tie tube A tube in a tie assembly joined to a scaffold. It may also be a transom.

Toe board A board on edge around a platform to prevent materials falling from its edge. The board may be fastened in place by toe-board clips and end toe-board clips.

Tower A scaffold in the form of a tower, which may be static or mobile on castors.

Tower scaffold An assembly of towers joined at intervals with a continuous platform across the top.

Tower-and-beam scaffold, tower-and-ledger scaffold A scaffold formed of towers spaced so far apart as to require beams to join them at various levels and at the top to provide a continuous platform.

Transom A tube at right angles to the ledgers and the building.

Board-bearing transom A transom that supports platform boards but that can be moved when the boards are moved.

Board-end transom A board-bearing transom placed to support the ends of the boards in a fleet of boards.

Structural transom A transom that forms part of the structural framework of a scaffold. It must not be taken away when the boards are moved.

Sway transom A transom projecting inwards around the end of a building façade or into the reveals in the face to prevent lateral sway of the scaffold by being fixed tightly to the wall end or reveals.

Trapeze A U-shaped assembly of tubes lowered down as the first stage in the construction of a slung scaffold.

Truss-out A scaffold framework anchored in or to a building to act as the starting framework for a scaffold to be built outside the building.

Truss-out scaffold Assembly of the truss-outs and the scaffold founded thereon.

Wall plate A vertical timber plank to protect a wall from damage by butting tubes and to transfer forces to shores.

Weather protection scaffold A scaffold, temporary building or canopy that is sheeted over to protect a site from the weather.

Working place A platform on which workmen carry out building work.

1.4 Units

Scaffolds are usually used for supporting persons, plant and materials, the weight of which is known in kilogrammes. The self-weight of the scaffold decking and framework is known in kilogrammes. Accordingly, all weights are given in kilogrammes in this book. Pounds and tonnes may be added in brackets.

Distributed forces that have to be assessed by a workman on a job or given to him to enable him to build a suitable structure are also given in kilogrammes per square metre. Newtons per square metre may be included where the information is for designers. Pounds per square foot may also be quoted.

Areas are given in square metres, with square feet added in brackets where relevant. Lengths are given in metres, with feet added in brackets where relevant.

2 Types of scaffold, choice of systems and standardisation

2.1 Types of scaffold

Access platforms can be categorised into four main types:

(a) Platforms on scaffolding tied to the face of a building.
(b) Platforms hanging on wire ropes down the face of a building.
(c) Broad-area platforms either built up from the floor or hanging from the roof to service a ceiling or the roof.
(d) Moveable gangways, towers and masts.

These four main types can be further subdivided. Each subtype has special advantages or disadvantages; economy or high cost; simplicity or impracticability.

Sketches of common types of scaffold are shown in Figure 2.1 and these are also listed in Tables 2.1 to 2.5.

2.2 Choice of system

The list given is sufficiently comprehensive to indicate that some guidance is desirable to assist with the choice. There are further subdivisions of the types, and some systems incorporate more than one type.

The major consideration must be the safety of work people using the platform.

The second consideration is that the platform and its supporting scaffold must comply with the statutory regulations applicable to the job and location in question, e.g. works of engineering construction, building operations, work in docks and harbours, and work in mines.

The third consideration is that the platform and its supporting structure should comply with the appropriate codes of practice, which includes not only the British Standard Code of Practice but also any code specifically prepared for the industry or location of the scaffold, e.g. for work connected with the railways, outside television sites, or on motorways and public highways.

The Construction (Working Places) Regulations 1966 include most means of access within the definition of a working place. A ladder for one man with a pot of paint and a heavy-duty masonry platform for a gang of men and storage of stone are both scaffolds and are subject to the regulations. Both should be in accordance with the relevant British Standard codes of practice and specifications.

The choice of the most suitable system should be made having regard to the following:

(a) How safely and well the job in hand can be done from the platform provided. This item heads the list because work carried out by workmen in an unsafe or awkward posture is rarely done well or economically.
(b) Hidden traps for workmen, not only when performing the work in hand but also when gaining access to the place of work, or when handling and temporarily storing materials.
(c) Danger to others from the scaffolding itself and from materials falling from it.
(d) Susceptibility of the equipment to displacement or overturning by other activities and traffic.
(e) Safety in erection and dismantling. Regard must be paid to the work people available to erect, dismantle and periodically modify the scaffold. If the user intends to subcontract the scaffolding, then he will be less concerned with the skills required; but if he intends to build the scaffold with his own work people, then he must ensure that they have the ability and training to build the chosen type of scaffold correctly and to understand the requirements of the law and the codes. Regard must also be paid to the periods and number of times that workmen are exposed to risk in assembling any particular type of scaffold.
(f) Where there are several trades or several different types of work to be performed, the scaffold system must be satisfactory for all of them. For instance, a scaffold suitable for painting the trusses in a roof might be suitable for rewiring the lighting system but quite unsuitable for temporarily storing and fixing the sprinkler mains.

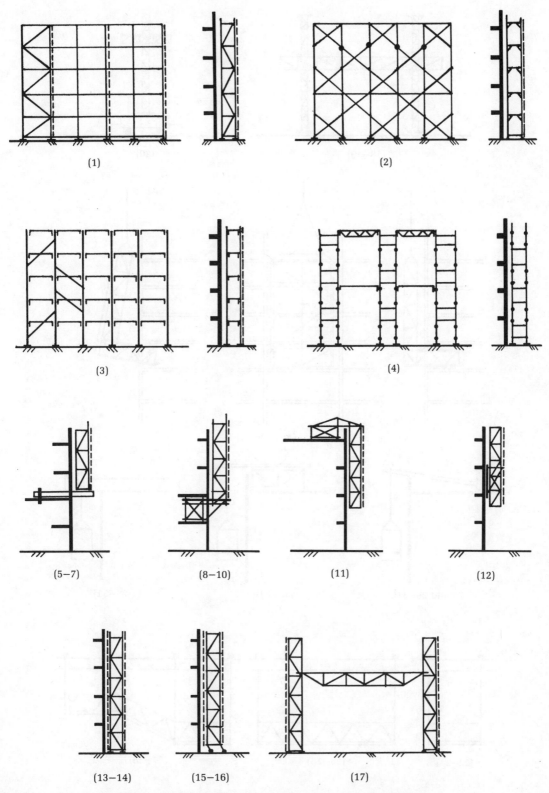

Figure 2.1 The more commonly used types of scaffold. See Sections 2.2 to 2.5 for the descriptions.

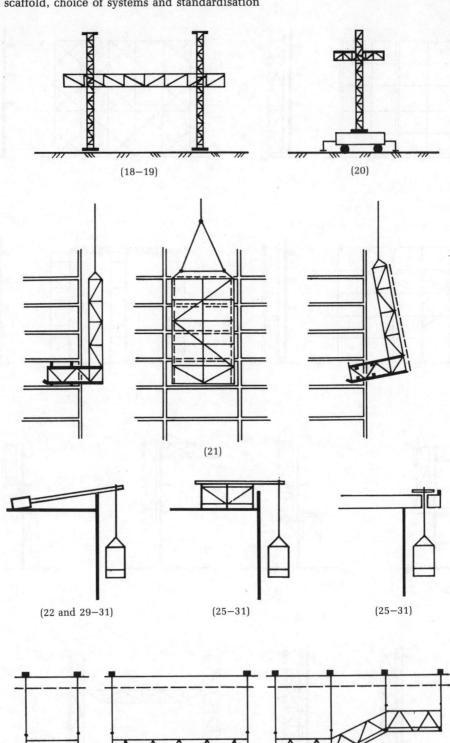

(18—19)   (20)

(21)

(22 and 29—31)   (25—31)   (25—31)

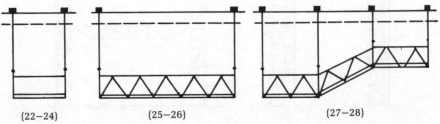

(22—24)   (25—26)   (27—28)

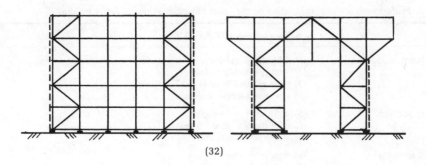

(32)

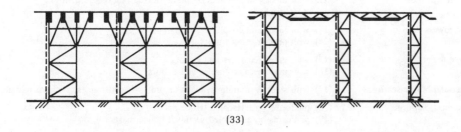

(33)

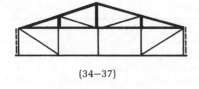

(34—37)

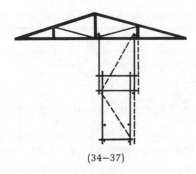

(34—37)

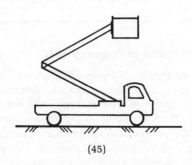

(45)

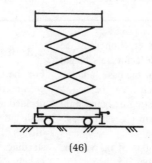

(46)

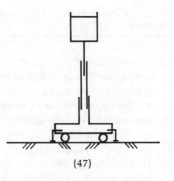

(47)

Table 2.1. Platforms on scaffolding standing on the ground and tied to the face of a building

Main types	Sub-types (shown in Figure 2.1)
Ground-based scaffolds	(1) with tubes and couplers (2) with prefabricated frames (3) with scaffold systems (4) tower-and-ledger scaffolds
Cantilever scaffolds	(5) with tubes and couplers (6) with prefabricated frames (7) with scaffold systems
Truss-out scaffolds	(8) with tubes and couplers (9) with prefabricated frames (10) with scaffold systems
Drop scaffolds	(11) with tubes and couplers
Bracket scaffolds	(12) with tubes and couplers on prefabricated brackets, e.g. for steeplejacks
Static towers	(13) with tubes and couplers (14) with prefabricated frames
Mobile towers	(15) with tubes, couplers and castors (16) with prefabricated frames and castors
Tower-and-bridge scaffolds	(17) with steel or aluminium tubes or prefabricated frames, static or mobile
Static-mast platforms	(18) platforms on telescopic lattice masts (19) platforms on assembled sectional lattice masts
Mobile-mast platforms	(20) rack-and-pinion scaffolds on masts on lorries or trolleys
Prefabricated assemblies	(21) scaffolds assembled in units moved by tower cranes

Table 2.2. Platforms hanging on tubes or wire ropes down the face of a building

Main types	Sub-types (shown in Figure 2.1)
Cradles capable of being raised and lowered	(22) painters' cradles (23) window-cleaning cradles (24) one-man work cages, power-operated or manual
Suspended independent scaffolds capable of being raised and lowered	(25) single decks with stop ends (26) multiple decks with stop ends, power-operated or manual
Suspended continuous platforms capable of being raised and lowered	(27) single decks in articulated hinged units (28) multiple decks in hinged units, power-operated or manual
Bosun's chairs capable of being raised and lowered	(29) manual on fibre rope (30) power-operated on wire rope
Work cages either capable of being raised and lowered or suspended on fixed-length wires	(31) tube-and-coupler platforms on wire rope suspensions

(g) The stability of the system after erection must be considered. This is related to two factors, its height in relation to the area available for the base, and the availability of firm places to which it can be tied.

(h) The complexity of the façade existing or being built and the availability of suitably firm places for ties.

(i) The loads to be sustained.

(j) The maximum recommended heights of the various access systems, as detailed in Table 2.6.

2.3 Standardisation of imposed loads

For the sake of uniformity throughout the building and construction industry, five load classifications have been decided upon. The advantage of this standardisation is that both the buying offices of customers and the tendering offices of scaffold contractors can work with a full understanding, in any part of the country and on any site, of what is being asked for and what will be built in response.

Table 2.3. Broad-area platforms either built up or hanging down

Main types	Sub-types (shown in Figure 2.1)
Ground-based birdcages	(32) single working platforms extending in both directions
	(33) multiple piers with bridges between them
Slung scaffolds not capable of being raised or lowered	(34) single working platforms on ropes to service soffits
	(35) single working platforms on tubes to service soffits
	(36) single or multiple platforms to service end or side walls
	(37) drop-link or chain scaffolds capable of being rigged out of plumb to service the sides of ships, etc.

Table 2.4. Moveable platforms, towers and masts

Main types	Sub-types (shown in Figure 2.1)
Timber elements	(38) lightweight stagings between roof trusses or between towers
	(39) prefabricated cat walks
	(40) pedestrian bridges
	(41) ramps and roads
Towers	(42) static tube-and-coupler towers
	(43) mobile tube-and-coupler towers
	(44) prefabricated frame towers, usually mobile
Mobile-mast scaffolds	(45) hydraulic arms and platforms
	(46) scissor lifts
	(47) hydraulic columns and platforms
	(48) extendable lattice masts and platforms

Table 2.5. Other types of scaffold not classified above

Main types	Sub-types (shown in Figure 2.1)
	(49) ladder towers
	(50) stairway towers
	(51) ladders
	(52) trestle scaffolds
	(53) ladder brackets
	(54) roof crawling boards

Normalisation of the number of working platforms will be helpful in avoiding overloading.

Fixing the maximum length of bays for each loading will also be helpful in ensuring safe scaffolds in the normal run of the industry.

Table 2.7 is taken from the Code of Practice BS 5973 1990 and gives the five basic classifications of loads.

Tables 2.8 to 2.10 give additional distributed and horizontal loadings.

If the scaffold conforms to the requirements of Table 2.7 and is built in accordance with the code requirements, no special calculations need be done.

Other loadings and bay lengths are sometimes required. These should be the subject of special design.

Above a distributed load of 3 kN/m², the scaffold will be for an unusual or special purpose, so its loading capacity should be negotiated between the scaffolder and the user. Some larger loadings are detailed in Table 2.7.

2.4 Concentrated loads

In Table 2.7, load ratings have been detailed as distributed loads, and the platforms and scaffold structure must be capable of carrying these loads over the whole area of each platform in use. However, in practice concentrated loads can occur, so the platforms and scaffold structure must be capable of carrying the loads detailed in Table 2.9, or other values agreed between the user and the scaffolder.

These values may be adjusted according to the practice in other countries. They should be given special consideration in those parts of the access scaffold between the material hoists and the building.

2.5 Horizontal imposed loads

It should be ensured that where there is a vertical imposed load consequent on a platform being loaded, there should also be a horizontal load of 3% of the vertical imposed load. This will be at varying places on the scaffold according to the height of the working platform at any given time.

The lateral strength of the scaffold at any level derived from its bracing and its ties should be such that it can

Table 2.6. Types of scaffold and recommended maximum heights

Scaffold type	Recommended maximum height
Platforms for continuous access founded on the ground	
Tower and ledger or beam scaffolds	3–12 m
Tube-and-coupler birdcages, frame and system birdcages	2–20 m
Tube-and-coupler, frame and system orthodox access scaffolds	3–75 m
Platforms for continuous access not founded on the ground	
Cantilever scaffolds and truss-out scaffolds	Can be started at any level and finished about 20 m higher
Single decks slung under soffits or fixed in roof trusses	Any height
Suspended platforms	Any height, but require restraint when higher than 30 m
Prefabricated frames or brackets sometimes incorporating shuttering and moved by tower cranes	Any height
Platforms for a few operations	
Timber steps, telescopic trestles and splitheads	3–4 m
Pole ladders and extending ladders, smaller steel and aluminium sectional towers	4–12 m
Larger demountable towers, motorised mechanical and hydraulic towers	10–25 m
Larger motorised towers, platform mast scaffolds, larger lorry-mounted hydraulic platforms	25–60 m

provide a horizontal restraint to the vertical members of 3% of the total of the self-weight and the imposed loads above that level.

Any known horizontal forces such as wind loads and from movements caused by plant or other constructing activities should be catered for as a special feature. Some horizontal imposed loads are given in Table 2.11.

2.6 Dynamic loads

When loads are handled by crane, the weight of items being placed on or removed from a scaffold should be increased by 25%.

When loads are handled manually or with the use of manual winches, the weight of items being placed or removed should be increased by 10%.

The horizontal force required to drag materials along a scaffold on wheels should be taken as 10% of the moved weight. If the materials are dragged along without wheels, the horizontal friction should be taken as 40% of the weight, and the vertical weight being moved should be increased by 10%.

The above values are in general use, but some countries and business operations may have their own special codes of practice dealing with dynamic loads.

Loading from helicopters and from ships needs special consideration according to the circumstances.

2.7 Deflections

The foregoing loads are recommended for standardisation to facilitate design and to regulate site use. In addition to considering the loads for the design of scaffolds, consideration should also be given to the elastic movement of scaffolding and the components of scaffolds when under load.

A platform should not deflect or oscillate more than about 25 mm under normal use. This does not apply to beams forming bridges, where some deflection will have been anticipated by the user and will not create a hazard.

Guard rails should not deform outwards by more than 35 mm under the load for which they are specifically designed, nor more than 25 mm if they are the normal edge protection for platforms and roofs.

Table 2.7. From BS 5973 1990   Loads on normal access and working scaffolds of tubes and couplers

Duty	Use of platform	Distributed load on platforms	Max. number of platforms	Commonly used widths using 225 mm boards	Max. bay length
Inspection and very light duty	Inspection, painting, stone cleaning, light cleaning and access	0.75 kN/m^2 75 kg/m^2 15 lb/ft^2	1 working platform	3 boards	2.7 m
Light duty	Plastering, painting, stone cleaning, glazing and pointing	1.50 kN/m^2 150 kg/m^2 30 lb/ft^2	2 working platforms	4 boards	2.4 m
General purpose	General building work, including brickwork, window and mullion fixing, rendering, plastering	2.00 kN/m^2 200 kg/m^2 40 lb/ft^2	2 working platforms + 1 at very light duty	5 boards or 4 boards + 1 inside	2.1 m
Heavy duty	Blockwork, brickwork, heavy cladding	2.50 kN/m^2 250 kg/m^2 50 lb/ft^2	2 working platforms + 1 at very light duty	5 boards or 5 boards + 1 inside or 4 boards + 1 inside	2.0 m
Masonry or special duty	Masonry work, concrete blockwork, very heavy cladding	3.00 kN/m^2 300 kg/m^2 60 lb/ft^2	1 working platform + 1 at very light duty	6 to 8 boards	1.8 m
Masonry or blockwork with storage capacity	Masonry and blockwork where large amounts of materials have to be stored on the platform	4.50 kN/m^2 450 kg/m^2 90 lb/ft^2	1 working platform + 1 at very light duty	8–10 boards	1.6 m
Special duties defined between the user and the scaffolder		6.00 kN/m^2 600 kg/m^2 120 lb/ft^2	By special arrangement and design accounting also for point loads	By special arrangement	To be calculated
Heavy construction units, shuttering units or equipment, plant, hoists, etc. Impact forces must be taken into consideration. Loads must be clearly stated to the designer		stated to the	1 special working platform + 1 at very light duty		To be calculated. Consider double standards

Table 2.8. Loads on other scaffolds of tubes and couplers

Type	Duty	Distributed load on platforms	Notes
Putlog scaffold	Brickwork	2 kN/m^2 200 kg/m^2 40 lb/ft^2	Multiply the loads by 2 for rough-terrain forklift operation
Birdcage scaffold	Inspection, cleaning and painting roofs and ceilings	0.75 kN/m^2 75 kg/m^2 15 lb/ft^2	Side scaffolds can be incorporated to service walls, as detailed in Table 2.8
Birdcage scaffold	Roof work, sprinkler fitting, storage of materials	1.5 kN/m^2 150 kg/m^2 30 lb/ft^2	This scaffold should be calculated and knee bracing in two directions incorporated

Gangways and stairs for persons	Dense crowds
	
	Normal crowds
		5 kN/m^2
		500 kg/m^2
		100 lb/ft^2
		4 kN/m^2
		400 kg/m^2
		80 lb/ft^2
			Special provision should be made for lateral sway, e.g.
			  1.0 kN/m^2 on gangways
			  1.5 kN/m^2 on stairs
			Add wind forces

Sitting entertainment stands	Shows and exhibitions	4 kN/m^2
		400 kg/m^2
		80 lb/ft^2
			Special provision should be made for lateral sway, e.g.
			  1.0 kN/m^2 on gangways
			  1.5 kN/m^2 on stairs
			Add wind forces

Standing entertainment stands	Shows and exhibitions	6 kN/m^2
		600 kg/m^2
		120 lb/ft^2
			Special provision should be made for lateral sway, e.g.
			  1.0 kN/m^2 on gangways
			  1.5 kN/m^2 on stairs
			Add wind forces

Camera towers	Outside broadcasting	4 kN/m^2 400 kg/m^2 80 lb/ft^2	Special attention to wind force displacement and rotation
Stages and rostrums	Shows and exhibitions	Agreement with the producer	Lateral sway 1 kN/m^2 Add wind forces
Ramps for vehicles	On sites	Agree loads with the user	Allow 50% of vertical load for horizontal sway
Ramps and bridges for pedestrians	On sites	4 kN/m^2 400 kg/m^2 80 lb/ft^2	Allow for one side loaded and the other side empty as a special condition
Mezzanines	Storage	Agree loads with the user	Lateral force should be 7.5% of the vertical force + wind load
Racking	Storage	Agree loads with the user	Lateral force should be 5% of vertical force + wind load
Roof loads	Wind forces snow load	Design as per codes	Agree winter/summer conditions Agree access or no access
Birdcages to support prefabricated stands	Entertainment stands	Agree with the stand designer and producer	Lateral force should be 15% of the vertical weight
Temporary support for plant	Supporting contractors equipment	Agree with the user	Ensure that plant sway, prestressing forces, etc. are catered for

Table 2.9.

Scaffold designed for the distributed loads given below (kN/m²)	Alternative concentrated loads at positions on the platform (kN)		
	On a 1000 × 1000 mm area	On a 500 × 500 mm area	On a 200 × 100 mm area (which can be considered as a point)
0.75	2.0	1.5	1.0
1.50	2.5	2.0	1.0
2.00	3.0	2.5	1.0
3.00	3.5	3.0	1.0
4.00	3.5	3.0	1.0
6.00	3.5	3.0	1.0

Table 2.10. Environmental loads for non-designed structures

Wind loads	0.25 kN/m²	25 kg/m²	5 lb/ft²
Snow loads	0.75 kN/m²	75 kg/m²	15 lb/ft²
Icing	Normal: add 0.75 kg/m of tube		
	Severe: add 2 kg/m of tube		

Table 2.11. Horizontal imposed loads

On guard rails	25 kg/m
On roof edge protection scaffolds	50 kg/m + 25 kg/m if used for access platforms
On traffic fenders	250–500 kg/m
Crowd control barriers	200 kg/m
Crowd guidance frames	50 kg/m

3 Basic independent tied scaffolds and putlog scaffolds

3.1 General description of an independent tied scaffold

This scaffold has been chosen for description first because it is the basic and most commonly used type. It will serve to establish the definitions and general principles of scaffolding.

With regard to the inclusion of the word 'tied' in the type name, it must be stated at the outset that the majority of access scaffolds assembled on the face of a building are not safe free-standing structures because they are so narrow, are often of considerable height and are not maintained upright by buttresses of guys and anchors. They are in fact tied to the building. If the building were to be taken away, the scaffold would collapse. If the ties were to be taken away, leaving the building in place, the scaffold would collapse. Originally, the term 'independent scaffold' was introduced to distinguish it from the 'putlog scaffold', which has its board-bearing putlogs embedded in the building. The use of the description 'independent scaffold' induced both scaffolders and builders to believe that the scaffold could stand as an independent structure. This led to collapses, and the term 'tied' was introduced to the type name to remind both scaffolder and user of the necessity for ties. The rules for ties are given in Chapter 6.

The independent tied scaffold consists of two lines of uprights or standards set parallel to the building, with the standards placed opposite each other in pairs to form frames at right angles to the building as well as parallel to it. The standards are joined by horizontal tubes in both longitudinal and tranverse directions, termed lifts, at specific height intervals to create a three-dimensional rectangular structure. This is subsequently braced diagonally across some of the rectangular bays to give the framework resistance against lateral forces and to maintain the structural nodes in their proper places, thus achieving stability in the whole structure. This braced framework is then tied to the building

that it serves in order to transfer the stiffness of the building to the scaffold and so ensure its load-bearing capacity and its stability.

Base plates and sole plates are fixed at the bottom of the standards on the ground to form the foundations of the structure.

Where working platforms are required, extra board-bearing transoms are added to the structure and scaffolding boards placed on these. Guard rails and toe boards are fixed around the working platforms.

The vertical loads from the weight of the tubular framework and the timber-boarded decks, and the imposed loads on the platforms, are carried to the ground through the uprights. The lateral forces derived from any displacement of the structure, from surges due to working on the platforms and from wind forces, are transferred to the ground by the bending resistance of the uprights, by diagonal bracing maintaining the framework square, by the friction of the bases on the ground, by the lateral resistance that is generated by the ties to the building, and by any sway transoms installed.

The dimensions of the structure are fixed to meet the functional requirements of the job to be done and to match the features of the ground and the wall to which the scaffold is to be tied. Lift heights are fixed by the nature of the work on the platform. The spacing of uprights along the face of a building, i.e. the bay length, is fixed so that the ledgers are adequately strong and there are enough pairs of uprights to carry the self-weight, the imposed loads and the leg loads resulting from the horizontal forces.

The lift heights are usually 1.35 m for brickwork construction and 1.8 m for other trades. For concrete frame construction, these heights may be at the storey height of the building and can be modified subsequently for other trades.

The spacing of uprights away from the building and the width of the scaffold measured at right angles to the building are fixed by the nature of the building surface, by the nature of the work being done from the platform, by the statutory width required for the various duties of the scaffold and, in some cases, by the type of couplers being used.

The frequency and nature of the bracing is fixed by the degree of lateral force to be dealt with, the nature of the work being carried out on the platforms, which are frequently partly obstructed by the bracing, by the requirements of the codes of practice, and by any special requirements to deal with the use of the pavement and access to the building at lower levels. Chapter 5 deals with bracing.

The nature and frequency of ties is fixed by the size of the scaffold, its function, its location, the nature of the building surface and the requirements of the codes of practice. Chapter 6 deals with ties.

All dimensions can be calculated structurally. Unfortunately, it is impracticable to build a structure in scaffolding materials sufficiently accurately or free from joints to comply with these calculations. Empirical rules must be used. Alternatively, having carried out rigorous structural calculations, a modification factor can be applied to represent all the deficiencies of scaffolding construction and use. The empirical rules are suitably embodied in a series of tables that take into account the shortcomings of the materials, the system, the construction and the user.

Ledgers are joined to the uprights by right-angle couplers. In some systems, transoms may also be attached to the uprights with right-angle couplers and be at the same height as transoms fixed to the ledgers. These transoms are then fixed to the ledgers using putlog couplers.

If there are transoms at intervals of about 1.2 m along every lift of ledgers, all the couplers attaching them to the ledgers can be putlog couplers. If there are no boards to be placed on the lift, and there is only one transom per pair

of uprights, then this should be attached with right-angle couplers.

Braces should preferably be joined to the horizontals with right-angle couplers or brace couplers. They can be attached to the uprights with swivel couplers.

Tie tubes and the whole of the tie assembly should be coupled using right-angle couplers.

Coaxial joints in uprights can be made with either expanding spigot pins or sleeve couplers. Coaxial joints in ledgers should preferably be made with sleeve couplers, and coaxial joints in longitudinal bracing should also be made with sleeve couplers.

3.2 Hazards in independent tied scaffolds

- Absence of ties.
- Absence of bracing.
- Coaxial joints in tubes being situated in the same lift height or bay width.
- Undermined foundations.
- Attachment of tarpaulins.
- Too many working lifts.
- Overloading.
- When the same scaffold is built a second time or repeatedly during a long-term building project, the quality of the scaffolding becomes worse in stages as a result of inattention to the original details.

3.3 Empirical design tables

The empirical rules referred to in Table 2.8 are elaborated in Part 4. They are for general use when there is nothing special to take into account, and they are conservative. The scaffold must contain the correct bracing quantity and pattern, be tied in the orthodox manner and be used correctly without overloading.

3.4 The standard form of independent tied scaffold

Figure 3.1 gives the normal form of an independent tied scaffold. The sole plates can be reduced in size if the scaffold is on a pavement used by the public.

3.5 Joints in uprights and ledgers

The joints in uprights should be staggered into different lifts, preferably as shown in Figure 3.2(a). The arrangement shown in Figure 3.2(b) is inferior but acceptable.

Joints in ledgers should be staggered into different bays. A guard rail may be considered a supplementary ledger.

To correct a scaffold in which the joints have not been staggered, the joints can be lapped with a short tube and

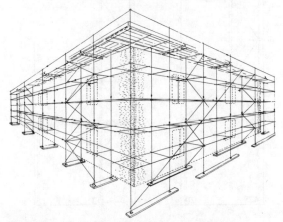

Figure 3.1 The normal form of independent tied access scaffold showing permissible variations in sole plates, bracing patterns, ties and typical joint locations. On hard surfaces the sole plates can be omitted.

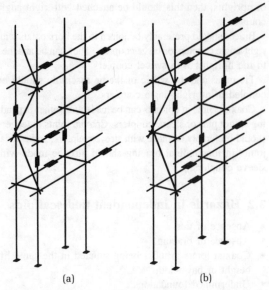

Figure 3.2 Methods of staggering the joints in standards and ledgers: (a) Preferred method with two joints per lift and one per bay. (b) Acceptable minimum with three joints per lift and two per bay, with an unjointed guard rail.

two fittings or by introducing a supplementary tube across the lift or bay.

3.6 Unorthodox independent tied scaffolds

If there are duties to be performed not given in Table 2.8, or loads greater than the values given in that table, then special calculations should be made and height limitation factors applied as detailed in Part 4.

If the regular pattern of bracing in either direction or the regular pattern of ties is not practical, then modifications should be made to overcome the problem. Extra ties can overcome deficiency in bracing and extra bracing can improve the stiffness between more widely spaced ties, provided that these are adequate.

Figure 3.3 shows four variants of an independent tied scaffold that could be considered unorthodox. These are the types of scaffold that need engineering calculations.

3.7 Independent tied scaffolds with tubes other than the British Standard type

A variety of tubes in various parts of the world are used in scaffolding. Strength in bending governs the adequacy of both the ledger and the uprights, hence spacing of the uprights can be reduced if a weaker tube is used. The strength of various tubes is given in Part 4.

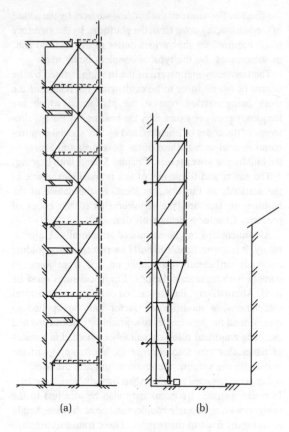

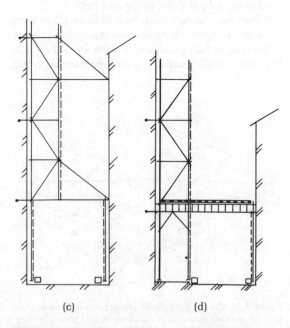

Figure 3.3 Examples of unorthodox independent tied scaffolds.

3.8 Independent tied scaffolds with prefabricated joints

These systems generally do away with loose scaffolding couplers and have instead a half joint on each tubular element. When the two elements are assembled, e.g. a ledger on a standard, the two half joints unite to make a complete joint.

Such proprietory systems are not dealt with in British Standard Code of Practice 5973 1990 in sufficient detail, so the user is advised to obtain guidance from the supplier. He should ascertain the strength in bending of tubes used in the system.

By way of general guidance, the rules for bracing and the rules for ties given in this book for tube-and-coupler scaffolds apply to system scaffolds.

3.9 Access for traffic through independent tied scaffolds

Traffic access is frequently needed through a scaffold, and uprights have to be omitted. The general rule is that, for each standard removed, an additional one is placed at the side of the opening. An A frame or a V frame is inserted over the opening to transfer the load from the centre to the replaced standards at the edge.

Figure 3.4 shows five ways of making a safe access. Consideration should be given to using supplementary couplers on the puncheons over the opening.

It should be noted that the standards are doubled at the sides of the openings.

Figure 3.4(a) uses A-shaped frames of tubes and couplers over the entrances. Figure 3.4(b) uses V-shaped frames. Both types of frame must pick up all the puncheoned uprights and there must be sufficient A or V frames to transfer all the vertical loads away from the opening.

Figure 3.4(c) uses unit beams with a calculated number of couplers supporting the ends. Figure 3.4(d) is a traditional span made entirely with tubes and couplers arranged as a scaffolding tube beam. Figure 3.9(c) uses ladder beams.

Traffic fenders of 300 mm × 300 mm timbers can be incorporated where necessary and practical, and all the beam arrangements shown should be repeated on the inside and outside lines of standards.

3.10 Independent tied scaffolds with prefabricated frames

Scaffolding frames can be fabricated from standard scaffolding tubes or from tubes with a different diameter, wall thickness or metal strength. The rules for tube-and-coupler scaffolding need only slight modification to take account of the fact that there will be a multiplicity of joints in the uprights, all in the same plane. Modifications will be needed to take into account different metal cross-sections and metal quality. The degree of modification is a question for a structural engineer, bearing in mind that one variation may compensate for another.

The frame manufacturer's advice should be requested, especially with regard to his rules for ties and bracing. The applicability of his rules to the particular problem being studied should be carefully considered.

In this circumstance, attention is drawn to Section 6(1) of the Health and Safety at Work Act 1982, which requires the supplier to have adequate information about the equipment available.

Attention must also be paid to Section 6(2) of the same Act, which requires the user to ensure safe use of the equipment.

3.11 Hazards in prefabricated frame scaffolds and system scaffolds

- Because a frame scaffold or a system scaffold does not require separate couplers, it cannot be assumed that the tubes are any better than the normal tubes used in tube-and-coupler scaffolds, or that the joint formed by two prefabricated parts offers better lateral stability than a scaffolding coupler. Therefore, the belief that prefabrication reduces the amount of bracing and tying may create a hazard.
- In this connection, it must be borne in mind that the contribution that the stiffness of joints makes to the lateral stability of a scaffold is only about one-tenth of that afforded by a diagonal brace.

Other hazards in prefabrication are:

- failure to use locking pins or wedges to fix the upper parts of the scaffold to the lower parts;
- lack of longitudinal stiffness;
- dangers in dismantling due to the projection of components.

3.12 General description of a putlog scaffold

A putlog scaffold consists of a single row of standards parallel to the face of a building and set as far away from it as is necessary to accommodate a platform of four or five boards with a gap of about 100 mm between the inner edge of the platform and the wall. The wall gap should be the minimum required for plumbing the brickwork and for the fall pipes.

Sole plates and base plates should be used under the standards, which should be laced together with ledgers fixed with right-angle couplers. The putlogs can be fixed to the ledgers with right-angle or putlog couplers, and the blade

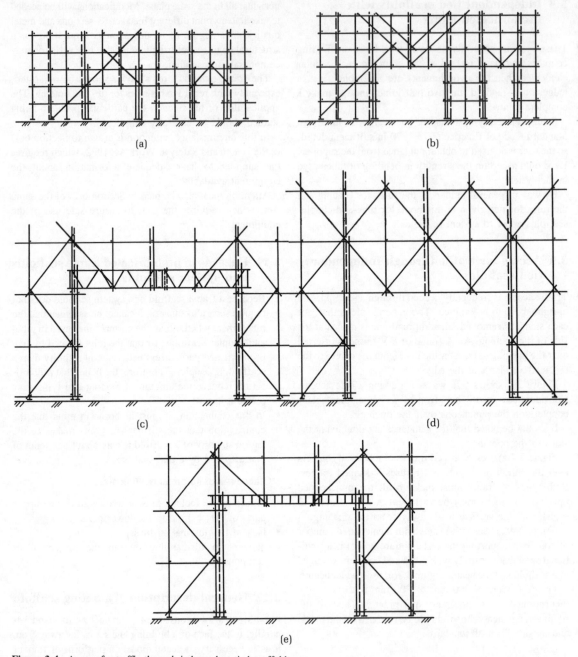

(a)

(b)

(c)

(d)

(e)

Figure 3.4 Access for traffic through independent tied scaffolds.

end of the putlog tube or putlog adaptor should preferably rest flat on the brickwork being built.

The scaffold should be tied to the building using the same number of ties as for an independent tied scaffold, as detailed in Chapter 6.

Special precautions should be taken to achieve adequate

fixing when a putlog scaffold is started and has only one or two lifts in place. This can be achieved by using temporary raking tubes assembled on the outside of the scaffold and bearing on firm ground, while their bottom ends are foot-tied back to the standards of the scaffold.

Putlogs should be spaced not more than 1.5 m apart for

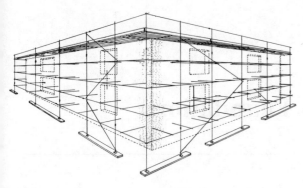

Figure 3.5 The normal form of putlog scaffold showing various bracing patterns, ties, typical joint locations and the use of bridle tubes. Ledger bracing or rakers may be necessary to start off the lower lifts.

38 mm thick boards, or at the appropriate spacing for other thicknesses of board as set out in Chapter 7.

Where a putlog is required as a board support, and it is opposite an opening in the building such as a window or doorway, the inside end of the putlog should be supported on an underslung bridle tube spanning adjacent putlogs, as shown in Figure 3.5, which also shows a method of tying through these openings.

Reveal ties, screw-in anchor ties and wire ties can also be used.

In cases where the brickwork is green, or if there is doubt about its adequacy to support the blade ends of the putlogs, continuous underslung bridle tubes can be used.

Where a putlog scaffold is erected against an existing brick wall for repointing, the old putlog holes can be reused or others raked out. In this case, the putlog blades can be inserted vertically.

Guard rails and toe boards should be provided on working platforms.

Longitudinal bracing is required at intervals of 30 m, but ledger bracing is not required.

For brickwork, the lift height should be about 1.3 m but, for certain types of masonry work, a lower lift height may be necessary.

3.13 Hazards in putlog scaffolds

- Reliance solely on the grip of the putlog ends in the brickwork for the attachment of a scaffold to a building has frequently resulted in a collapse, especially when the scaffold has reached its second lift level. Slight side sway in the wind loosens the putlog grip.
- Dislodgement of the lower lift by forklift truck loading.
- When the first lift is in use, removal of the guard rail leaves the joint in the ledger unassisted, and lateral failure may occur following impacts.

3.14 Masonry walls

In many stone and artificial stone-clad walls, there are no regular courses and, in this case, the putlog blades cannot be placed flat. The putlogs should then be positioned in a convenient place with the blades vertical between the ends of pieces of masonry. This arrangement is more susceptible to loosening with wind oscillation than when the blades are flat, and care must be taken to ensure good tying through the windows and adequate longitudinal bracing.

3.15 Bay width

Putlog scaffolds are frequently associated with brickwork jobs and the mechanical handling of bricks. There should be close liaison between the scaffolding contractor and the main contractor about the width of bays, with special attention to the width of brick pallets or packs. For forklift handling on a rough site, a clearance of at least 200 mm should be left at each side of a pallet.

The scaffold should be fixed with very firm and frequent ties as there is a tendency for forklift operators to pull scaffolding off the wall when the forks are being withdrawn.

3.16 Ledger joints

Traditionally, platforms were loaded with piles of bricks placed beside the standards but, with mechanical handling, loads are often bigger and are centrally placed in a bay. If there is a joint in a ledger at mid-span, the weakness may produce a collapse. It is important to arrange the ledger joints close to the uprights. Alternatively, joints should be spliced with short tubes and parallel couplers.

4 Foundations

4.1 Classification

Foundations in scaffolding have two critical differences from building foundations. They are (a) on the surface, and (b) not calculated. Empirical rules must be applied. Fortunately, the tubular framework of scaffolding is relatively flexible, even when correctly braced. It can accommodate a multiplicity of small movements from one cause or another and, in particular, from one standard settling downwards on its foundation.

The contact pressure underneath the base plate or sole plate of a standard is dissipated down into the earth in the same way as that of any other foundation, i.e. the pressure falls away quite rapidly with increasing depth. This fall off in pressure approximately follows a decreasing square law, so the uppermost layer, i.e. that immediately below the base plate or sole plate, is the most critical. Hence, classification of the surface needs careful consideration.

Three subdivisions form a useful basis.

(a) Surfaces that are hard and will support the lower end of an upright tube without question, e.g. thick concrete or masonry paving, steel in ships, and floors in factories and industrial buildings.
(b) Materials with a hardness between the two extremes or that will be damaged if loaded without distribution pads, or crack or break up if loaded over too small an area.
(c) Soft soils and poorly compacted surfaces that will consolidate under load.

4.2 Hard foundations

In the case of rigid materials, e.g. thick concrete, thick masonry paving, steel floor tiles or sheets and granolithic screeding on concrete, it is satisfactory to set up vertical tubes directly on the bearing surface. There is no need to spread the load when the foundation material will spread it adequately.

4.3 Medium hard foundations

In this class are grouped asphalt, internal floor surfaces, terrazzo flooring, indented and embossed sheeting and flooring, screeds, brick paving and similar surfaces that, even if tough enough, must not be damaged.

The standards should be set on steel base plates. These in turn can be set on hardboard if rust staining and scratching must be avoided.

4.4 Soft materials

The majority of cases fall into this category, including all soils, compacted ash, clinker and hardcore. Hard materials not fully bedded such as tiles and pavement slabs are also included.

In these cases, the standard is pitched on a steel base plate, which in turn rests on a sole plate or sole board of timber to achieve a further spreading of the load.

In the case of internal floor boarding, the load should be spread over several joists by a sole plate parallel to the floor boards and thus across the joists. A double thickness of scaffolding board may be necessary to provide adequate strength.

When founding on a concrete floor in buildings under construction, slabs may be reinforced strongly in one direction only. The sole plates should cross this direction to bring into play several elements of the reinforced concrete. The location of hollow pots and waffle profiles should be ascertained and appropriate measures taken to spread the scaffold load.

4.5 Sole plates

Taking the ordinary soil on a building site as the norm, it is desirable to have a sole plate area of at least 2000 cm², i.e. about a metre length of scaffolding board beneath each standard for each tonne of load in the standard.

In the case of a very high scaffold, this rule might result in an impracticably long sole plate and, for this case, it is preferable to harden the ground artificially either by compacted hardcore or concrete slabs about 500 m wide. The improved ground and increased width of the footing will enable a shorter length of sole plate to be used.

If the load in the standard comes down within the middle third of the length of the board, the whole length of the board will remain in contact with the ground and contribute to the resistance. If the standard is on one or other end third of the board, the opposite end will tend to lift under the eccentric load and so reduce the area of ground supporting the load. Hence, the standard should always be within the middle third of the length of the sole plate. Similarly, it should be within the middle third of the width of the board to prevent a tendency to tip sideways.

When two standards are placed on one continuous board, the combined load will have its centre of force near the centre of the length of the board, and a better foundation will result. Thus, the general practice has developed of placing a half-length scaffolding board (about 2 m long) on the ground at right angles to the building under each pair of standards.

Figure 4.1 gives some technical points to bear in mind with foundations on the surface.

Figure 4.1(a) is a typical pressure dissipation diagram for a load on a sole plate. The contact pressure is 1.0 p, equal to the load divided by the area. The curves are plotted for 0.9, 0.7, 0.5, 0.3, 0.2 and 0.1 values of p. Figure 4.1(b) shows the mode of failure in an overloaded soil. Figure 4.1(c) shows the load alongside the middle third of the width, when the opposite edge of the board supports no load. Figure 4.1(d) shows the load outside the middle third of the width, when the contact area is reduced and the board tips up sideways. Figure 4.1(e) and (f) show one or two loads correctly central or symmetrically placed along the length. Figure 4.1(g) shows the load at the end of the middle third of the length, when the opposite end supports no load. Figure 4.1(h) shows the load outside the middle third of the length, when the contact area is reduced and the board tips up lengthwise.

4.6 Hazards with sole plates

- Using rotten boards.
- Laying down continuous runs of boards so that some standards bear on the ends of boards or across the joints between them.

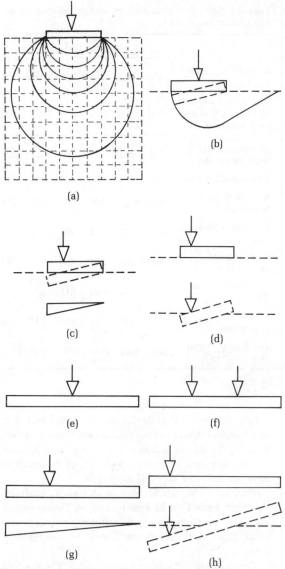

Figure 4.1 Sole plate behaviour.

- Placing boards directly on waterlogged ground.
- Rain or site surface water washing away the supporting soil.
- Excavation below or too near the boards.
- Boards knocked away by road or site traffic.
- Frost in the ground lifting the boards and the scaffold, leaving the scaffold unsupported during a thaw.
- Boards levelled using intermittent bits of brick or blocks.

4.7 Foundations across trenches

Trenches on building sites frequently need to be spanned by sole plates.

Table 4.1. Safe centre point loads on scaffold boards used as sole plates across trenches

	Load (kg)					
Width of the trench (mm)	250	500	750	1000	1250	1500
Span to be considered by adding 150 mm at each side (mm)	550	800	1050	1300	1530	1800
A single scaffold board 38 mm thick	345	238	181	146	124	106
Two scaffold boards	690	476	362	292	248	212
A single board 50 mm thick	589	323	309	249	212	180
Two boards 50 mm thick	1178	646	618	498	424	360
A single board 63 mm thick	960	660	501	406	345	293
Two boards 63 mm thick	1920	1320	1002	812	690	586
A single board 75 mm thick	1383	951	724	585	497	556
Two boards 75 mm thick	2766	1902	1448	1170	994	1112

Table 4.1 gives the loads that may be placed mid-span on a trench bridge of various thicknesses of board. When assessing the span of boards across the trench, a distance of 150 mm should be added at each side of the trench to allow for defective edges of the trench.

From the table, it will be seen that single scaffolding boards should not be used because of the very small loads they can carry. Doubled scaffolding boards across a half-metre wide trench are limited to about half a tonne.

When a standard is sited on a plank bridge or is observed to rest on disturbed or weak ground, its load should be transferred to its neighbours by connecting an A-frame with raking legs from the first lift node to the bottom of the two adjacent legs. This A-frame need not be in the lowest lift but can be higher up the scaffold.

4.8 Foundations on steelwork

When joists are cantilevered out from a building to form the starting level for a cantilever scaffold, when the flanges of a bridge beam are used to support a scaffold, when broad surfaces such as the deck or hold of a ship have to be used for the base, the scaffold must be prevented from skidding sideways off the steelwork.

In the case of the top flange of a joist, a base plate with a spigot can be welded on to the flange A forkhead used in falsework can be inverted straddling the flange. If none of these is possible, then the standard can be pitched on the steelwork and its bottom end box-tied around the steel section in two directions with tubes and couplers.

When bearing on the bottom flange of a bridge beam, the top flange gets in the way. Rakers can be used tucked into each side of the web and coupled together into a V-frame over the top of the beam. In this case, the scaffold is set on puncheons on the ledgers of this V-frame. Supplementary couplers may be necessary on the puncheons if the load exceeds 635 kg.

In the case of pipe gantries, where it is possible or necessary to found on the pipes, a timber sole plate should be box-tied on to the top of the run of pipes and the scaffold founded on this.

Ships' decks and the bottom of ships' holds usually have enough places where a foot lift can be anchored in place near the surface, and the scaffold can be raised from this. The plates on ships are often thinner than one might expect, so the owner should be consulted.

4.9 Founding on sloping surfaces such as a steep gradient in a street

In hilly towns, gradients of up to 10% are sometimes encountered that have a surface hard enough to permit the whole scaffold to creep downhill. Sometimes only the bottom horns of the legs creep down.

The sole plates will lie on the ground and thus be at an angle to the end cross-cuts of the scaffold tubes. This circumstance does not give trouble in practice but will need consideration if the scaffold is high or heavily loaded — say, more than 2 tonnes per leg. The limit of inclination of the street before special precautions are necessary is the angle at which a base plate may be skewed in the end of a tube. This is possible because the spigot on the base plate is not a true fit in the bore of the tube, and it permits assembly at an angle. For this purpose, the base plates should not have a shank larger than 25 mm in diameter nor longer than 50 mm.

The bottoms of all standards should be joined in both directions by a base lift placed about 150 mm above street level. The base lift can follow the ground contours to avoid long horns at the bottom, but high lifts may also have to be avoided. On a sloping site, a stepped bottom lift with horizontal tubes uses more tubes and couplers than a sloping one.

On streets steeper than described above, timber wedges should be used to level the surface on which the base plate is placed.

4.10 Founding on base lift tubes

There are several circumstances where it is neither desirable nor practical to place base plates and sole plates under the standards.

A common case is when the scaffold has to be built on a steep embankment, say to service the face of a building adjoining a railway cutting. The slope of the cutting may be too great to put sole plates along the slope on the surface. The standards will slip off, and wedges will slip down. If the lines of uprights are marked out along the slope, a chase can be cut into the slope for a board to be placed longitudinally and flat for each line.

Even if permission can be obtained, this arrangement is unsatisfactory because one side of the sole plate is cut deeply into the slope, while the other is only just on the surface. When such a sole plate is loaded, the edge near the surface depresses into the slope and the board tends to roll over, letting the standard slide off.

This problem can be avoided by basing the standards directly on the soil with neither base plates nor sole plates. The scaffold is partly built, the couplers on the standards are loosened and the standards are driven down into the soil with a sledgehammer. Before they firm up too strongly but when they have penetrated about 25 cm, the base lift is assembled with ledgers only about 10 cm above ground level. The base transoms can be steeply sloping. The standards are then driven strongly down into the ground until the ledgers rest on the ground and can be driven no further. The scaffold is then built upwards as usual without any tendency to slide down the slope or to settle further.

The couplers are then re-tightened with the ledgers bearing hard on the ground. Figure 4.2(a) is an example.

On very soft slopes, the transoms can be put below the ledgers and spaced every half metre, and extended a metre either side of the scaffold. In this way, a very substantial mattress is formed on the ground. Figure 4.2(b) shows this, and also the longitudinal bracing that has been driven into the ground.

In cases where there could be some lateral force down the slope and the scaffold might tend to turn over downwards, a longer length of transoms can be placed on the ground, as shown in Figure 4.2(c). External driven rakers are added.

In these three ways, gangways, rostrums and show stands can be constructed safely on the poorest of sloping ground.

Taller towers for cameras and radio transmission masts may be required on poor sloping ground but need to retain precise verticality. Figure 4.2(d), (e) and (f) are examples of how these can be built. The base lift can be modified into the form of Figure 4.2(b) or (c) if desired.

For some applications, it is desirable to place sole plates on the ground surface and still retain the driven-in tubes.

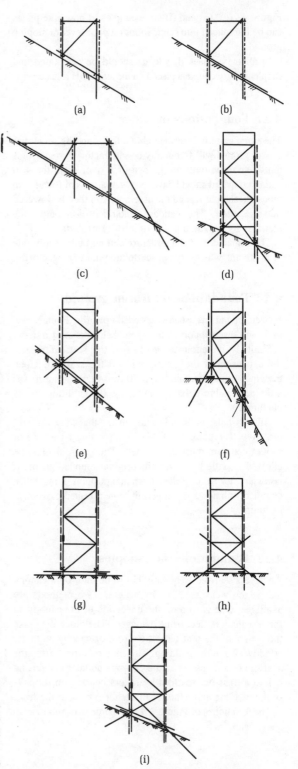

Figure 4.2 Foundations on sloping and weak ground.

Figure 4.2(g), (h) and (i) are examples of how sole plates can be introduced into foundations on poor level or sloping ground.

In all these cases, leg loads should be calculated and supplementary couplers placed on the standards if necessary.

4.11 Foundations in water

The same system described above for founding on slopes or soft ground will frequently be satisfactory for scaffolds that have to be built on the bed of a river or in the sea.

Each standard should have its own mattress of foundation tubes attached to it, and the assembly can then be lowered into the water. The standards should project below the mattress by an amount determined by probing.

The remainder of the structure can be built at low tide or by divers, who carry out scaffolding quite satisfactorily.

4.12 Foundations on frozen ground

Frozen ground has usually expanded upwards. Foundations on it will lose support as the ground shrinks after a thaw.

Similarly, if the foundation of an existing scaffold freezes, the expansion of the ground will lift the scaffold. Upon thawing, it may remain up to 50 mm higher than it was built and collapse later, damaging the building or its cladding.

All ties must be checked to ensure that they are still effective. Temporary and flexible ties using wire rope or swivel couplers must be inserted. The scaffold must be allowed to settle back on to its original foundations or, if necessary, loaded or jacked down into position. It must then be retied with ties that will permit some repeated movement in future.

4.13 Foundations on canopies

An access scaffold frequently has to be built on a canopy that is too weak to carry its weight. Two methods are available. One is to prop the underside of the canopy so that it cannot deflect or be damaged. The other is to take the weight of the scaffold off the roof or canopy on to the window sills of the building. This may require starting the scaffold on sole plates on the canopy and later converting it into a truss-out scaffold by rakers bearing on the sills of the building and with ties to anchor it back to the face.

The techniques of Figure 3.4 may come in useful to span short canopies.

4.14 Foundations over services and other places where access is needed

Very frequently, some of the standards of a scaffold occur over manhole covers, on wooden pavement covers or over hydrants and stopcocks. The simplest way of dealing with this type of problem is to move the standard along the ledger until it does not rest on the place to be avoided. A movement of half a metre one way or another will not make a significant difference to the strength of an access scaffold. If this technique is not suitable, then the bottom lift of the standard can be left out and the load distributed through an A-frame to the neighbouring standards, which may require doubling as a consequence.

4.15 Foundations that have to resist lateral forces

Figure 4.2 shows scaffolds on which both the standards and the bracing tubes of the bottom lift have been driven into the ground through loose couplers. This results in a foundation that can resist lateral displacement. The diagonal bracing when driven in makes a very good anchor against uplift and overturning.

These diagrams include some examples of the use of sole plates under the foot ties instead of under the standards. Supplementary couplers may be necessary on the standards to prevent them being forced down through the foot ties. Bracing couplers can supplement the foot tie couplers.

Arrangements such as these can resist considerable uplift, so the foundations can be used to give extra stability to exposed camera towers, cable support towers and ladder towers.

4.16 Foundation on brackets

Scaffolds are frequently not founded on the ground but must rest on brackets attached to the building.

Generally, there should be one bracket for each pair of standards, but the system results in large eccentricities of load applied to the building. The rotational forces applied to the building must be calculated and catered for with a safety factor of two after taking into account the effects of impact, thermal expansion, distortion and possible alterations. Fixing ties to every pair of standards higher up the scaffold will relieve the bending moment on the building.

Bracketed scaffolds are frequently used by steeplejacks working on old buildings. The strength of the building and the means of anchoring the brackets to it must be assessed very accurately.

5 Bracing

5.1 The inadequacy of the couplers to resist sway

Without bracing, a scaffold is a rectangular framework of uprights and horizontals. The couplers at the intersection points, while termed 'right-angle couplers', cannot be considered structurally to provide a fully restrained 90° angle between the tubes.

Firstly, consider the joint between the ledger and the upright as viewed in front elevation. Even if the coupler is a very accurate forging, its grip on the tubes is over a distance not greater than 40 mm under the flap of the coupler. The flap itself is joined to the body by a hinge at one side and a single bolt at the other, and does not press the flap on to a bearing surface. The tube gripped by the flap does not have great sideways crushing strength − in fact, it deforms under the bolt tension before any external load is put on it. Therefore, it is inevitable that the joint will have a relatively low angular rigidity.

Secondly, consider a pressed coupler, customarily weaker than a forged one in rotation. This will have less angular rigidity. Some couplers have no angular rigidity at all, being pin joints up to a certain angular movement.

Scaffolds are structures that are both tall and wide in front elevation, so a small angular movement of the couplers produces a large displacement at the top of the structure. The rigidity that might be hoped for in the coupling would have to be applied instantly to prevent deformation, whereas in fact it takes a considerable degree of deformation to become effective. Its resistance is low when the lateral displacement of the structure is low, and it only gains resistance when the displacement has become hazardous.

For the above reasons, it is not advisable to assume any rigidity in the couplers, although the structural continuity of the uprights and ledgers through the fittings is not affected.

Now consider the scaffold in end elevation. If it is a putlog scaffold, it will have the full ridigity of the wall on one side, and it cannot distort away from the wall unless something slips. If it is an independent and infrequently tied scaffold, the couplers tending to prevent lateral deformation towards or away from the wall are those fixing the angles between the transoms and the uprights.

If these couplers join the transoms directly to the uprights and are right-angle couplers, the weaknesses described above apply. Usually, however, the transoms are coupled to the ledgers, and thus the deformation of the scaffold towards or from the wall is determined by the rotational strength of the coupler, i.e. its resistance to twisting around a tube; and the coupler may well be a weaker putlog coupler.

There is therefore no justification for assuming that a scaffold, either a normal one or a birdcage, gains useful lateral stability when loaded from the rigidity of its joints. Rigidity at low deformation merely helps the scaffold to be built without its intended applied load.

5.2 Bracing

Consequent upon this argument is the necessity for all lateral effects to be catered for by bracing. This does not mean that every rectangular space in the structure must contain a diagonal. The framework can be allowed sufficient internal and local deformation to collect the lateral forces imposed on it in various places and transfer them to intermittent points, from where they are transferred to the ground by bracing.

Two requirements must be met. First, the collecting points for the imposed lateral forces must not be so far apart that the structure cannot transfer the imposed forces to these points. Second, the bracing from the collecting points must successfully transfer the accumulated forces to the ground without imposing hazardous loads on any other part of the structure.

LEDGER BRACING

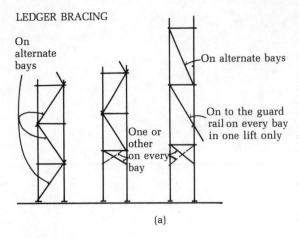

On
alternate
bays

On alternate bays

On to the guard
rail on every bay
in one lift only

One or
other
on every
bay

(a)

FACE BRACING

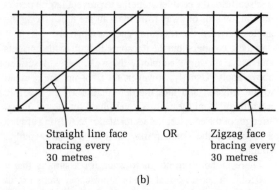

Straight line face   OR   Zigzag face
bracing every   bracing every
30 metres   30 metres

(b)

Figure 5.1   The basic rules for bracing independent tied scaffolds.

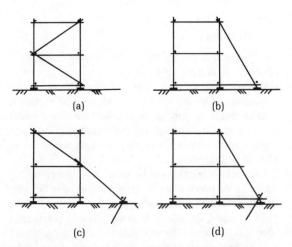

(a)   (b)

(c)   (d)

Figure 5.2   Four methods of bracing a two-lift structure.

Because the ground is fixed, and the imposed lateral forces are above it and sometimes at a very high elevation, a large distorting moment is applied to the structure. There are two ways of dealing with these lateral forces. One is to provide raking tubes from the location of the forces to the ground and so direct these forces from the structure and transfer them directly to the ground. The other is to build into the structure sufficient lateral rigidity, by the addition of diagonals, to withstand the distorting forces applied to it. Either method is satisfactory, provided that the analysis is done correctly. The forces can be dealt with partially by both methods.

Figure 5.1 shows the basic rules for bracing independent tied scaffolds. Scaffolds with large lateral loads may need calculating.

5.3 Lateral stability

Figure 5.2 shows four methods of bracing a two-lift structure.

(a) By internal bracing, which keeps each lift square.
(b) By an external raker, which keeps the whole structure square. This method does not eliminate the tendency of the uprights to buckle under load.
(c) By a combination of an internal brace and an external raker.
(d) By an external raker coupled to both lifts and the foot tie.

In case (a), the limiting lateral force would not only have to cause the couplers on the lower lift to slip, but also the couplers on the top lift.

In case (b), the maximum safe load that can be transferred along the raker is 635 kg, being the safe slip load of the right-angle coupler at one or other end. The maximum allowable load along the raker may be less than 635 kg if its strength as a strut is less than this because it is long.

In case (c), the lateral force would have to overcome the couplers on the top lift brace and also those on the external raker.

In case (d), the lateral force would have to overcome the coupler on the top or bottom of the raker and deform the centre horizontal tube.

Thus, system (b) is the weakest. Systems (a), (c) and (d) are stronger.

To understand this mechanism, it must be remembered that a scaffold is not a true pin-jointed structure, i.e. with all the short lengths having pin-jointed ends. It is a trellis, i.e. one in which the members are continuous through the joints but pinned together at the joints.

Theoretically, such a structure consisting of continuous vertical members pinned to continuous horizontal members will fall sideways if it has no bracing in it, but it can be

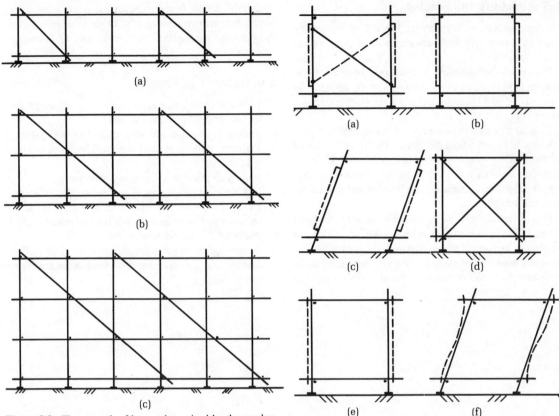

Figure 5.3 The strength of braces determined by the number of couplers on the braces.

Figure 5.4 Methods of attaching bracing.

stabilised by a single diagonal placed anywhere in the structure, provided that this diagonal and its end joints are strong enough, and the vertical and horizontal continuous members are also strong enough.

In practice, there are joints in the horizontal members that can fail, and there are joints in the vertical members that can fail to resist uplift. Therefore, the best practice in scaffolding is to ensure that every lift has a brace in it.

In long access scaffolds, there should be a longitudinal brace in every lift for every 30 m of length. There should be a transverse bracing system in every lift on alternate pairs of standards.

When the bracing is called on to resist horizontal forces, horizontal forces are generated in the ledgers and vertical forces in the standards. This emphasises the need to stagger the joints in both horizontal and vertical tubes.

5.4 The lateral restraint achieved by bracing

The brace across a single bay is as strong along its axis as the weakest coupler in it, i.e. 1.25 tonnes at failure and

0.625 tonnes safe. Figure 5.3(a) shows a structure with two bays braced with two tubes and coupled with four couplers. The safe horizontal resistance for each of these two braces is approximately 0.5 tonnes, giving a total horizontal resistance of 1 tonne.

From this comes the traditional calculation of horizontal resistance by counting all the couplers on all the bracing in one direction, then halving the number and multiplying the answer by 0.5. Thus, in Figure 5.3(c), there are eight couplers on the two braces (8 ÷ 2 = 4, 4 × ½ = 2. Thus, the structure would offer a horizontal resistance of about 2 tonnes). If the structure tended to deform sideways, two of the couplers at the bottom of the braces might hold firm while the top two slipped, confirming the rule.

Figure 5.3(b) has an odd number of couplers per brace, but the rule can be applied. There are six couplers. Three are slipping and three are holding fast, giving a lateral resistance of 3 × ½ tonnes = 1½ tonnes.

Thus, the efficiency of a bracing system is proportional to the number of couplers connecting it to the basic framework. The same calculation must also be made for stability in the direction at right angles to the first calculation.

5.5 Attaching the bracing

The best practice is to attach the bracing by right-angle couplers to the horizontals rather than by swivel couplers to the uprights.

Figure 5.4 explains this. In cases (a), (b) and (c), the bracing is attached by swivel couplers to the uprights. In cases (d), (e) and (f), it is attached by right-angle couplers to the horizontals.

Case (a) is braced on four sides and is subjected to a force from the left-hand side of the figure. This force is resisted by the bracing, which can be seen as a cross. If this is removed, as in case (b), the structure distorts as in case (c). The side braces make no contribution and move over with the uprights.

Case (d) is braced on four sides and subjected to the same force, which is resisted mainly by the bracing, which can be seen as a cross. If this is removed, as in case (e), the structure tries to distort as in case (f), but meets some resistance from the couplers of the side braces and the strength of the side brace tubes, i.e. the side bracing makes a contribution to the resistance of the structure to distortion in a plane at right angles to itself. Thus, case (d) is stronger than case (a).

5.6 Hazards in bracing

- Failing to make strong in-line joints by using spigot pins or not tightening sleeve couplers.
- Failing to attach all the lifts to the bracing system.
- Failing to start the bracing at ground level and continue it to the top of the scaffold.
- Failing to place bracing in two directions.
- Not realising that forces in braces are always accompanied by forces in the verticals, which might be upwards or downwards, and in the horizontals, which might stretch the scaffold.
- Removing bracing to improve the convenience of the scaffold and not replacing it by some other stabilising means.

6 Ties

6.1 Fundamental requirement

An access scaffold is not a very rigid structure in itself, even if it is braced in the orthodox manner. The reasons for this are:

- There is some flexibility in the couplers.
- There is some flexibility in the tubes.
- There are joints in the tubes.
- The centre lines of the members do not come together at node points but are separated by distances of up to 300 mm.
- The structures may be very long or very high in relation to their width. For instance, a 16-storey block will have a height to base ratio of about 40 to 1, and a terrace of eight houses will have a length to width ratio of about 40 to 1. Considering the braced frames as lattice beams, they will undergo very high deformation at such high ratios.
- Alternate vertical scaffold tube frames are not braced as structural lattices, and the horizontal frames are hardly braced at all.

The consequence of these weaknesses is that the assembled scaffolding would collapse if it did not rely on the building to give it rigidity.

Practical experience is that collapses by buckling frequently occur due to the absence of ties in very well-braced scaffolds exceeding 20 m in height. Badly built scaffolds may collapse when the height exceeds 10 m.

Figure 6.1 shows some deformed scaffolds. Figure 6.1(a) shows a scaffold attached only once near half height, leaving the top to fall away from the building. Figure 6.1(b) shows a scaffold tied only at the top, leaving the middle to bulge away from the building. Figure 6.1(c) shows a scaffold tied twice, providing a much greater strength because before it can fail it will have to buckle as three short columns rather than one long column. Reference to Part IV, Chapter 31,

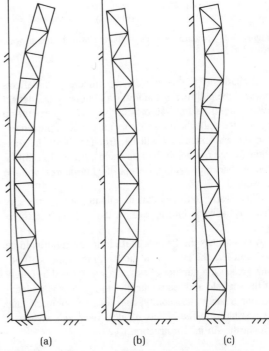

(a)   (b)   (c)

Figure 6.1 The deformation of scaffolds is controlled by both the ties and the butting of transoms on to the building walls.

will indicate how a long single-tube column or strut is much weaker than a short one. The same is true for the whole scaffold, even for the braced frames.

Figure 6.1(a) shows that because the top portion leans away from the building, the lower part deflects towards the building. If the transoms butt up against the wall, this will be prevented, and the amount of the top part that will deform is reduced, making this whole scaffold safer.

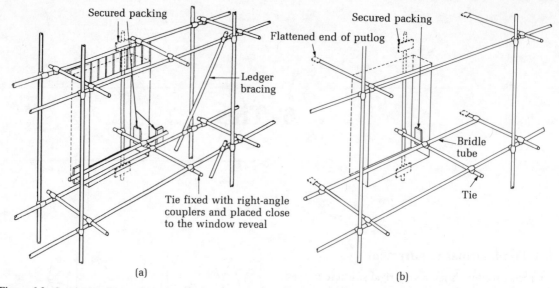

Secured packing

Ledger bracing

Tie fixed with right-angle couplers and placed close to the window reveal

(a)

Secured packing

Flattened end of putlog

Bridle tube

Tie

(b)

Figure 6.2 Standard methods of fixing 'through' ties in window openings: (a) For independent tied scaffolds. (b) For putlog scaffolds.

Similarly in Figure 6.1(c), the middle part deflects towards the building as the lower and upper parts deflect away from it. If the transoms in the middle part butt up against the wall, the inwards curvature will be prevented, and the whole scaffold will be stronger. If all the transoms for the full height butt up against the wall, a failure due to buckling with the opposite curvatures will be prevented.

This clearly indicates that it is as important to ensure that the transoms butt up against the wall as it is to fix ties.

A tie attaches the flexible scaffold to the rigid building and transfers the rigidity of the building to the scaffold. The greater the number of ties, the greater is the rigidity of the scaffold and, consequently, the greater is its load-bearing capacity, its stability, its resistance to impacts, and its resistance to the effects of local overloads and local irreglarities in its construction.

6.2 The traditional tying system

Before any Code of Practice recommendations were made, the traditional tying system was to insert an extended transom tube through every alternate window on every alternate floor. Because rooms were about 12 ft wide and 8 ft 4 in high, the face area of the building containing each window was about 100 ft². This area was known as a 'square' of wall or a 'square' of scaffolding. The tie spacing was thus one tie per four 'squares' of building, equivalent to one tie per 400 ft² or 1 tie per 37 m² of building surface. This was a very safe procedure and it allowed any

tie tube to be taken out while the sashes or casements were glazed. The tie was temporarily inserted in the adjacent window opening and later returned to its original place through an open transom light, sash or casement.

Figure 6.2, taken from the British Standard Code of Practice, shows the method of fixing through ties for an independent tied scaffold and for a putlog scaffold.

6.3 The basic rules for the spacing of ties

In recent years, the industry has pressed for the use of anchors drilled into the surface of the building wall. The ties shown in Figure 6.2 are removable and so must be inserted at a greater frequency than anchor ties, which are non-removable.

A further and most important consideration governing tie spacing is the presence of vertical protection sheeting on the outside face of the scaffold. This sheeting increases the wind forces on the scaffold, so more ties are required.

Four different tie spacings have been codified, with the following basic requirements:

- Non-removable ties on unsheeted scaffolds   40 m²/tie
- Removable or through ties on unsheeted scaffolds   32 m²/tie
- Non-removable ties on sheeted scaffolds   32 m²/tie
- Removable or through ties on sheeted scaffolds   25 m²/tie

These spacings are satisfactory for right-angle couplers of the normal strength.

6.4 Tie spacing in windy areas

Some areas of every country are exposed to high winds, e.g. mountains and marine areas. In these circumstances, calculations should be made. Special attention must also be paid to the accuracy of the site construction to ensure that the attachment of the tie tubes to the scaffold is as strong as the tie fixing. Tests may be necessary if the building wall is old or in bad condition.

6.5 Box ties and lip ties

Figure 6.3 shows the general arrangements of box ties and lip ties. Both must be considered as movable ties and spaced as such. It should be observed that box ties usually have two couplers on the inside end of the tie because there are two tie tubes, whereas lip ties usually have only one coupler inside the building. This difference in strength should be taken into account in cases where the tie frequency has to be calculated.

6.6 Tie patterns

The ground can be assumed to fix the position of the bottom of the scaffold adequately.

The tie pattern must be adapted to suit the architecture of the building. Figure 6.4(a) shows a spacing of three bays horizontally and three lifts vertically, giving a tie area of 36–40 m². It is suitable for non-removable ties. Figure 6.4(b) shows a spacing of four bays horizontally and two lifts vertically, which would give a tie area of 32–35 m² and be suitable for movable ties.

Figure 6.4(c) and (d) are for the same scaffold built against a structure where the cladding cannot be penetrated so frequently. Here the spacing is 8 m or 8.8 m horizontally and 6 m vertically, giving 48–52.8 m² per tie. These areas are larger than the Code recommendations, so the scaffold has been stiffened by two levels of plan bracing fixed on alternate tie levels.

It is recognised that these four diagrams are idealised arrangements and that building façades do not conform to any set model. In general terms, the designer should aim for non-removable ties in a staggered pattern if possible.

6.7 Abnormal façades

A normal façade is a flat-faced building with window openings at about the usual spacings or a plain surface without openings into which anchors can be placed in any pattern required.

Sometimes, the façade of a building does not permit the tying patterns given in Section 6.3. The storey heights may be large, or there may be cladding panels or a glass surface.

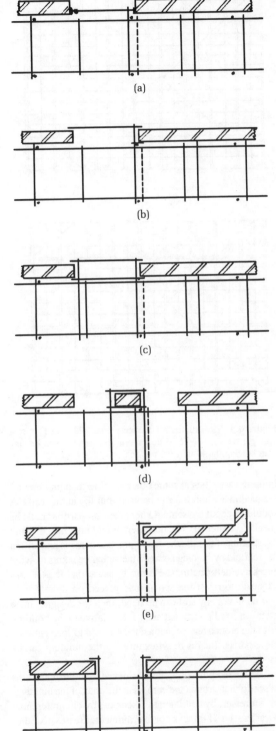

(a)

(b)

(c)

(d)

(e)

(f)

Figure 6.3 Box ties and lip ties.

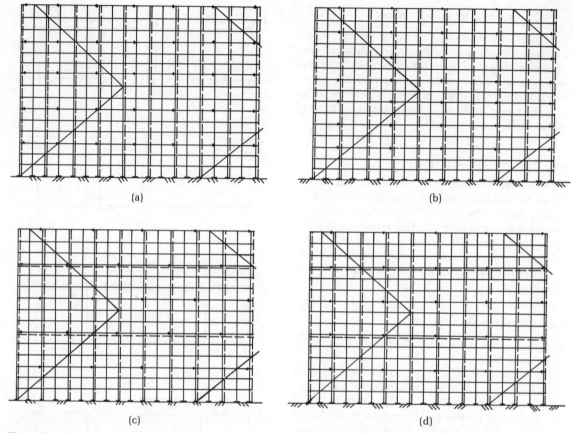

Figure 6.4 Tie patterns for unsheeted scaffolds: (a) Non-removable ties at 36 m^2 area per tie. (b) Removable ties at 36 m^2 area per tie. (c) Staggered tie pattern, assisted by plan bracing, at 48 m^2 area per tie. (d) Rectangular tie pattern, assisted by plan bracing, at 48 m^2 area per tie.

In these cases, ties at some other spacing than that recommended in Section 6.3 can be used, but the number of ties should be about the same. Otherwise, the scaffold must be specially designed. The designer should take into account the spacing of ties that he can in practice achieve and stiffen the scaffold by bracing between the actual tie points to avoid excessive deformations. He must ensure that the ties are adequate to resist the actual load placed on them.

If no ties of any sort can be fixed, then the designer must establish fixed points that are stable relative to the building or to the ground and strengthen the scaffold to span between these points. Beams or wires through the building can be used.

These fixed points might be achieved by utilising returns of scaffolding around the corners at the ends of the building, by buttresses, by drilling into balconies, by circumferential wires, as on chimneys, or by flying buttresses to other buildings.

Having assumed these fixed points in the design, the designer must ensure that he achieves them on the job.

6.8 Sheeted scaffolds

Scaffolds that have vertical protection sheeting, other than those that are so fully sheeted they become complete buildings, are subject to wind forces in excess of those applied to an unsheeted scaffold, but not as great as the pressure on the building. They are open at the ends and the top, and frequently the bottom as well, and have considerable pressure within the sheeting.

The pressure distribution along the surface of the sheeting and at the corners will form a similar pattern to what occurs on a building, i.e. a reduction along the centre of the walls and an increase at the corners.

It has been customary to increase the allowable loads in couplers, when these are due solely to wind, by a factor of 1.25. The allowable slip load in a single tie coupler is therefore $625 \times 1.25 = 781$ kg. Adopting the recommended area of 25 m^2 per tie for sheeted scaffolds with movable ties, the wind pressure that can be resisted is 31.2 kg/m^2. This corresponds to a design wind speed of

22.5 m/s (50 mph). The use of movable single tie couplers is therefore restricted to the southeast of England and to buildings less than 30 m high.

If the single tie coupler is non-removable, it will be expected to support 32 m² of area, i.e. for a pressure of 24.4 kg/m² over the larger area. This corresponds to a design wind speed of about 20 m/s.

From this, it will be seen that in nearly every location other than the southeast of England, each tie should be fastened with two couplers or a calculated bolt fixing. Thus, the tie tube should be coupled to the inside and outside ledgers or to the inside and outside standards. The wall end of the tie tube will require a supplementary coupler.

6.9 The tying value of the ground and of scaffold returns around corners

It is common practice to assume that the ground prevents the base of the scaffold being blown away from the building, and the width of the base gives the lower levels stability. It is normal to assume that this effect deals with the lateral stability of the lowest lift of the scaffold, so calculations of the area of the building to receive the tying system can start at 2 m above ground level.

When the face scaffold is connected at its end to a similar scaffold around the end of the building, this return scaffold can be considered to stabilise the front scaffold for two bays along the front. This may occur at both ends. Calculation of the area of the building to receive the tying system can thus start 4 m in from any end where there is a return scaffold. A centre return scaffold or centre buttress can be considered to provide the equivalent of 4 m of tied scaffold, i.e. 2 m on each side of the buttressed area.

6.10 Rectangular buildings fully surrounded by scaffolding

Rectangular buildings less than 8 m in width and depth can be scaffolded without ties, provided that the four face scaffolds fully butt up against the walls.

An internal scaffold to a rectangular building less than 8 m wide and deep can also be scaffolded without ties, provided that the internal scaffold fully butts up against the inside walls.

6.11 Circular buildings fully surrounded by scaffolding

Circular towers must never be scaffolded either inside or outside without ties because the shape permits a spiral type of failure, even if the transoms fully butt up against the wall surface.

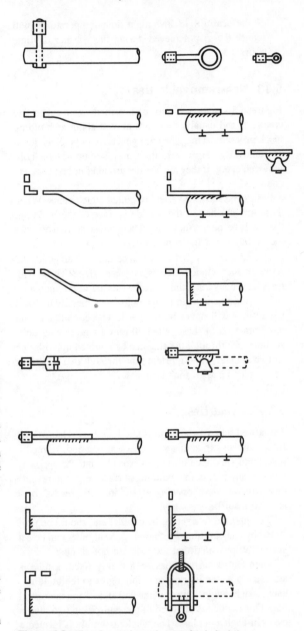

Figure 6.5 Tie tubes and attachments for bolted ties.

6.12 Hazard with sway transoms considered as a tie

• The projection of transoms to nip into window reveals, even in opposing pairs, should not be considered equivalent to a tie. This is good practice as a measure against displacement sideways, particularly on the inside of the scaffold, but it will have no restraining effect against the scaffold falling way from the building. Contraction of the scaffold in low night-time temperatures will reduce the grip achieved during the warmer daytime hours.

6.13 Non-removable ties

Figure 6.3 shows a variety of ties made with tubes and couplers, and usually these are moved about as building work proceeds. Thus, they are generally to be considered as removable. However, they may sometimes be non-removable, e.g. if they are box ties around finished concrete columns.

Figure 6.5 shows a variety of bolted attachments. When these are attached to the outside surface of the building, they will be non-removable. There must be liaison with the architect and the contractor.

The attachments that are cranked can be removed and replaced after window frames have been fixed. Those with transverse bolts for fixing on to reveals are also removable.

The attachments with in-line bolts can generally be fixed where they will never be removed. The commonest and most practical of these is the 50 mm diameter ring bolt. A short tube through the ring can be coupled up with one or two couplers, and a bridle tube through two rings can be coupled up with four couplers.

6.14 Reveal ties

A scaffold tube of the appropriate length is selected and fitted with an expanding screw jack at one end, termed a reveal pin. This assembly is then expanded into the opposing faces of any recess in a building surface, e.g. the reveals of a window, door opening or architectural recess, or a sill and a soffit.

The opposing faces must be parallel and should be protected by pads of thin, dry timber. Reveal tubes can be set vertically, provided that the sills are not sloping.

When the reveal tube has been firmly fixed, a tube is attached to it standing out at right angles to the building, using a right-angle coupler at the end away from the reveal pin. This tie tube is attached to the scaffold in two places by right-angle couplers. The whole assembly is termed a reveal tie and has the strength of the single right-angle coupler joining the tie tube to the reveal tube, i.e. 635 kg.

The screw jack should be inspected and retightened every few weeks.

The reveal tie relies on the friction of its end pads on the window reveals and as such has a variable pull-out strength. Accordingly, it is not advisable to rely only on reveal ties. It is better to have some ties of a more positive nature.

On some buildings, it is not possible to allow open windows or drilled-in anchors, and reveal ties are the only

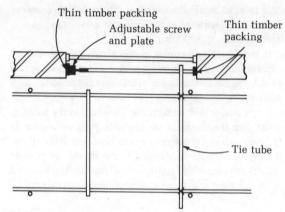

Figure 6.6 Reveal tie.

method possible. In this case, it is advisable to double the number.

Figures 6.3(a) and 6.6 show reveal ties.

6.15 Push-and-pull ties

The word 'tie' implies that the scaffold is tied to the building, but it must be remembered that the scaffold must not only be prevented from falling away from the building but also from buckling towards it. Reference to Figure 6.1 will make this clear. Figure 6.3 shows how box ties act as push-and-pull ties. In the case of lip ties, an extra short butt should be fixed, as in Figure 6.3(b). Alternatively, the transoms can be extended to bear on the building surface, protected by plastic caps, so as to convert the lip tie into a push-and-pull tie.

In the direction parallel to the wall, box ties act both ways. Lip ties can be arranged to do so, as in Figure 6.3(f), or by the use of sway transoms, as in Figure 6.3(e). The timber in both box ties and lip ties can be wedged off the building surface by wooden wedges.

6.16 Anchor ring or bolted ties

The tube anchor ring shown in Figure 6.7(a) should preferably have three features:

(i) The anchor that is set into the wall should be thin so that the hole to be drilled is as small as possible for the bolt diameter required, and so that lateral movement of the shank of the ring bolt is small.

(ii) The shank of the bolt should have a shoulder near to the ring that fits snugly into the hole drilled so that lateral movement is reduced to a minimum.

(iii) The ring should have a minimum internal diameter of about 50 mm so that a scaffold tube can pass through it without too much play in the direction of the shank of the bolt.

When the anchor rings are set into the wall, there should be enough space between the wall and the inside of the ring to enable a scaffold coupler to be assembled around the tube and through the ring. Its nut must also be accessible.

The best arrangement is to insert two rings about 3 m apart on a horizontal line. The plane of each ring is set vertically so that a 4 m bridle tube can be passed through both rings. Two tie tubes are then joined to this bridle and to the inside and outside ledgers of the scaffold, or to the inside and outside standards.

The attachment of the two tie tubes to the bridle tube should be both outside the rings or both between the rings so that lateral movement of the bridle is prevented.

Each of the tie tubes, while having two fittings attaching it to the scaffold, has only one fitting on the bridle, so its safe load is 635 kg. The whole assembly can support 1270 kg.

If special-quality forged couplers are used with a safe load of 1 tonne each, the wind resistance of the double assembly can be assumed to be 2 tonnes after the strength of the anchor fixing has been checked. In normal circumstances, this will be adequate for 33 m² of building surface with a sheeted scaffold.

It is sometimes impracticable to have two anchor rings in the wall. Where only one can be used, a short butt tube can be passed through the ring with a tie tube on each side of it. A butt tube with a single tie tube is an arrangement to be avoided because it offers no resistance to side sway in one direction. It has the same restricted strength as a lip tie, i.e. 625 kg.

The wire-tie anchor ring shown in Figure 6.7(b) can be used for wire or steel-banding ties. With such an arrangement, butting transoms must be used to hold the scaffold out from the wall. However, they offer no lateral stability, so side sway of the scaffold must be prevented by some other means and the longitudinal bracing must be adequate.

A wide choice of expanding anchor bolts is available. These can be fixed in the outer face of the building or in the side face of a column or division wall inside the structure. Figure 6.5 shows a variety of tube ends and tube attachments that can be used with ordinary expanding anchor bolts.

6.17 The outward slope of ties

The reason for the requirement in the British Standard Code for ties to be horizontal or to slope downwards from the building is so that, if the scaffold settles for any reason or collapses, it will be pulled towards the wall during the movement.

If the tie tubes slope upwards away from the building, they will rotate about their fixings and push the scaffold

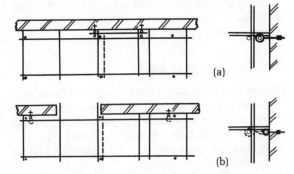

Figure 6.7 Anchor ring ties: (a) To engage scaffold tube. (b) For wire lashing.

away from the wall if any subsidence occurs. The value of the butting transoms will then be lost.

In a collapse where the minimum tie frequency is one tie per 40 m² of surface, the weight of the scaffold will probably exceed the strength of the ties. This area per tie is typical of ties on alternate storeys, i.e. 5.5 m apart and spaced every alternate window, and 8 m horizontally or on every fourth pair of standards. This is equivalent to four bays of scaffolding three lifts high, for which Table 37.6 gives a self-weight of 12×75 kg, or 900 kg. If there is one working lift, this will give an extra weight, as shown in Table 37.7, of 4×450 kg, or 1800 kg, at full load. The total weight of this portion of the scaffold will then be 2700 kg. This can probably be catered for by the likely failure strength of one special 3 tonne coupler, or two standard couplers with a combined minimum failure load of 2×1270 kg, or 2540 kg.

If the tie tubes are originally horizontal or sloping downwards, the transoms will press hard against the building wall and exert considerable frictional resistance to downward collapse, so relieving the weight applied to the tie in the collapse condition.

Tie tubes sloping upwards from the building are of no value against buckling of the standards due to local overloading, while tie tubes sloping downwards might prevent progressive collapse.

There are historical examples of both the inadequacy of ties sloping upwards and of the ability of horizontal or downward-sloping ties to prevent total collapse of a scaffold to the ground.

6.18 Oscillations

Scaffolds are subjected to oscillations from normal use and from wind forces. These small movements have a dangerous effect on the efficiency of the ties. Wedges may work loose and lip ties may flex slightly. The packing of reveal ties is particularly badly affected by oscillations.

Special attention must be paid to making all types of tie close-fitting and firm during installation. Subsequent inspections are essential.

6.19 Inspection of ties

Because the tying system is a critical feature in the stability and safety of a scaffold, the tie elements must be inspected very closely when the scaffold is inspected. The tightening torque of the tie fittings should be checked regularly, particularly the tightness of the reveal ties which bear on to wooden packs that may shrink with time.

A tie count should be made and a tie chart drawn during the handover. This should be referred to at each scaffold inspection.

At the end of the construction period of a building, it is important that the last trades do not remove ties. Glaziers and painters should be supervised most strictly.

When the scaffold is being taken down, its ties should be removed only under a controlled programme to avoid having a large expanse of untied scaffolding during dismantling.

7 Working platforms

7.1 General

The provision of the working platform on an access scaffold is the object of the whole structure but, regrettably, it too frequently receives inadequate consideration.

The features requiring attention are the width, the arrangement of boards, the length, proximity to the work, the slope, the guard rails, the toe boards, the means of access, the access holes, correct support to secure the boards against dislodgement, gaps and imperfections, and the final surface.

7.2 The width of the platform

Because the boards are usually 228 mm wide, the width of the whole platform is conveniently specified as a whole number of boards.

A single board is frequently seen as a catwalk across a trench or a barrow run from ground level up to the ground floor slab. A person cannot fall more than 2 m from such a platform, so there is no regulation to prevent its use, but it is not desirable and can be dangerous, especially if the single board is not firmly bedded.

A single board between two trestles less than 2 m high is frequently used to gain access to a ceiling. This is undesirable and dangerous. If a workman sits on a single board to attend to a wall surface, it is not so dangerous at the place of work, but he has to 'walk the plank' to get there, and this is undesirable.

A scaffolder can frequently be seen erecting a scaffold from a single board taken up lift by lift. This is undesirable and dangerous and is also frequently seen to span too great a distance.

Two boards (456 mm) should be used as a gangway, as a platform for light work without materials or for inspection.

Three boards (684 mm) can be used for work where tools are involved, but not for depositing materials.

Four boards (912 mm) are required when materials have to be put down on the platform, utilising 456 mm for the materials and 456 mm as a free walking width.

Five boards (1140 mm) are needed if materials occupy 456 mm and the remaining 684 mm is needed for a barrow run.

Six boards (1368 mm) are required if a trestle or higher platform is to be erected on the platform.

Widths required by the Regulations

The widths given above are practical widths, being multiples of the width of one board. The Regulations give some tolerance on these values.

7.3 The width of the scaffold

The scaffold width depends on whether there is an inside board or not, and on whether the toe board is placed on top of the platform or at the side.

Figure 7.1 shows four different widths of scaffold for a five-board platform.

7.4 Inside boards

For several reasons, it is sometimes desirable or necessary to place the inner line of standards well away from the building surface. For instance, they may have to avoid overhanging eaves and gutters, or the ledger may have to be placed outside projecting mullions. The gap between the boards and the building surface may then be 300–400 mm. In such cases, it is customary to extend the transoms inwards

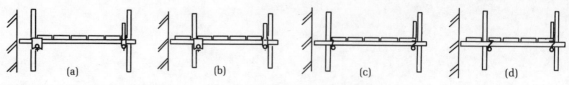

Figure 7.1   The widths and positioning of scaffolds for various arrangements of boards.

and to place thereon a board between the inside line of uprights and the building surface.

In brickwork and pointing jobs, it is sometimes desirable to have the working platform nearly touching the building surface to prevent droppings of mortar. An inside board enables this to be achieved.

Two inside boards can be used on extended transoms but, for three or more, consideration of the strength of the cantilevered transoms is necessary.

When an inside board is used, a gap of 50 mm occurs in the platform where the standards pass through it. In most cases this causes no hazard, but if there is a danger of debris falling through, e.g. old dismantled roof tiles, the gap should be covered with a plywood or hardboard strip.

7.5 Hop-up brackets

In tube-and-coupler scaffolds and in prefabricated frames and systems, a variety of hop-up brackets are available to support boards between the inside line of standards and the building.

Usually these brackets fix to the standards, so there is a problem obtaining two supports 300 mm apart for the ends of the boards where they butt end to end. For this circumstance, two puncheons can be inserted in the scaffold at an appropriate place and joined to the ledgers at and below the level where the inside platform is needed. The hop-up brackets are fixed to these puncheons.

7.6 Hazard with inside boards

• Where the façade of a building is not flat but includes bay windows or recesses, inside boards are a convenient means of placing a platform near to the wall in the recess. Special-length boards are often needed. There is a great danger of improperly supported and trap-ended boards in this circumstance.

7.7 The length of the platform

The length of the platform has to match the length of the job, but it should extend half a metre or so beyond each end to facilitate the work at the end of the job and to start off the work around the corner of the building if this has to be done subsequently.

In any event, the framework supporting the working platform can with advantage project beyond the ends of the building façade so that the transoms at the end of the scaffolding can be projected down the sides of the building to act as sway transoms.

7.8 The space between the working platform and the work

Reference has already been made to the use of inside boards. In general-purpose brickwork scaffolds, if the boards lie outside the fall pipes, i.e. about 150 mm from the surface, most trades can work satisfactorily. It is best if the scaffolding contractor and the main contractor agree on this matter before erection commences.

The Regulations require that if a person is liable to fall more than 2 m, guard rails and toe boards are required. This applies to the inside of a scaffold as well as to the outside if there is a gap through which a person can fall more than 2 m.

7.9 The slope of platforms

Platforms should preferably be horizontal, but it is not against the Regulations to have them sloping, except as detailed below. A cross fall should be avoided. A longitudinal slope, if required functionally, should be agreed between the scaffolding contractor and the main contractor before the job begins.

In the case of ramps steeper than 1 in 4, the Regulations require the fixing of stepping laths, with a 100 mm gap to enable barrow wheels to pass. The Regulations require that no platform steeper than 1 in 1½ be used.

Difficulties occur in platforms between roof trusses when these are not on the same level. A steeply sloping platform may tend to slip downwards across its supports.

Vehicle ramps such as those to enable dumper trucks and wagons to pass down into building excavations must be on slopes specifically agreed with the excavation contractor.

Pedestrian ramps for use by the public should not be set at a gradient steeper than 1 in 6. A ramp for building workers to pass from one floor level to another outside the building but within a scaffold framework can be at 1 in 4.

7.10 Guard rails and toe boards

Protection against materials falling from the platform is provided mainly by the toe boards and against persons falling mainly by the guard rails. If loose materials are being stacked higher than the toe boards and are likely to fall over the edge, then some additional safeguard such as a brick guard is desirable.

These protective methods are necessary at the ends of a platform as well as on the outside. They are also required when there are holes in the platform through which materials or persons can fall.

If part of a scaffold is boarded but is not in a condition for use, a notice must be posted on the usable part of the platform to warn users not to go beyond the safe limit. Such a notice should also be posted to prevent persons going on to a scaffold from the building if the scaffold is unfit for use.

The Regulations are very specific about the heights of toe boards and guard rails, and about the space between them. They require:

(i)   Guard rails and toe boards must be fixed along every side of a working platform, including the edges of holes in the platform from which a person can fall more than 2 m.

(ii) The height of the guard rails must be between 914 mm and 1143 mm above the platform and above any raised standing place on the platform.

(iii) The height of the toe boards must be not less than 150 mm.

(iv) The distance between a toe board and the lowest guard rail above it must not exceed 762 mm.

(v)   The height of the inside guard rail of a suspended scaffold need not be more than 686 mm.

(vi) In suspended scaffolds, there need not be an inside guard rail or toe board if the workers sit at the edge of the platform and a rope or chain handhold is provided.

(vii) Guard rails and toe boards are not necessary on ladder scaffolds, trestle scaffolds or platforms under a roof, provided that a firm handhold is available over the full length of the platform.

The Regulations permit guard rails, toe boards and barriers to be removed or remain unerected for the time and to the extent necessary for the access of persons or the movement of materials or other purposes of the work, provided that they are replaced or erected as soon as is practicable.

The codes of practice of various countries differ in their requirements for the edge protection of platforms. Some codes require intermediate guard rails spaced at 300 mm or 500 mm intervals.

7.11 Lapping guard rails and toe boards

Guard rails can be lapped one above the other. Both tubes should be coupled to the platform side of the uprights.

Toe boards can be lapped so that there are two thicknesses. Both toe boards should be on the platform side of the uprights and both should be firmly clipped in place.

7.12 Hazards with guard rails and toe boards

• The most common hazard with guard rails and toe boards is that they are not fixed at the ends of the scaffold.

• Guard rails and toe boards may have to be omitted where there is an access to the platform, but the gap should not be wide enough to allow a person to fall from the platform at a distance from the access point. Chapter 8 gives details on ladders.

• Where a ladder passes through a hole within the area of a platform, guard rails and toe boards must be placed to prevent a fall by persons or materials.

7.13 The correct method of supporting the boards

The standard scaffold board is 3.96 m long and needs supporting 150 mm from each end. This leaves a length of 3.66 m between the end supports, which is clearly an unsafe span. A single transom or putlog in the middle results in two spans of 1.83 m, but these are too long to be safe for boards 38 mm thick. Two intermediate supports are needed between the end supports, which results in three spans of 1.22 m. This is the standard method of supporting a normal length of 38 mm scaffold board.

The Regulations deal with boards of other thicknesses and require a distance between supports not greater than:

991 mm for boards 31.75 mm thick
1524 mm for boards 38.1 mm thick
2591 mm for boards 50.8 mm thick

To this list can be added three more figures:

3500 mm for boards 63.5 mm thick
4500 mm for boards 76.2 mm thick
3000 mm for two boards (on top of each other)
   38.1 mm thick

The Regulations require that a board does not project more than four times its thickness beyond its end support unless it is secured to prevent tipping.

Every board must rest on three supports if it is susceptible to sagging, and it must rest evenly on its supports.

7.14 Short boards and trap-ended boards

At the end of a platform, it is frequently necessary to use a board shorter than the standard length of 3.96 m. Short boards are also necessary where there are ladder holes in platforms.

The weight of a 3.96 m board is sufficient to prevent it tipping up if its end, when projecting 150 mm past its end support, is trodden on. A board 2.5 m long is the shortest that will not tip up. Therefore, any board shorter than 2.5 m must be prevented from tipping up if its ends can be trodden on. In a tower, the ends of the boards cannot be trodden on and no precautions need be taken. In a scaffold, if the short board is at the end, the end toe board will hold it down.

If the short board is in the middle of a run or adjacent to a ladder hole, it will need tying down. The neatest way is to tie it to the transoms at each end using nylon cord. Cross battens of timber 48 mm thick can be used as transoms under the ends of short boards and the boards nailed to these.

A standard British scaffold board 3.96 m long (13 ft), 228 mm (9 in) wide and 38 mm (1.5 in) thick weighs 23.6 kg (52 lb). A person stepping on to the end of a board will apply 76 kg (168 lb) to the overhanging end.

Working out the tipping moment and the righting moment will show that the correctly supported board in Figure 7.2(a) has a factor of safety of 3.8 against tipping up.

The shortest board that will just balance a person's weight works out at 2.11 m (7 ft), with no factor of safety. This is shown in Figure 7.2(b).

The largest overhang at the end of a standard board that will just be balanced by the remainder of the length works out at 47 cm (18.5 in). This is shown in Figure 7.2(c).

7.15 Lapped boards

The Regulations require that boards shall not be lapped, i.e. placed on top of each other. If such an arrangement is required, e.g. at the corner of a scaffold, a bevelled piece must be inserted or some other suitable means taken to remove the step.

An exception is made for platforms serving curved surfaces in engineering construction.

7.16 Gaps and imperfections in the final surface

Scaffold boards are raw sawn along their edges and after some time may warp as viewed in plan, making the edge curve by up to 20 mm in a length of 3.96 m. Two such curvatures opposing each other will give a 40 mm gap between two boards. This can be reduced by rearranging

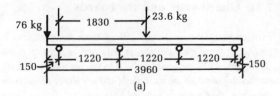

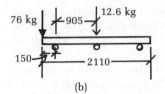

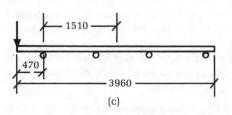

Figure 7.2   The correct method of supporting a 6.96 m long board (a). The hazards of the short board (b) and the trap-ended board (c).

the boards, but a gap of 25–30 mm at various places is unavoidable.

The 50 mm gap in the platform when a standard rises through it has already been mentioned and this can be tolerated, except when there is an observable danger.

At the ends of each fleet of boards, the tolerances in the length of the boards may give a gap of 50 mm, which cannot always be reduced by rearranging the boards.

Boards that are twisted along their length when viewed at the end inevitably cause trips in the levelness of the platform. A twist of 25 mm is not a hazard in general building operations, but is unsatisfactory for visitors such as TV cameramen, actors and civic dignatories.

Where these hazards occur, they should be made good by any suitable means. A plywood sheet is probably the most practical.

7.17 Brick guards

Fencing along the edge of a platform or a roof should be designed in accordance with local codes of practice or according to the duty required.

Brick guards should have mesh no longer than 50 by 50 mm.

Slate and tile guards for roof edge protection must have a continuous toe board along the bottom.

8 Ladders

8.1 Types considered

The ladders referred to in this chapter are those for general access by building tradesmen, namely timber pole ladders and timber and aluminium extension ladders. Roof ladders, crawling boards, stepladders and trestles are not included.

There is no doubt that pole ladders are the most serviceable on building sites. They are firm to use and have the same characteristics from site to site. Unfortunately, they are heavy, weighing 10–12 kg per metre length and are, in consequence, difficult to raise.

Single lengths of extending ladders are lighter but more flexible than pole ladders. However, the user does not feel so secure on them. Aluminium extending ladders are very light and do not give the same sense of stability as wooden types.

8.2 Hazards with single lengths of extending ladders

- Timber sections may be rigged with the tensioning wires on top, which is the weaker alternative.
- The running guides, particularly of the lower section, stick out at the top and may catch workmen's clothes.
- The tensioning wires on the underside of the sections may stand proud of the surface of the timber. This results in the section having very little sideways friction when used against scaffolding or steel roof trusses. This also applies to multi-section and aluminium ladders.

8.3 Rigging ladders for construction work

Ladders are included in the definition of the term 'scaffold' in the Construction (Working Places) Regulations. Their use either as a place of work or means of access in building operations or works of engineering construction is governed by these Regulations.

The Regulations require that the ladder has a firm footing.

This is essential even if the top is securely tied. Even if the top cannot fall sideways, it can twist if its base is not firm, and this may be dangerous. The Regulations forbid the use of loose bricks or other loose packing. The intention of this is clear, but it is unfortunately worded because it rules out the use of timber pads under one stile when the ladder has to be set up on a sloping surface.

Despite the wording of the Regulations, on a sloping pavement it is preferable to use a wooden wedge under one leg to prevent the ladder twisting. This wedge should be about 100 mm wide and 200–250 mm long, and should have an angle of 7–15°, preferably not greater than 10° and certainly not greater than 15°. The ladder should be bounced on its top resting place as it is wedged to ensure that its base is correctly placed to give it a firm contact at the top (see Figure 8.1).

The Regulations require that the top end of a ladder be securely fixed. It should be fixed against slipping sideways and against being blown off its support by the wind. It is essential that its top fixing is secure enough to prevent the ladder falling to the ground if its bottom end support fails or is knocked away accidentally.

The top fixing should be by fibre rope or wire lashing around the stiles to a firm place on the structure. Proprietary clips and lashings are available for this. Care must be taken to ensure that the rope or lashing does not cross behind or in front of the ladder and itself cause a trap. The uppermost rung on which a person will stop should be on a level with or slightly higher than the landing place so that there is no obstruction to the feet being placed on this rung.

Where there is no suitable place on the structure or building for the attachment of rope ties, the ladder must be prevented from falling by other means. It is preferable to fix two guys, one leading down each side of the ladder to the nearest firm fixing or ground anchor available. This is a suitable arrangement on a building site, but is not practical in the High Street.

The Regulations state that where a top fixing is imprac-

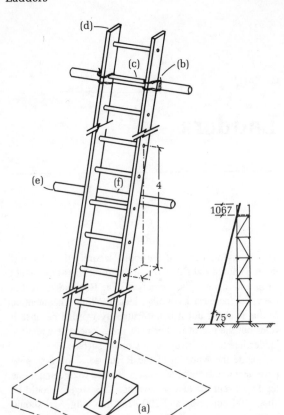

Figure 8.1 Ladder rigging.

ticable, the ladder must be fixed securely at its lower end. This enables it to be tied into the structure to prevent its base slipping outwards and to be guyed sideways below the top. This again is practical on a building site, where the object of securing the ladder near the bottom can be achieved, but it is not practical in the High Street.

The Regulations also state that where a ladder cannot be fixed at all to the structure, a person must be stationed at the foot of the ladder to prevent it slipping. This alternative deals with the High Street job and any other job where a ladder has to be moved so frequently that proper tying off is not reasonably practical. 'Footing the ladder', as it is called, has the further advantage that it protects the bottom of the ladder from accidental impacts by persons or vehicles.

The Regulations require that, where necessary, the ladder be prevented from undue swaying or sagging and be equally supported on each stile. The intention of this requirement is that the middle of the ladder cannot sway laterally or bounce up and down. It is preferable to ensure that the ladder is strong enough to reach from its bottom to its top without undue sway or sag rather than requiring a suppport at its middle.

The good features shown in Figure 8.1 are:

- The use of a wedge on sloping ground.
- The use of two separate soft rope or wire rope lashings at the top to avoid having a length of rope crossing the ladder and impeding the foothold.
- The scaffold tube behind the top of the ladder does not impede the foothold.

The hazards shown in this figure are:

- There is no projection of the ladder at the top to provide a handhold for the top landing.
- The ladder is shown resting on a tube behind it at about half height, which is discussed in the following section.
- The tube at half height obstructs the foothold.

8.4 Hazard in the case of a ladder supported at mid-span

- There is a danger in fixing a mid-support to a ladder to reduce sagging and bounce. Lashing or guying the ladder at its mid-point to prevent lateral displacement is satisfactory but, if the sag is taken out, the contact force on the wall at the top is reduced, which reduces sideways friction.

A ladder relies on its friction on the ground to prevent it slipping out at the bottom, and on its friction on the wall to prevent it slipping sideways at the top. These friction forces are proportional to the contact forces. Figure 8.2 shows how these contact forces are generated. As a person climbs a ladder, the force on the wall at the top increases and, consequently, the sideways resistance by friction also increases.

A ladder sags according to its rigidity, and the forces at the top and bottom vary according to the position of the workman on the ladder.

If the sag is taken out of the ladder by a support near its middle, the force between the ladder and the wall will be seriously reduced or even removed. Resistance to side-slipping will be seriously reduced or even vanish. Thus, if the Regulations requiring the sag to be reduced are carried out with a ladder that is only 'footed', a hazard results. A typical case is when a ladder is inclined and touches a boundary wall or the side of an outhouse at about one-third of its full height.

Other dangerous bottom conditions to be carefully considered are ladders based on footpaths with a steep slope, on cobbles, on terraced gardens and on oily factory floors. Dangerous top conditions result from wet painted surfaces, glass, steel scaffold tubes, structural steelwork and guttering.

8.5 Access at the top of the ladder

Getting off a ladder or getting on to it at the top are the

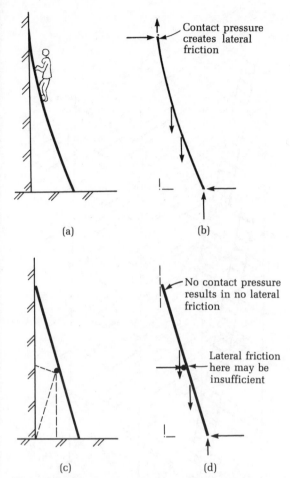

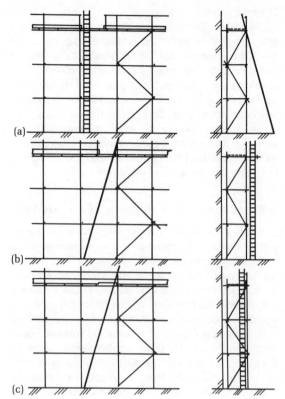

Figure 8.3   Three arrangements of ladders for access scaffolds.

Figure 8.2   The forces on a ladder in the normal position [(a) and (b)] and propped near the centre [(c) and (d)].

most hazardous parts of the journey. This is only partly dealt with by the Regulations, where the requirement is that the ladder should extend at least 1067 mm above the landing place or the highest rung to be reached by the feet. Alternatively, an adequate handhold is required. The Regulations also require sufficient space at each rung to provide adequate footholds.

The location and nature of the guard rails on a working platform adjacent to a ladder are not dealt with in the Regulations. There are two circumstances to be considered. One is when the ladder is set at right angles to the edge of the landing it serves, and the other is when the ladder is set parallel to the edge of the landing (see Section 7.12 and Figure 8.3).

8.6 Ladders at right angles to the edge of the landing platform

The preferred arrangement is that the ladder rests on the outside edge of the landing platform so that its 1067 mm extension above this level is over the platform. In this way, the guard rails and toe boards of the platform can be fixed right up to one side of the ladder and the access gap left at the other side. This gap should be at least 600 mm and not greater than 750 mm. The guard rail on the side of the ladder where the gap is should terminate with a guard rail post so that it is firm to hold. An advantage of having the top extension of the ladder over the platform area is that it alerts a person on the platform to the fact that there will be a gap in the guard rails and toe boards.

A ladder fixed to a guard rail instead of to the outside edge of the platform is not as satisfactory. A slope of 4 in 1, which is typical for a ladder, results in the top rung being 230 mm from the edge of the landing place, and this gap has to be stepped over by the user.

8.7 Ladders parallel to the edge of the landing platform

For a ladder set parallel to the edge of the landing, the centre line of the ladder should be 300−400 mm from the edge of the landing to permit the shoulder of the user to pass clear of the landing. The gap to step across is then about

200 mm, which is satisfactory.

The guard rail behind the ladder should terminate in a post immediately behind the ladder, leaving a space of 50–100 mm for the hand. At the front of the ladder, the gap to the end of the guard rail should be 600–750 mm from the place where the ladder passes the landing edge, i.e. the total gap in the guard rail should be 900–1100 mm.

A ladder arranged parallel to the landing edge should rest on an extended transom and should be prevented from sliding off it by a scaffold fitting or a fitting and a butt tube on the outside. The ladder should be tied on to the transom or to both the transom and the ledger.

8.8 Ladders within the area of a scaffold

In some circumstances, a ladder may be fixed within the area of a scaffold. Problems arise with the necessary hole in the landing platform or in several holes if the scaffold has more than one working platform. Difficulties also arise in fixing suitable guard rails to the holes in the platforms.

Reference must be made to Section 7.1 for fixing the short boards that frequently occur at ladder holes in the platform.

It is generally convenient to make the hole two boards wide, i.e. 456 mm, but this is inconveniently narrow for a big man. A better width is 684 mm, but the extra width of the hole reduces the width of the rest of the platform.

Figure 8.3 shows these various circumstances.

8.9 Ladder towers

For long jobs or where the full width of the working platform is required, a ladder tower can be built outside the scaffold. Figure 8.4 shows such a tower. Special board lengths are needed at every lift and the securing of short boards must receive special attention.

8.10 The inclination of ladders

There are no regulations concerning the angle at which a ladder must be fixed, but there is the practical rule that the slope should be as near to 75°, or 4 in 1, as possible. This is a comfortable slope. It enables a person to climb the ladder holding on to one stile with one hand while carrying something in the other hand.

Vertical ladders are to be avoided wherever possible, for the simple reason that they require both hands and nothing can be carried safely.

In mine shafts, excavations, docks, ship repairs, skeleton structural steelwork, dams, grain bins and many other large structures, vertical ladders are unavoidable. In these circumstances, the ladders are often wet and dirty. Special attention is needed in these cases to ensure that there are no extra hazards in passing through holes in working platforms and in access to and from the ladder at various levels.

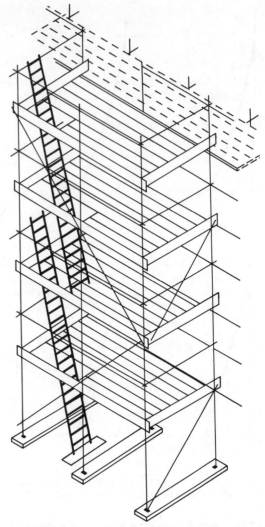

Figure 8.4 A ladder tower coupled to a scaffold.

Guard rings on the front of a vertical ladder are advisable above 2 m from the ground. In a permanent access this can easily be achieved, but in a temporary situation it may be difficult, such as in the case of a vertical ladder fixed to the outside of a static or mobile tower. In a mine shaft or excavation, where the vertical ladder will be in place for many months, a scaffold surround can be built.

The clearance inside the guard rings should preferably be 600–650 mm.

8.11 Suspended ladders

In some circumstances, ladders have to be slung below the main access point and suspended clear of the ground. The Regulations require that slung ladders be securely and evenly suspended, and prevented from swinging.

This can be achieved by attaching the stiles to a firm point by wire lashings. Two separate lashings on each stile at two different rung levels are preferable.

The Regulations say nothing about the hazard of climbing down off and falling from the bottom of the ladder. If there is no independent deck at the bottom, a platform and guard rail surround should be built on to the ladder at its lower end to prevent a person walking off the end and to ensure that he descends into a cage.

Rope ladders, with or without wooden rungs, should be avoided on building and construction sites. They are sometimes unavoidable, but it must be remembered that it is impossible to climb a rope ladder unless the bottom is tied off to maintain it in tension. In any event, it is not safe to carry anything up in one hand when using either type of rope ladder.

8.12 Leaning out to work outside the side of a ladder

When giving instructions to work people on the width of work they can attempt from a ladder, regard must be paid to its narrow base.

Consider an aluminium ladder 6 m long for wall painting. A movement of the shoulders of a person on the ladder of only 300 mm to one side will bring his centre of gravity to the edge of the base width. If he moves his feet to one side of the rungs, the situation will be worse. He can reach out with one hand to paint at a distance of only 750 mm, at which point he is at risk because the ladder will tend to slide sideways.

The tendency to lean out from the side of a ladder is greater when one is not at the top because the work is further away. Sideways friction between the ladder and the wall decreases as the worker descends the ladder, so the most dangerous position is not when he is at the top but when he is working at the half height and about 25 m up the ladder. This is just that situation when the person 'footing' the ladder may walk away, thinking the workman is safe.

8.13 The use of ladders by scaffolders

In current practice, ladders are not provided for scaffolders to use as construction aids and scaffolders do not ask for them, except in cases where the bottom of the scaffold is not on the ground. The scaffolder can climb up and down the braced frames in the structure quite safely to gain access to and leave the rapidly changing place of his work. If there are any difficulties with this, tubes and couplers are available on the job to make a safe climbing frame.

The place at which the scaffolder works not only gets higher or lower as the work proceeds, but also moves horizontally at a rate of 5–6 m every five minutes. A ladder would never be in the right place. It would have to be re-rigged every few minutes and, in many cases, attached to scaffold members that had not yet been made firm by the extension of the scaffold beyond that point.

In the case of a High Street scaffold, the ladder may have to be rigged at right angles to the scaffold, and this will result in interference to and by the traffic. It cannot be rigged inside the scaffold because it is not practical to re-rig it continually inside the framework. It cannot be rigged parallel to the scaffold because each such occasion requires special extended transoms. These would have to be taken out at a later date from below the scaffolders' erection boards. All this re-rigging exposes the scaffolders to risk many more times than is desirable.

A ladder cannot be used for carrying materials to the place of work because couplers are too numerous to be handled a few at a time and too heavy to be carried when bagged. Tubes are too heavy to be carried up the side of a ladder one-handed, and in any event the scaffolder would be faced at the top with a no-handed stance to swing the tubes into the framework, and he could not adjust his own centre of gravity to deal with any movement of the tubes.

If a ladder were to be rigged at one point in a job and not moved along with the progress of the job, the scaffolder would have to climb along the horizontal members at the top of the unstabilised scaffold. This could be more hazardous than climbing up and down the scaffold. The scaffolder would have to rig a run of boards at a suitable level. This would involve the fitting of transoms which, with the boards, would have to be moved up or down as the work was extended. The transoms would have to be recovered without the benefit of the run of boards.

A pole ladder is about three times heavier than a scaffold tube of equal length, so it is a two-man job to rig or re-rig a 6 m ladder.

The ladder would be resting on a steel tube at the top of the partly built scaffold and be subject to slight movements. It would always require tying off at the top and, in practice, this would never be properly done.

8.14 Further hazards with ladders at the end of a platform

- The ladder may be resting on the projecting ends of the scaffold boards forming the platform. The lateral force from the top of the ladder may push the boards along the scaffold until they become unseated from their end transoms.
- When the access ladder is at the end of a boarded platform, a person stepping on to the platform from the ladder may step on to a projecting end of a board and dislodge it, especially if it is short.

9 Prefabricated beams

9.1 Background and economics

The first prefabricated elements in tubular scaffolding were roof trusses of various sizes and shapes. The necessity for these was that when trusses were made of tubes and couplers, the ends of the tubes and the bolts on the couplers spoiled the top surface of the truss for receiving the sheeting. It was soon realised that these trusses could be used as beams either the correct way up or upside down. The next stage in development was to cut off the apex of the roof truss, leaving a very useful trapezoidal beam which could be used either way up. When used upside down, it became the normal fish-bellied beam still used to the present day, but losing favour to the ladder beam and the unit beam.

By the 1970s, the use of prefabricated beams had increased considerably. They are now used in circumstances that previously would have been dealt with by the scaffolding gang using ordinary tubes and couplers. Figure 9.1(c), (d) and (e) show examples. The traditional method of making an opening wider than 6.4 m is to bracket out from each side, stabilising the projecting ledgers by the V bracing or temporary rakers. The scaffold is completed in the normal manner and additional bracing or cross bracing is inserted to stabilise the spanning ledgers.

The cost in man hours, depot hours, maintenance hours and capital lock-up is much greater in cases (c) and (e) than in (d).

In fans, canopies and temporary roofs and buildings, the prefabricated unit is more advantageous.

9.2 Advantages of prefabrication

An expert scaffolder can build a lattice beam in tubes and couplers that is the equivalent of any prefabricated beam. He will cantilever out from each side of the gap to be bridged, supporting the cantilevers on rakers and continuing to do this until the gap can be spanned by a 6 m tube. The rakers form knee braces to the beam. If it is undesirable to have knee braces across the corners at the ends of the span, he will cantilever out higher up to form the top chord first. The rakers will then form part of the diagonal bracing of the finished beam. He may add V rakers over the beam to assist it.

Tube-and-coupler beams will be deeper than prefabricated ones because couplers are not as strong as welds. They will have tube ends projecting, and these may have supplementary couplers which may spoil the structure for other users.

Prefabrication has the advantage up to 10 m that beams can be transported and raised into position in a short time. They can be quickly assembled in families according to the loading requirements, and the top and bottom surfaces are clear of projecting tubes and couplers. Beams can be spliced to increase the span, and they can be supplemented with tubes and couplers more easily when the gap has been spanned.

9.3 Behaviour of prefabricated beams

There are so many failures of scaffolds incorporating prefabricated beams that special training must be given to scaffolders required to fix them. These problems have not usually been due to deficiencies in the beams, but in the rigging of them. A thorough understanding of the behaviour of beams is essential for the design engineer and the scaffolder.

When a beam fails, there is a likelihood of a complete or partial collapse, with serious consequences, because a beam carries a considerable load and spans a space intended to be kept free.

The fish-bellied shape has an advantage over the rectangular shape. Consider a rectangular beam 6 m long and 600 mm deep, as in Figure 9.1(a). It will weigh approximately 80 kg and require two men to handle it. When it

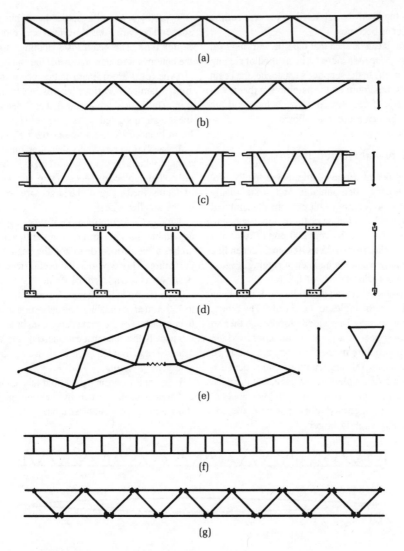

Figure 9.1 Types of prefabricated beam.

is placed across the gap it has to span, resting on the ledgers of two scaffolds or on two brick or concrete piers or parapets, it will be resting on the ends of its bottom chord. It will have to be held upright by one man at each end while it is laced into position with tubes and couplers. The scaffolders cannot leave their stations at each end.

Now consider a fish-bellied truss that has the bottom chord turned up towards the top chord, as in Figure 9.1(b). This beam can be rested on its supports, then left without any further support. It will swing into its final vertical position, and the scaffolders will have their hands free to lace it into position with tubes and couplers. The enhanced

safety of this is obvious.

The same advantage occurs with longer spliced beams, but these will probably be lifted by crane or tackle and maintained in position until coupled up.

When used as a simple beam between two supports, the top chord of the beam is a compression member and the bottom chord is in tension. The top chord behaves like any other scaffold strut. At its full length it is weak, but when stiffened at intermediate intervals it gains strength. Reference to Part IV will show the extent of this gain.

Thus, a beam without lateral restraint will be weak. It will roll over sideways and cease to act as a beam. The more frequently the beam is stabilised to prevent rolling over, the stronger it becomes.

From the above, it can be seen that the safe working load of a beam assembly depends more on its method of rigging than on the beam itself. In the supplier's catalogue, the beam will be rated at its maximum on the assumption that it will be rigged correctly. Herein lies an obvious hazard if the rules for use of the beam are not followed.

9.4 Types of prefabricated beam

Many proprietary beams are available, some totally prefabricated, some partly demountable and some totally demountable into short lengths of tubes. When assembled, the final product is either a lattice beam or a ladder beam. Beams are available with depths from 200 mm to 2000 mm and in supporting ability from 0.25 to 10 tonnes, depending on the depth, the length and the manner of rigging. Figure 9.1(a) to (g) shows various types of beam.

Type (a) is an ordinary zigzag lattice beam. It may have cross battens in it at right angles to the chords. The braces in between the chords may be welded closely together so that their axis lines meet at a point on the chord axis in the classic tubular structure method, or they may be fixed so that there is a clear 100 mm length of chord between them, enabling scaffold couplers to be attached at the node point. This latter type is more useful in scaffolding applications, particularly in temporary roofs, but its deflection under load is usually slightly larger.

Type (b) is a similar lattice beam but is fish-bellied in shape. The variations described above may also be made in this type. The beam loses nothing in strength because the bottom chord extremities and the end battens are left out.

Type (c) is an ordinary lattice beam, but it is made up from a family of various lengths, e.g. 1 m, 2 m, 3 m and 4 m. This has the advantage that each section is lighter than the longer unjointed types. The deflection under load may be influenced by the slope of the bolt holes.

Type (d) is a completely demountable type consisting of a chord with node plates welded on to it. Battens and braces are bolted in place as necessary. A beam may have a series of depths, the battens of the larger size becoming the braces of the smaller size.

Type (e) is a combination of types (b) and (c) in which adjustable links are inserted in the lower chord to enable arches to be formed or to hog the beam up to the desired curvature. This type may be prefabricated either as a flat beam or as a triple-chord frame.

Type (f) is a ladder beam. Various depths from 300 mm to 750 mm are available. The deflection under load is higher than in beams of the same depth with braces, but in practice it is usually less than that calculated by the Virendeel theory. The strength is also greater than calculated and is best assessed by testing.

Type (g) uses ordinary scaffold tube chords, with battens and braces prefabricated by welding half couplers on to short lengths of structural tube.

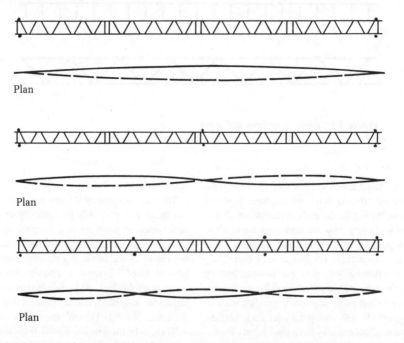

Plan

Plan

Plan

Figure 9.2 Lateral deformation of the top chord of a beam under vertical load for various restraints.

9.5 Rigging prefabricated beams into a scaffold

The simplest way of rigging a beam between two parts of a scaffold structure is to fix it at its ends with scaffold couplers. If the couplers are attached to only one chord, the beam will roll over when loaded. If they are fixed to both the top and bottom chords, each end of the beam will be held vertically, but the centre will still tend to roll over. The top chord is a compression member, so it will tend to buckle sideways.

Figure 9.2 shows the buckling characteristics of the top chord of a beam under load when prevented from rolling over at various intervals along its length.

The effective length of each part of the top chord is reduced as more couplers are attached. As a consequence, the chord is strengthened. For details, consult Tables 31.3 to 31.10.

From this, it will be seen that there is no single safe working load for a beam. The strength increases as the rigging becomes more thorough. Tables 34.1 to 34.25 give some values for various methods of rigging beams made from scaffold tubes.

When a working platform has to be fixed along a lift containing a prefabricated beam bridge, it is difficult to fix the transoms on the same level on both the bridge and the scaffold lift leading up to it.

Figure 9.3(a) shows beams resting at their ends on support transoms fixed to the uprights. Two such support transoms are shown to give double support and end stability to the beam. The board-bearing transoms across the bridge beams are underslung on right-angle couplers so that they are on a level with the board-bearing transoms on the scaffold lift.

In Figure 9.3(b), the board-bearing transoms are on top of the beams, resulting in the boards being higher than the transoms on the scaffold lift.

In Figure 9.3(c), the beam is coupled to the uprights and the board-bearing transoms are underslung on right-angle couplers.

Figure 9.3(d) is the same case of the beam being coupled to the uprights, but the board-bearing transoms are set on top of the beam, making it impractical to sit the scaffold lift transoms on the same level.

9.6 Hazards with prefabricated beams

Tables 34.1 to 34.25 reveal an unexpectedly large range of values for the supporting ability of each length of a prefabricated beam for the various methods by which it can be rigged. This presents three hazards.

- If a safe working load is given for a beam, it will probably be the largest value, on the assumption that

it will be fixed firmly every 1.2 m. If its moment of resistance is quoted, and its supporting ability is calculated from this, it will have been assumed that it will be fixed at 1.2 m intervals. If it is not fixed at this frequency, it will not support the intended load.
- If a designer has specified a close spacing of the fixings because he requires this to achieve a specific strength, and the scaffolder fails to put the fixings in at these points or at all, there will be a failure. An inspector may not know of or see the fixing details on the drawing or on the job. He may not have been informed of their importance.
- In Figure 9.1(b), the bottom chord is in tension and may thus have been fabricated in thinner tube than the top chord. This will give it a much lower load rating if it is used upside down.

9.7 Methods of preventing a beam rolling over

There are several ways to prevent the rolling over of a beam. The lateral force required to maintain it in a vertical plane is only about 3% of the thrust in the top chord, so the choice of method is not critical and can be made to suit the job. The lower chord must be stabilised as well as the top chord because it can rotate upwards as easily as the top can rotate downwards.

First, the ends must be fixed at the top and bottom chords. If the beam is a fish-bellied type, a raker must be added to achieve this, or the bottom chord can be extended with scaffold tubes.

Figure 9.4, which has eight parts, gives several means of achieving the required lateral stability. The left-hand half of each beam is wrongly and hazardously rigged, while the right-hand half is correctly rigged.

9.8 Lateral displacement

In addition to being prevented from rolling over, a beam must also be restrained from lateral displacement. Otherwise, a load applied near the middle will be eccentric to the line joining the end supports and the beam will have a greatly increased tendency to roll over.

Figure 9.5(a) shows the plan of a footbridge between two supports. Each beam is rigged so that its top chords are stabilised. Knee braces are fixed correctly at each end, but there is no plan bracing. The pair of beams has deflected sideways under the influence of the wind or other lateral forces.

Figure 9.5(b) shows the corrective plan bracing added.

9.9 Families of beams

A family of parallel beams is frequently required, e.g.

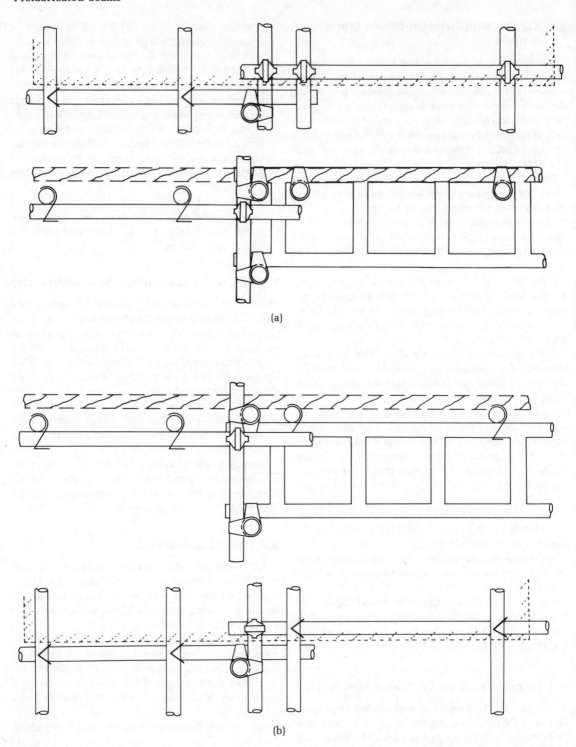

(a)

(b)

Figure 9.3 Rigging beams to obtain a level platform.

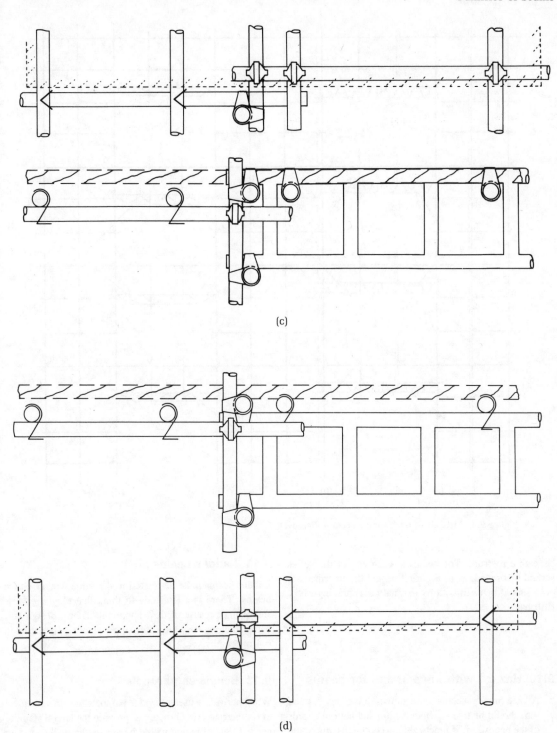

(c)

(d)

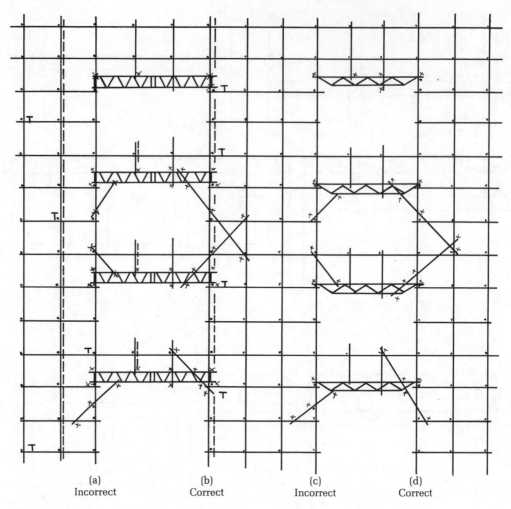

Figure 9.4 Methods of preventing a beam rolling over.

beneath a rostrum. The beams at each end of the series should be braced, as in Figure 9.6, and the intermediate ones should be dealt with by internal cross bracing every fifth bay.

9.10 Hazard with knee braces for beams

• A knee brace, such as those shown in Figures 9.4 and 9.5, should be fixed to both the top and bottom chords of the beam and, if practicable, to both uprights of the side scaffold. A short knee brace with only one coupler at each end is inadequate. Although it will act as a strut, it will not provide any resistance against the rotation or sideways displacement of the beam.

9.11 Ladder beams

Ladder beams must be treated in the same way as lattice beams. There is a tendency to think that they need less stabilising because of their apparently robust design but, being shallow, they are surprisingly slender when used in 6.4 m lengths.

9.12 Beams as parapets

When beams are used as bridge parapets, it is inconvenient to fix internal cross bracing to provide the lateral stability needed. Such bracing would be across the walkway over the bridge. In this case, the stiffness of the top chord could be achieved by external triangulation after the manner of the classic railway footbridge. Figure 9.7 shows three methods.

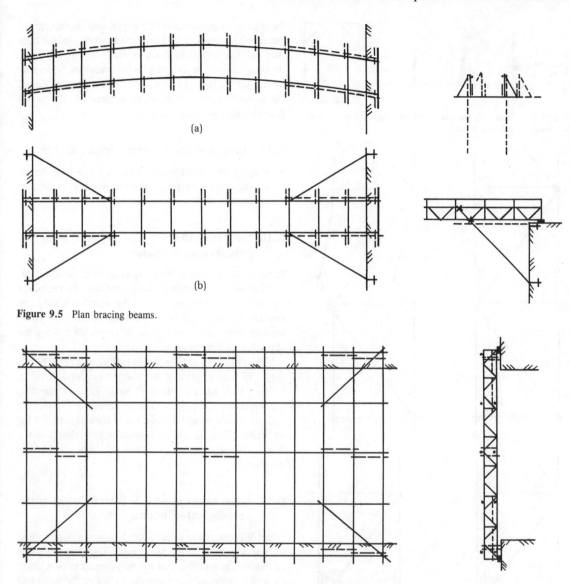

Figure 9.5 Plan bracing beams.

Figure 9.6 Correct rigging for a family of beams. If the top chords of the beams are required to support boards, the top chord stiffening can be placed below the top chords.

9.13 Hazard when starting scaffolds on beams

- Figure 9.8(a) and (b) shows how a beam can be induced to roll over if a scaffold is puncheoned up from the top chord only, e.g. over a traffic entrance to a building site. Figure 9.8(c) and (d) shows how the upper scaffold can stabilise the beam against rolling over if it is attached to both chords. This should be an inflexible rule.

9.14 Fish-bellied beams used upside down or as cantilevers

Some fish-bellied beams have different tube strengths in the two chords. The top chord, which is intended to be the compression chord, may be made of standard scaffold tube, while the bottom chord, which acts as a tension member in the normal use of the beam, may be made of smaller and weaker tube. If such a beam is rigged upside down,

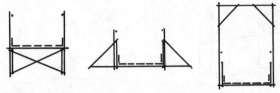

Figure 9.7 Beams as parapets, with bracing used to support the parapets and guard rails.

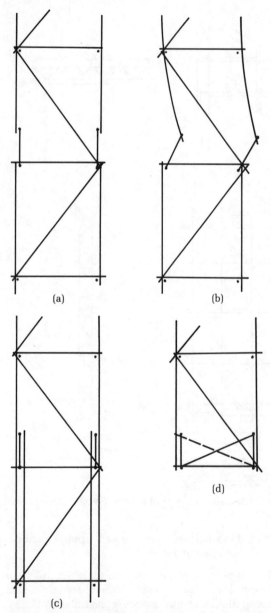

(a)

(b)

(c)

(d)

Figure 9.8 Beams induced to roll over by wrong rigging: (a) and (b) are incorrect. (c) and (d) are correct.

the weaker tube becomes the compression chord, and the beam will not support the same load as when used the right way up. The welded braces may also be of thinner tube, which is adequate in tension with the beam the correct way up but inadequate when the beam is used upside down. Beams with different chord strengths therefore require two tables of safe working loads.

9.15 Hazard with unsymmetrical beams

- Rigging an unsymmetrical beam as a cantilever forces it to bend the wrong way. For this assembly, a smaller moment of resistance will apply.

9.16 Supplementary couplers used with prefabricated beams

Because beams are used to span gaps in scaffolding, the intermediate standards supported by the beam are puncheons. If these are fixed to the beam with only one coupler, not only will they induce the beam to roll over, but they will also not support safely more than 635 kg, the slip value of one coupler.

If the puncheons are coupled to both chords of the beam, they will help to prevent it rolling over and will safely support 1270 kg. For standard loads greater than this, supplementary couplers will be required.

Extra standards will be required at the end supports for the beams. They should be coupled to both chords of the beam, and supplementary couplers should be fixed where necessary.

9.17 Knee bracing and V frames used with prefabricated beams

A and V frames fixed to both chords to stop a beam rolling over accept some of the beam load. The question arises as to how much of the load can be assumed to pass along these diagonals and so relieve the weight on the beam and also shorten its span.

There can be no exact calculation because scaffold couplers are friction devices and not positive fixings such as nuts and bolts. Overloaded couplers may not resist the overload until they fracture. They will slip and distort and may throw the load elsewhere. Bearing this in mind, the vertical load relief from one side of an A or V frame will not be more than 400 kg, and the effective reduction of the span will not be more than half of the distance from the beam end to the point of attachment of the A or V frame at each end. This will vary, depending on the number of couplers at each end of the braces of the members of the A or V frames.

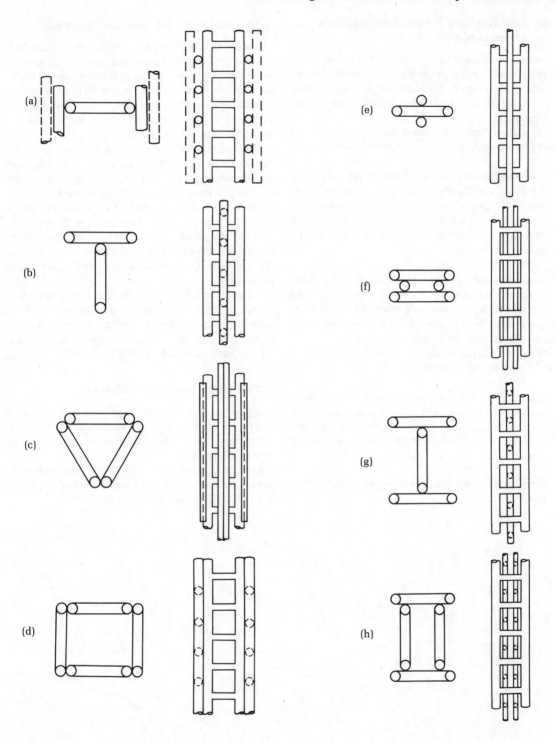

Figure 9.9 Ladder beams used as columns. The column strength depends not only on the number of vertical tubes but also on the vertical spacing between the places where the beams are coupled together.

9.18 Safe working loads: the supplier's responsibilities

From the above, it will be seen that there is no single safe working load for a prefabricated beam. It depends on the way the beam is rigged. The manufacturer must be asked to supply a table of safe working loads for each method of rigging.

The relevant clause in the Health and Safety at Work Act 1974 is Section 6(1), which can be paraphrased as follows:

It shall be the duty of any person who supplies any article for use at work to take such steps as are necessary to secure that adequate information will be available about the use for which it is designed and about any conditions necessary to ensure that when put to that use it will be safe.

Having obtained the rules for rigging any prefabricated beam from the manufacturer, the onus is on the scaffolding contractor and the user to rig the beam and load it in the proper manner and to inspect its installation and use to ensure that the rules have been carried out.

The relevant clause in this case is Section 6(3), which can be paraphrased as follows:

It shall be the duty of any person who erects or installs any article for use at work in any premises to ensure, so far as is reasonably practicable, that nothing about the way in which it is erected or installed makes it unsafe.

Tables of safe working loads of some beams will be found in Part IV, Chapter 34.

9.19 Ladder beams used as columns

Reference should be made to Chapter 34, Section 34.8.

If a single ladder beam is used as a column, it is strong in one direction but weak in the opposite direction. To make good this weakness, the column must be laced at intervals at right angles to the rungs. This vertical spacing can be calculated to give the vertical tubes the desired strut strengths using Tables 31.3 to 31.10.

It is clear from this that a lot of lacing is necessary. For this reason, it is better to stiffen the vertical ladder beam column with a second beam column at right angles to the first. This arrangement is shown in Figure 9.9(b).

The strength of the combination varies with the spacing of the couplers joining the two ladder beams together.

More than two ladder beams can be coupled together into compound columns. Figure 9.9 shows several such assemblies, while Table 34.30 gives the supporting strength of the column.

If the structure has to be higher than 6 m, joints will be needed and these will introduce weaknesses into the assembly if they are all at the same height. The joints should be staggered, and the joined beam should be coupled to the unjointed beam at both sides of the joint.

9.20 Lattice beams as columns

Similar arrangements of lattice beams to those shown in Figure 9.9 can be made. The same rules apply and the manufacturer must supply the safe working loads for various assemblies.

Some manufacturers supply special ridge-and-eaves fabrications that allow lattice beams to be used as columns and rafters for temporary buildings.

10 Pavement frames

10.1 Definition

A pavement frame is the lower part of an access scaffold, hoist tower or protection scaffold over a pavement that has had to be modified in design from the upper part to permit the pavement to be used, to provide a longitudinal access through the scaffold or to support contractors' cabins, loading bays or protection fans.

10.2 Basic types

(i) Access along and through the bottom lift of a scaffold on a pavement or walkway when the scaffold does not require any modification other than the omission of ledger bracing.

(ii) Walk-through pavement frames with increased width, as in Figure 10.1(a), (b) and (c).

(iii) When hoist towers or contractors' huts or protection decks have to be incorporated, as in Figure 10.1(d) and (e).

(iv) Frames above vehicular traffic or narrow streets similar to Figure 10.1(d) and (e).

(v) Walk-through frames associated with shoring, as in Figure 10.1(f) and (g).

10.3 Access through a normal independent tied scaffold

When access is required along or through an ordinary scaffold on a pavement, the ledger bracing is not put in place in the bottom lift. This weakens the scaffold because of the lateral instability of this lift. This deficiency should be made good by ties into the building at the first storey. These should be spaced about one every 6 m if the scaffold is higher than 20 m.

If it is impractical to tie the scaffold into the building at the first storey, the legs should be doubled on the transoms to increase the stiffness at right angles to the building.

For a scaffold not higher than 20 m, the lowest lift can remain without ledger bracing provided that it is no higher than 2.5 m. Stability is improved by fixing knee bracing from the first ledger down to a swivel coupler on the opposite upright 2.15 m above pavement level. These knee braces are fixed at every pair of standards but are inclined in alternate directions.

A guard rail inside the outside standard will assist longitudinal stability. Longitudinal bracing should start at ground level and the lowest bracing tube will improve transverse stability slightly.

For scaffolds taller than 20 m, consideration should be given to erecting a hoarding around a fully braced scaffold and providing a walkway around it that is fenced off from the traffic.

10.4 Pavement frames wider than the scaffold

The main difficulty in codifying pavement frames is that there are many ways of building them and they have many different duties to perform. Each job should be designed on the drawing board, with special attention to the height of any scaffold to be built on it, the weight of contractors' huts if these are to be incorporated, the weight of stored materials if it is to be used as a materials handling bay, and the weight and shape of falling objects to be arrested if it is only for protection.

Special attention should be given to the access into each end and to the desirability of preventing access from the face of the frame into the road. In some cases, it may be unsafe to allow pedestrians to move into the road through the side of the frame. In other cases, it may be essential to have access. Figure 10.2 is a typical front elevation.

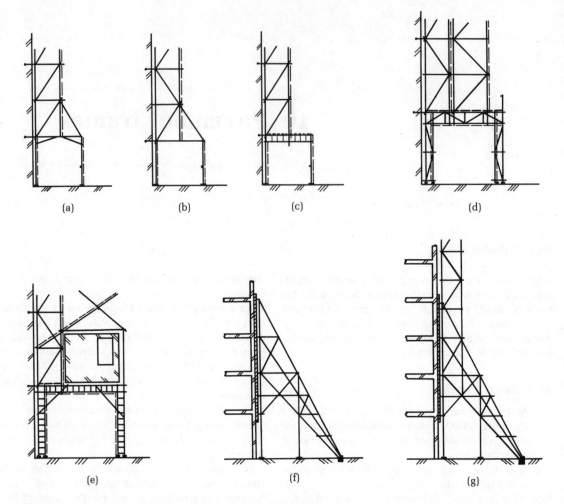

Figure 10.1 Basic types of pavement frame wider than the scaffold.

Traffic fenders and temporary lighting need to be considered. Approval must be obtained from the local authorities concerned with temporary works on pavements.

There is no rule concerning the imposed load for which a pavement frame should be designed in excess of its own weight and the weight and imposed load on the scaffold above. The design load on the pavement frame decking should be not less than 1 kN/m² (100 kg/m², 20 lb/ft²). Preferably, boards should be doubled and so arranged that no materials can fall through.

A typical construction incorporates double-tube piers along both edges of the pavement. Short prefabricated beams or tube-and-coupler beams span these and are so arranged that the decking is laid parallel to the building face.

Four or six-tube standards can be used at wider intervals, with prefabricated units or ladder beams in two directions. Special prefabricated frames with telescopic legs and

sometimes structural steel sections can be used. The structure must be stabilised in both directions by transverse and longitudinal bracing. Multi-tube columns must be well braced in both directions, and plan bracing is necessary if separate multi-tube columns have been used.

Care must be taken to use the correct lengths of tube because the public will be passing through the inside, and traffic will be passing the outside.

When hoardings are used, they should be securely attached either by joinery work, by butt tubes and couplers or by pipe clips and screws. On demolition jobs, the hoarding must be dustproof. On stone-cleaning jobs, the decking must be both waterproof and dustproof, which can be achieved by a layer of plastic sheet sandwiched between two layers of boards.

It is good practice to arrange the pavement frame so that members of the public are not unhappy to use it, that they

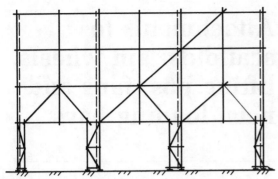

Figure 10.2 Front elevation of a pavement frame with entrances from the road and access to shop windows and doors.

are guided into it, that they are assisted by proper lighting and that they are provided with entrance ramps if the floor level is to be raised at any point.

10.5 Hazards with pavement frames

- Overloading long, unbraced bottom standards.
- Placing all joints in the standards at the same level, causing weakness at the beam level.
- Longitudinal instability due to the omission of longitudinal bracing in the bottom lift.
- Omission of one of the fittings required to join the standards to the two chords of the beams.
- Not providing an equivalent number of bottom standards to the upper standards, or a greater number if required.
- Lack of longitudinal stability when the scaffold is on a curve, necessitating short ledgers.
- Absence of supplementary fittings when required on puncheons.
- Omission of one of the fittings at the bottom of puncheons, which should be attached to both chords of the beams.
- Absence of adequate or correctly placed traffic fenders.
- Inadequate rearrangement of the standards opposite shop entrances and other doors.
- Not enough ties at the beam level to ensure rigidity.
- Failure to take into account the wind forces on contractors' cabins or sheeting.
- Overloading at the hoist tower and rubbish chute areas.
- Failure to counteract distortion due to eccentricities.
- In the case of a combined pavement frame and raking shore, starting the shore at the beam level instead of ground level.

11 Attachments to scaffolds: gin wheels, lifting jibs, fans and nets, loading bays

11.1 Gin wheel loads

Gin wheels are single pulleys fitted with fibre ropes. A person pulls one end to lift a load on the other. There are no brakes or ratchets, so their capacity is whatever a workman can pull hand over hand, perhaps 30—32 kg (66—70 lb).

Because the pull is down on one side of the rope and the weight pulls down on the other, the gin wheel must be supported strongly enough to carry twice the lifted load. Figures 11.1 and 36.4 show how this occurs. The same principle applies when pulley blocks are used to raise loads too heavy for a gin wheel.

11.2 The Lifting Regulations

The Construction (Lifting Operations) Regulations 1961 frequently apply in scaffolding. These list the statutory obligations with regard to equipment used for raising and lowering and as a means of suspension. They detail the requirements for inspecting and testing lifting tackle, and the frequency of such inspections and tests and their recording.

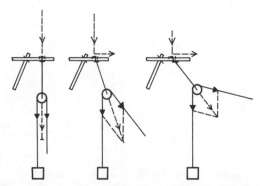

Figure 11.1 The forces exerted on a scaffold by a gin wheel.

Because gin wheels are pulleys to raise less than 1 tonne, they are not subject to the Regulations on construction sites, but they may be in factories.

Lifting gantries are defined in the Regulations by the term 'lifting plant' and are therefore subject to the Regulations, as are the joists, winches, ropes and pulley blocks that are fitted to these gantries.

11.3 Gin wheel attachment

A gin wheel for scaffolding should have a 55 mm (2.25 in) ring at the top so that it can be threaded on to a scaffold tube, where it should be retained by a scaffold coupler on each side of the ring.

The pulley should be fixed about 50 cm (18 in) from the scaffold on a horizontal tube, which is knee braced back to the scaffold. This propped cantilever should be fixed to one of the braced frames of the scaffold, and the scaffold should be tied to the building at the place and level of the gin wheel. Figure 11.2 shows some preferred arrangements.

11.4 Inspection of gin wheels

Inspections should be done at the depot for insurance reasons, general safety and to follow any regulations that apply at a site. It is preferable to mark a serial number and a safe working load on the wheel and enter these in a lifting plant register.

The wheel will usually not have special axle bearings and, if it seizes up through continued exposure on site, the axle will rotate in the side sheaves and wear away. Inspectors must watch carefully for this defect. The pulley should be kept well oiled.

11.5 Ropes for gin wheels

Gin wheels are grooved to receive a 20—25 mm (0.75—1 in)

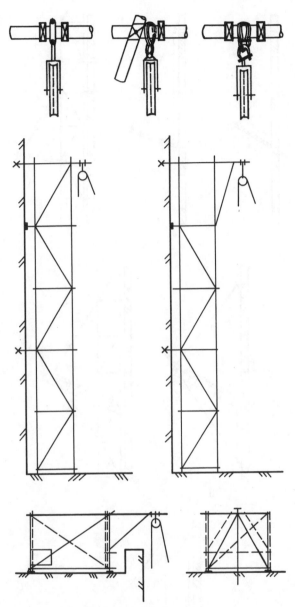

Figure 11.2 Gin wheel attachments.

fibre rope, which can be of natural or man-made fibre. The typical rough handling and continued exposure to which gin wheel ropes are subjected makes it preferable to replace them periodically, as follows: coir every year; sisal, hemp and manila every two years; and mono-filament polypropylene every three years.

To control this, it is necessary to have a serial number and a safe working load for the rope.

If physical damage is observed, the rope should be taken out of service. A 5% fraying of the fibre should be con-

sidered as physical damage, even if it is due to fair wear and tear.

Ropes of equal strength to those detailed above, but much thinner, can be obtained in some man-made fibres, but these are difficult to grip under load, and sometimes their surface is shiny, which may also lead to an accident.

11.6 Hazard with gin wheels and pulleys (see also Chapter 36)

- The pulling part of the rope must be reasonably vertical. Pulling at an angle exerts a horizontal force on the scaffold, which it might be unable to resist (Figure 11.1(b) and (c)). Two collapses are on record where camera towers were pulled over during attempts to lift a television camera on a gin wheel rope, which was attached to a motor vehicle that was driven away from the tower until the angle of the pull rope was about 45°.

11.7 Scaffold hoists

A scaffold hoist is a small crane that is attached to the outside standard of a scaffold. The bay to which it is attached should be one of the ledger-braced frames of the scaffold, with the brace omitted on the lift where the hoist is fixed. The scaffold must be tied to the building as near as possible to the hoist.

The hoist and its lifting gear are subject to the Construction (Lifting Operations) Regulations 1961 and so should be marked with a serial number and a safe working load. The equipment should be inspected, tested and certified according to the Regulations.

The hoist swivels on the upright of the scaffold to bring its load on to the scaffold platform. Guard rails are therefore not practical in the area of the hoist. The safest arrangement is for rotation to be in one direction only so that a guard rail can be fitted on the opposite side of the hoist to provide a safe place for the operator.

11.8 Protection fans and nets

A fan is a decking placed in or outside a scaffold to protect persons below from falling objects. If the protection is against falling materials, continuous sheeting is needed, but a netting is frequently used to catch falling persons. A vertical net can be rigged around exposed floor slab edges to prevent persons falling off.

Continuous sheeting or material protection fans can be corrugated steel, plywood, hardboard (if the duration is short) or scaffold boards.

Three grades are proposed in the Code (light, medium and heavy), but each case can be treated on its merits.

Light-duty fans are intended to prevent paint droppings,

stone-cleaning water spray, rubbing-down dust, pointing mortar, paint brushes and small tools reaching the ground. Corrugated iron sheeting is satisfactory for this.

Medium-duty fans are for heavier work and are expected to stop concreting aggregates and small droppings of concrete that have spilled over the sides of shutters, and nuts, bolts, small tools, bricks, blocks and the various bits and pieces that the wind might blow off the concrete slabs of a building during construction. For this purpose, a fan should be boarded out with 38 mm scaffold boards spanning no more than 1580 mm between supports, or a heavy gauge of corrugated iron.

The success of such a fan depends not only on the weight of the object falling on it, but also on the distance the object falls. About 10 m is the limit a medium-duty fan will stop the sort of object described. The intention behind including this figure in the Code was to draw the attention of builders to the need to move the fan upwards or downwards as the work requires.

Heavy-duty fans are for cases where construction work is being carried out on the face or at the edge of the slab of a building, and when heavier tools might be dropped or shuttering timber might fall. This type of fan will also stop a person's fall, but it is not specifically intended for this purpose. These fans should preferably be double-boarded, i.e. two layers of 38 mm scaffold boards supported at 1500 mm centres. Single boards supported at 750 mm centres can be considered if weight is a critical factor.

Pavement frames should perform the same duty as heavy-duty fans, but they should be capable of stopping objects falling a much greater distance because they are not raised as building work progresses.

Nets are specially for falling persons and should be constructed on a framework similar to that for a heavy-duty fan. They should be moved up to keep within two storeys of construction operations so that a person will not fall more than about 7 m before reaching the net. Nets usually have a mesh size of 100 mm and can be overlaid with a finer mesh if necessary.

Figure 11.3 shows six arrangements for a range of duties.

11.9 Design details

Designers and scaffolders need more guidance than the general description given above. In all cases, a static load should be defined for design purposes. This should be adequate to deal with typical dynamic loads encountered in the duties described above.

11.10 Light-duty fans

The nominal static rating given in the Code is 75 kg/m^2 (0.74 kN/m^2, 15 lb/ft^2).

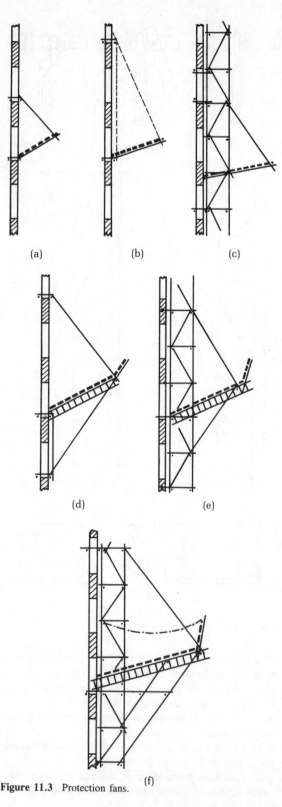

(a)   (b)   (c)

(d)   (e)

(f)

Figure 11.3 Protection fans.

A typical construction comprises a transom hanging from each higher window of a building with two wires, one down the face of the building and one inclined to support the outside end of the transom. These should be about 4 m apart. Two ledgers and an intermediate transom are then fixed, giving a series of 2 m bays.

The decking can be a thin gauge (24 g) of corrugated steel that is laid either along the building and resting on the transoms over the ledgers, or laid at right angles to the building on ledgers fixed over the transoms. The sheeting should provide a deck 1.2−1.8 m wide, with the inside edge tight to the wall and the outside edge about 300 mm above the inside edge. This incline inhibits material falling on the sheeting from bouncing outwards.

Each bay will have an area of 2 by 1.5 m, i.e. 3 m^2, giving a bay load of 225 kg on two wires, or 112.5 kg per wire. For this load, 6 mm wire is adequate.

The top fixing of the wires should be on to a tube projecting through a window and restrained by a bridle tube across the window inside the wall, or the wires can pass through the window opening and be attached to the bridle tube. A horizontal bridle is best suited for this purpose, with a cross butt at each end to prevent sideways movement.

The lower end of the wires should be fixed to the transom outside the ledger, or to the ledger beside the transom, in a way that prevents the knot slipping along the tube. A round turn and two half hitches is a suitable knot. The loose end can be used to tie a tube over the top of the sheeting to prevent uplift by the wind. The suspension wires will have to come round the edges of sheets that are adjacent to it, and the gap so caused should be protected by the tube directly beneath them. If close sheeting without any gaps is required, for instance in the case of protection against stone-cleaning water spray, the edges of the sheets will have to be notched for the wire.

The inside ledger should be tied to the building so that the fan cannot swing out from the building in a wind. This can be achieved by wire ties on to a bridle inside a lower window, on to reveal ties or by lashings jammed in the window framework.

Generally, this type of light-duty fan is hung on a building surface where there is no access scaffolding, for instance below a painter's cradle. In cases where an access scaffold is present, the same sort of sheeting can be fixed to one of the lifts without any suspension wires being used. It is good practice to ensure that the sheets are fixed closely to the wall and do not project irregularly over the road.

11.11 Hazards with fans

- There are gaps in the sheeting.
- The fan is not tied to the building at the bottom.
- The sheets are not secured against wind uplift.

11.12 Medium-duty fans

These can be fixed directly to the building where there is no scaffold, on the outside of an access scaffold or through from the building surface past the outside of the scaffold.

For all these applications, the static rating given in the Code is 100 kg/m^2 (1 kN/m^2, 20 lb/ft^2). This figure does appear to be large for a sheeted area that is not a working platform and not intended for the storage of materials, but it includes an allowance for wind turbulence.

This static loading is for the local design of the fan structure and the immediate scaffolding area supporting it. It is not intended that the whole area of the fan be regarded as fully loaded to this extent and the resulting total added to the leg load on the scaffold. The extra load on the scaffold should be taken as the dead weight of the fan and its decking plus an allowance of 33% for wind turbulence.

There will be exceptional cases where it is suspected that the wind force may be as much as 30 kg/m^2 (0.3 kN/m^2, 6 lb/ft^2), and these cases must be calculated from the data assumed. Regard must be paid to the leverage effect of both the dead weight and the wind force on the main scaffold, for example, when projecting tubes are threaded over the outside scaffold ledgers and under the inside ledgers.

It is appropriate here to remind designers that wind speed is exaggerated around the corners of buildings. Refer to Chapter 38 to find the degree of increase. The wind force can be upwards or downwards but, in general terms, if a fan is on the lower third of a building the wind pressure will be downwards and, if it is on the upper two-thirds, it will be upwards.

Where a medium-duty fan is attached directly to a building surface, e.g. outside a building that is being modified inside without changes to the outside wall, the inside end of the projecting tubes or beams can be rested on window sills and joined along the building face by a ledger outside and bridle tubes inside. The outside of the projecting tubes should be held up and down by tube rakers or guy wires secured at suitable angles through the window openings above or below the fan level. These securing points should be on ledgers or bridles both inside and outside the windows. They should rest on the sills and be prevented from lifting by vertical tubes bearing under the lintels.

The width of the fan at right angles to the building is a matter of judgement, taking into account the following factors: the vertical distance up to the construction work; the nature of the work; the width of the pavement; the height of the traffic; the camber of the road, which will cause vehicles to tilt towards the edge of the fan; the length of the sheeting elements; the strength of the building wall supporting the fan; and whether the top portion of this will be demolished below the level of the top tie-back. This list shows that some advance planning is needed to assess the best width and other design features.

Where the projection is up to 5 m, the projecting members can be tubes. For projections larger than this, prefabricated beams are more suitable, and the layout should be formally designed.

A typical construction is to fix a ledger along the outside of a building at the required height by projecting tubes through the window openings, and to repeat this process above and below the fan level for the tie-back rakers. The fan tubes or beams that will eventually be at right angles to the building are then raised from the ground vertically on wires and attached to the ledgers at the required level. They are then rotated outwards on the wires to the required angle, which should be about 1 in 3. The weight of beams must be carefully assessed, with special regard to the low angle of pull that may occur.

The fan tubes at right angles to the building should be laced parallel to the building with tubes at 1.2 m centres. Sheets across these ledgers can then be hook bolted on to the tubes.

The raking members that locate the outside edge of the fan tubes can be either guys or tubes but, in general, one of the pair should be a tube because this is a more rigid connection that will damp any wind oscillations. There must always be resistance to both upward and downward forces, although the emphasis on one or the other will vary with the location of the fan on the building.

If the fan tubes are 5 m long and spaced every 1.2 m, the area per bay will be 6 m^2 and the vertical force will be 600 kg per fan tube. If alternate tubes are anchored back by rakers, the vertical load per raker will be 1200 kg. The centre of this force will be about 2.5 m out from the wall and the vertical tie-back force at the end will be half this value, i.e. 600 kg. This means that the inclined tie-back force, if there is only one tube raker, will be about 850 kg. Allowing a 25% increase in the allowable load on a coupler supplying resistance to wind forces, one fitting will have a wind force safe load of $635 \times 1.25 = 794$ kg. This bay size of 5 by 2.4 m is therefore likely to overload a coupler on a single raker; hence, the diagrams in the Code show a medium-duty fan supported at the top and bottom. It is clear that these structures must be carefully designed and accurately constructed in accordance with the designer's recommendations.

The ability of the structure adjacent to windows to resist the inclined force of 850 kg from the raker must also be taken into account. The strength of a bridle spanning a window must also be considered. If the fan is 6 m wide, i.e. one long scaffold tube, and is raked back only at 4 m centres, i.e. in each window of a typical building, the area becomes 24 m^2, and the vertical load acting at the centre of the area will be 2400 kg. Normal long scaffold tubes used as rakers will not reach the end of this fan when set

at 45°. They will attach to the ledger about 4.25 m along the fan tube. The vertical force in a single raker thus becomes $3/4.25 \times 2400$, i.e. about 1700 kg, and the inclined force will be 2400 kg. This is why attention is drawn to this large raker force in the Code.

Clearly this raker force must be divided by two if there are inclined tubes above and below the fan. Even so, if the tubes are 4 m apart, doubled couplers will be required. The building will have to withstand a force of 1.2 tonnes at the point of attachment of the rakers, assuming that they both work evenly and neither is a guy.

As in the case of light-duty fans, this calculation is based on the recommended load and is to assist the design of the fan itself and its adjacent tie points. The assembly can be taken to impose a vertical force on the building or on the access scaffold, depending on how it is fixed. It will also impose horizontal forces at right angles to the wall, which may be inwards or outwards. All these forces must be calculated and the building's ability to resist them checked.

The decking can be of heavier-gauge corrugated sheet, e.g. 18 or 20 gauge, or single scaffold boards. Either type should be held down against local wind uplift, even if the fan is in the lower third of the building.

The minimum slope is 1 in 12, but 1 in 4 to 1 in 6 is preferred.

Some specifications require a parapet at the outside edge of the fan. When this is required, its weight must be assessed very carefully because it is situated at the outside edge of the fan, where it exerts maximum leverage for its weight.

In the case of a medium-duty fan attached to an access scaffold, construction is essentially the same as described above, except that the inside end is attached to the scaffold. The forces are the same, so the scaffold must be tied to the building opposite the raker positions in such a manner as to resist the raker forces. In this case, there will be no need to site the raker tubes at every window opening at 4 m centres, giving the large forces referred to. The rakers will normally be at every upright of the access scaffold, i.e. every 2 m. The forces for each of the rakers above and below will be 600 kg, which can be resisted satisfactorily by one coupler. This demonstrates that that maximum spacing of rakers for a fan 5 m wide should be about 2 m. The Code does not give this figure because of the wide variation in bay lengths of access scaffolds, but it alerts the designer to the consequences of exceeding it.

Good practice requires that the forces be properly calculated and that supplementary couplers be used where necessary. Long rakers must be stiffened where necessary. The load distribution between tube rakers and guy rakers is biased towards tube rakers, which will take up the load first because they are rigid.

11.13 Hazards in medium-duty fans

- Failures have resulted from building a fan by rule of thumb, without calculation. The building has sometimes been unable to resist the forces on it, especially in a demolition job where the wall has lost some of its internal support.
- Sheeting and boards have frequently blown off because of the wind flow up the building.
- Beams have been overloaded. They have sometimes been bent backwards as a result of being supported in their centres instead of at their ends.
- The ledgers of the access scaffold have been overloaded by the fan tubes or its rakers. The whole access scaffold has been distorted and sometimes pulled off the building if inadequately tied.
- There have been several accidents to scaffolders constructing or dismantling a fan because of its large projection to one side of the access scaffolding.
- With progressive movement of the fan up the building, the quality of its construction sometimes deteriorates.

11.14 Heavy-duty fans

These are usually attached to an access scaffold. They should have an applied load of 100 kg/m^2, as for medium-duty fans, but their weight will be much greater because of the greater strength of their decking. The weight of double boarding will be twice that of medium-duty fans, i.e. about 50 kg/m^2. The weight of the supporting structure must be increased to deal with this increase and to give a greater allowance for impacts. Because heavy-duty fans are frequently specified for use over pavements and below construction work at the edge of a building, an edge parapet is often required, perhaps four boards high. This and its tube-and-coupler framework is heavier than its counterpart in the medium-duty range.

In general, the rakers should be tubes above and below the fan at centres not exceeding 1.2 m. Calculations should be made to ascertain the raker loads and the correct number of couplers to use at each end.

The decking should be of doubled boards if there is a danger of large falling objects. Some designers believe that it is an advantage to separate the layers with timber battens to protect the lower layer from the maximum shock of a falling weight and to ensure that the full breaking strength of one layer must be exceeded before the second layer comes into play. This can sometimes be achieved by placing one layer across the ledgers with the top surface of the boards 10 mm below the top of the transoms. The second layer of boards is at right angles to the transoms, leaving the required gap.

If the two layers of boards are in close contact, they should preferably be at right angles, but this may require cutting the boards, which is expensive. If the boards are placed parallel, the joints should be staggered so that each upper board rests on two lower ones. Thus, three boards will come into play if an object drops on the fan, giving a more or less point load.

Two layers of boards in contact offer the opportunity to insert a plastic or plywood sheet between them and so make the fan drip and dustproof.

Hazards with heavy-duty fans are the same as those for medium-duty fans.

11.15 Obstruction of daylight by fans

When a protection deck is placed within a scaffold, there is always obstruction of daylight. The protection deck will work as satisfactorily at a high level as on the first lift above the pavement, and obstruction of daylight will be reduced. If there is a fan, the obstruction will be worse.

Figure 11.4(a) and (b) shows a city centre location where the fan would be a nuisance to pedestrians and to shopkeepers because of the reduced light. Where all the fan cannot be put at a higher level, it can be set on two levels, as shown in Figure 11.4(c) and (d).

This arrangement is shown in elevation in Figure 11.4(e) and (f). It will be seen that no part of the street level is deprived of direct daylight. The length of each separate section of the fan should be the length of a half scaffold board.

The pavement is fully protected against objects falling within 5° off vertical. If this is inadequate for a particular circumstance, the gaps in the top level can be netted.

Negotiations between the person ordering the scaffold and fan and the property owners, particularly if they are shops, should take place before the job is started.

11.16 Safety nets

The relevant British Standard is BS 3913, and the Code is CP93. This details the net and gives a test in which a sack weighing 140 kg is dropped 6.1 m. This is approximately equivalent to a person falling into the net from a height of 8–10 m. Unfortunately, it is not practical in most construction jobs to fix the net within 6 m of the construction work. For instance, if a person is fixing column shuttering clamps 2 m above the last concrete slab poured, the net would have to be bracketed off the slab below to comply with the 6.1 m rule. This is not practical because shuttering materials are being taken out of the side of the building at this level. The net could be fixed one further storey below this, i.e. nearly three storeys, or 7–8 m,

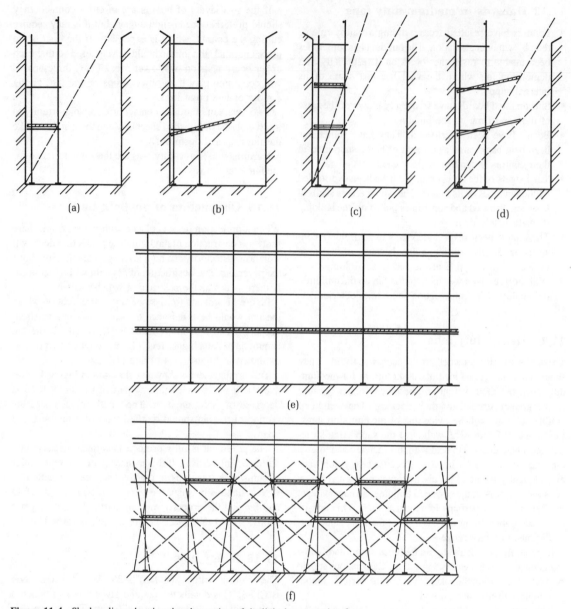

Figure 11.4 Shadow lines showing the obstruction of daylight by protection fans.

below the working place. In practice, this would probably not make much difference to the performance of the net, but it would not be in accordance with the net manufacturer's recommendations, with all its consequent problems.

Safety nets can be overlaid with a finer-mesh net to catch debris. This is good practice and can frequently be seen to have been useful on jobs.

The rigging of nets presents a problem because of the need for periodic movement up and down the building. When there is no scaffolding, and several movements are required, the best arrangement is to make or purchase prefabricated tubular frames to clip on to the slab edge. If there is a scaffold, similar prefabricated frames can be designed to be coupled on to the uprights or to be hooked on to the ledgers. These frames can be handled by a crane without removing the net. They should be angled upwards, as in the case of fans, so that the outside of the net is about 1 m higher than the inside.

When the net is to be in place for a long time, it can be rigged on the type of framework used for a medium-duty

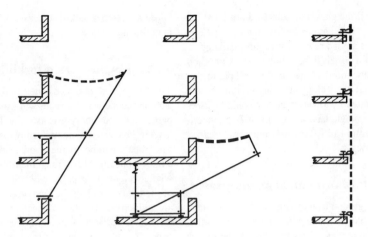

Figure 11.5 Safety nets.

fan. Care must be taken to rig the net as high off the tubular framework as possible because, on falling into a net, a person passes several feet below its nominal level before he is stopped. Figure 11.5 shows some net assemblies.

As far as design is concerned, the weights of nets must be ascertained because they are heavier than would be expected. Windage can be calculated on the actual area of the threads and knots multiplied by a factor of 1.5, with a further allowance of 10% for flapping.

When practicable, safety nets can be rigged vertically within the confines of the building on scaffold frames locked on to the inside of the exterior columns.

The durability of nets is good. Excluding abuse, they will last ten years, but they must be discarded if they have performed their function. The test cords supplied with each net can be tested every two years to give an indication of the state of the net.

It is good practice to rig the net simply so that it can be moved easily, i.e. to rig it in the manner set down by the manufacturer, with the four edges tied.

11.17 Hazards with safety nets

- Trying to manhandle the net, thereby putting men at risk because of its unexpected weight.
- Getting the net caught on projecting tube ends and thus causing damage to the net or the scaffold.
- Not re-rigging the net at a higher level as the building grows.

11.18 Loading bays

The Code distinguishes between two types of loading bay: those founded on the ground and those cantilevered or trussed out from the scaffold at a higher level.

The former can be subdivided into loading bays that are separate towers and those that are strengthened parts of an access scaffold.

11.19 Loads

There are three loadings to be considered for each type:

(i) A distributed load that is intended for general design purposes.
(ii) An impact load on placing materials to be stored.
(iii) The largest load that the handling device is authorised for or is likely to place on the platform.

The Code makes the general recommendations that cantilevered platforms be limited to distributed loads of less than 400 kg/m^2 and, for loads heavier than this, that the bay has its own foundations on the ground. This does not rule out special cantilevered platforms being designed for heavier loads; nor is it intended that platforms with smaller loads do not need proper calculation. On the contrary, loading towers, in which should be included strengthened parts of access scaffolds, are generally regarded as requiring calculation.

The Code is specific about the impact to be added. The recommendation is 25% of the largest lifted load. This is intended to apply to mechanically handled loads, and the 25% is in addition to the load lifted and subsequently deposited. It does not apply to the total load stored on the platform.

If the loads are handled manually or by manually operated tackle, an allowance of 10% is adequate.

The Code makes no comment on item (iii) above because the mechanical handling devices used on each site are different. The designer must be informed of the maximum package load. For instance, it is little use designing a platform for 400 kg/m² if the packages delivered are 2000 kg on pallets, with only two battens below them.

This example gives a good reason for requiring dialogue between the user and the designer, but there is another anomaly that needs attention. While cantilevered platforms are generally limited to 400 kg/m², the ground based towers are not so limited. A tower crane that could be used for loading the former might have a capacity of 2–5 tonnes, whereas a rough-terrain forklift truck is unlikely to carry more than 1–2 tonnes.

11.20 Loading towers founded on the ground

Having regard to the ground irregularities on building sites, it is not practical to have a loading bay detached from the access scaffold that it serves. The gap between the tower deck and the working platform must be properly and particularly firmly boarded. Hence, it must be rigidly spanned so that the boards cannot be dislodged. This requirement results in the transmission of vertical impacts and accidental horizontal impacts being transferred to the working scaffold, which must therefore be specially braced in two directions in the vicinity of the loading tower. Ledger bracing on every pair of standards behind the tower is good practice. Longitudinal bracing in the lifts up to the height of the loading tower should be fixed either behind or adjacent to the tower.

As far as general construction is concerned, the tower can be a scaffold system or made of tube and couplers. For scaffold systems, adjustable legs are desirable so that the deck level can be fixed at the exact height of the access scaffold working platform. With tubes and couplers, this presents no difficulty if the transoms in the access platform are extended outwards to form part of the deck of the tower. This arrangements results in the loading bay boards being parallel to the access working platform boards, which has two disadvantages:

(i) There will be a 50 mm gap between the two decks where the scaffold standards pass through.
(ii) The adoption of a single board length for the loading platform, i.e. 3.96 m.

If this is divided in two by an upright at the front extended to form a centre guard rail post, the front loading gap on each side of this post becomes no more than about 1.75 m. This is adequate for most brick packs, but may be too small for block packs.

Therefore, the centre front standard of the tower should be offset from the centre line of the tower to give a larger and a smaller gap with dimensions discussed by the designer and the user. Removable guard rails should be fitted.

The front face of the tower must be braced longitudinally because sideways impacts can be severe.

The ground in front of the tower should be levelled and

hardened so that forklift trucks can use it accurately and have turning space.

11.21 Rough-terrain forklift truck operation

The designer should study the lifted loads with particular attention to the way packages will load the scaffold. Normal pallets will inevitably give two line loads on the scaffold, even if the materials are uniformly distributed on the pallet. The impact will be on one or both sides of the pallet. With a tipped-back forklift arrangement, on many occasions the whole weight of the pallet will come on to the front ledger of the tower, i.e. half the load on each side. Packed bricks are unlikely to be stacked near the uprights, as happens after their redistribution along the scaffold. They will be deposited in the middle of the bay.

Regard must be paid to the location of the transoms along the ledger spans because this affects the bending stresses in the ledgers. For instance, if there is a transom in the middle of a ledger span, the weight of the decking and the loads stored on it pass through this transom as a centre point load on to the ledger. Two transoms near the ends of the span receiving the load would be much more suitable.

For example, consider the loading gap of 1.75 m referred to above, into which is placed a 1 tonne pallet. With its own weight and an additional 25% impact, this will exert a downward force of 1250 kg, distributed between two ledgers, i.e. 625 kg on each ledger.

The safe centre point ledger load for these circumstances will be about 160 kg. This means that the ledger will be loaded to four times its safe load if all the pallet weight is transferred to the centre transom. Failure would be inevitable if the decking boards did not come to the rescue as structural members.

If transoms are placed 200 mm from each end of the 1.75 m opening, the safe load at each of the two points on the ledger will be 350 kg, and there will be only 320 kg at each of these points. There will be no failure, and the strength of the boards will not come into play.

A further precaution that can be taken to supplement the strength of the ledgers in a loading tower is to double them, either by slinging the additional ledger below the transom or by attaching it to the uprights below the main ledger and ensuring that the ledger loads are shared by the use of parallel couplers or butt tubes.

11.22 Hazards with loading towers

• The user has a responsibility to distribute the loads on the loading tower evenly and to refrain from overloading it. This latter requires some planning because the loading tower can be fed with materials more quickly by mechanical means than it can be unloaded by manual movements of the stored materials along the working platform.

- Loading towers are often filled up after two or three journeys of a forklift truck and can then take no more. Subsequently, the adjacent working platform is loaded by the forklift truck and is overloaded.
- This operation also results in the working platform being stripped of its guard rails and toe boards.
- Moving the toe boards and guard rails on the front of a loading tower requires supervision to ensure that a safe working place is always afforded to the person distributing the delivered loads from the tower to the working platform. Falls have occurred from loading platforms.

11.23 Loading bays forming part of the working platform

It is sometimes difficult or inconvenient to build a separate loading tower, and the working platform of the access scaffold has to be used as the receiving bay for delivered materials. When this happens, the same recommendations concerning extra bracing apply. Because the scaffold will not have the buttressing afforded by the width of a separate loading tower, it will require additional tying to the building at the loading points. The tying should be to firm points and not to green brickwork.

Building work at the loading points will probably be impossible, and access will be restricted. Therefore, regard must be paid to the requirements of the Construction (Working Places) Regulations concerning the width of the platform to be left unimpeded.

Attention must be paid to the replacement of toe boards and guard rails after materials have been deposited on the scaffold.

Doubling of the front outside ledger and the ledger couplers will almost certainly be necessary. The additional tubes and couplers can be moved up or down as the working lift is changed.

11.24 Hazards with forklift trucks

As a general recommendation, it is not good practice to use forklift trucks loading directly on to the working platform, even if an extra one or two boards width has been provided to facilitate this operation. The pitfalls are:

- Impact damage, particularly if the ground has not been prepared.
- The removal of guard rails and toe boards.
- Overloading.
- Obstructing the working place and creating hazards for the work people.

11.25 Cantilevered loading bays

The Code proposes maximum distributed loads of 400 kg/m^2. The same impact factors and point load considerations described above for ground-based loading towers apply to the cantilevered design. The Code gives a diagram with a deck 1.8 m wide and two scaffold bays wide, or 4 m. Such a platform will have an area of 7.2 m^2. When loaded to the maximum rating, the total load will be 2880 kg. Half of this is additional load down the outside standard of the access scaffold behind the middle of the cantilevered loading bay, so this standard carries an additional 1440 kg.

Usually, a cantilevered loading platform is used to service a tall building, and this extra load may be critical. It can be dealt with by doubling the standards at this point. The loads on the couplers must also be calculated, as with ground-based loading towers.

If there are two cantilevered platforms above one another, the standards are double loaded. If the loading bays are staggered without a gap of a bay between them, the standards opposite their ends will be loaded from the edges of two platforms, and this may be critical. For this reason, the Code requires a clear bay to be left between cantilevered loading bays. If such an arrangement is not possible, then design calculations must be done.

Impacts from tower crane operations are unlikely to be as great as those from forklift trucks, but the 25% allowance must still be made. Building operations that require a cantilevered loading platform must be studied carefully because construction plant such as shuttering units may be very heavy.

There should be extra bracing of the access scaffolding adjacent to the cantilevered loading platform. Extra tying and butting of the access scaffold are also critical. The main deck transoms can be loaded up to 750 kg in tension away from the building, so the tying method and couplers must be designed to accommodate the forces calculated.

11.26 Hazards with cantilevered loading bays

- Bad design in providing a size and shape that is unsatisfactory for the materials used, leading to dangerous stacking and the removal of guard rails.
- Overloading the access scaffold standards by the addition of fans or hoist towers in the vicinity of cantilevered loading bays.

11.27 Hoist towers

Each site and each hoist must be considered on its merits with regard to the storage area and loads involved.

The size of the tower and the positioning of the hoist mast within it should be agreed with the main contractor.

The purpose of the hoist will govern whether its stopping levels coincide with the general scaffold lift heights or with the floor levels of the building.

Special consideration must be given to the position of the edge of the hoist platform and the edge of the general scaffolding. There must be a continuously boarded and level access across the gap between the hoist platform and the scaffold decking. The requirement for side boards within the tower must be considered because workmen may step on to the hoist platform to bring off materials. Side meshing may also be required.

Safety gates should be incorporated at each access level.

11.28 Ties

When a hoist tower is attached to an ordinary access scaffold, extra ties must be fixed adjacent to the hoist. One extra tie every two storeys is recommended, and these should be distributed systematically within the ordinary tie pattern of the access scaffold so that there is one tie on every storey next to the hoist.

11.29 Wind forces

Wind forces may be considerable on the hoist tower and mast. The sides of the tower will probably be covered with mesh, and three sides will be braced. Two sides will be wind braced. The platform and its load will present 1–4 m^2 to obstruct the wind, while the mast may be effectively a solid face. Toe boards may be fitted at stopping levels. A passenger hoist will need special wind calculations for the cage.

These factors added together for an ordinary goods hoist and tower give a face area of 5 m^2 (59 ft^2) per storey. This gives a typical additional wind force of 200 kg (441 lb).

The centre of this force will be about 1.5 m (5 ft) outside the face of the access scaffold and 2.9 m (9.5 ft) away from the building, so it may exert considerable leverage on the ties.

11.30 Bracing

Three sides of the tower must be face braced, but the fourth side containing the entrances and exits cannot be braced.

The access scaffold uprights adjacent to each side of the hoist tower should be ledger braced continuously from top to bottom.

Horizontal wind bracing, i.e. external plan bracing, should be fixed at every other lift on alternate sides of the hoist tower.

The access scaffold should have façade bracing in the vicinity of the hoist tower.

11.31 Rigging the hoist and lifting gear

This is a matter for clear understanding between the main contractor and the scaffolding contractor because it is governed by the Construction (Lifting Operations) Regulations 1961. Responsibility under these regulations lies with the user, so it is preferable for all lifting gear to be rigged, inspected and tested by the main contractor.

11.32 Hazards with hoists

- Operatives putting their heads inside the tower into the hoist platform zone.
- Materials falling from the platforms.
- Inadequate maintenance and inspection of mechanical, electrical, structural and safety features.
- Impact damage.

PART II
Special scaffolds

12 Access birdcages

12.1 Multi-lift birdcages: general description

The term 'birdcage' is used in the scaffolding industry to mean a structure several bays long and several bays wide that can be boarded over the whole or a large part of the top lift. Such a structure is required to decorate a ceiling in a large building. It can also provide lifts for working platforms on its sides to permit treatment of the walls of the building.

The uprights are spaced out in both directions, depending on the loads to be supported on the top. These loads may not be uniform over the whole area. For instance, the fixing of distribution pipes in a sprinkler system in a roof gives a relatively light loading, but the fixing of the large supply mains to the system may result in very heavy local loads. The spacing of the uprights must be varied to accommodate differences of this nature in the same birdcage.

If there are no heavily loaded areas and only general access for painting and cleaning is to be provided, the load on which the design should be based is 0.75 kN/m² (15 lb/ft²). This is quite small, and the user must be aware of this design figure and arrange his work accordingly.

12.2 Design

The top lift of a birdcage scaffold consists of ledgers joining the top of the standards with transoms fixed on the ledgers at a closer spacing than the ledgers so that they can support the boards lying parallel to the ledgers.

The Code of Practice standard load for a birdcage platform is 0.75 kN/m² (15 lb/ft²). The combined weight of the top lift of tubes in both directions and the boards is 0.25 kN/m² (5 lb/ft²), so the total load on the platform is 1 kN/m² (20 lb/ft²).

Each upright supports an area extending halfway across to the next upright in each of the four directions, so it supports an area equal to that between any four uprights, i.e. the whole bay area.

Four limiting strengths must be considered: the strength of the couplers joining the ledgers to the uprights; the strength of the top lift transoms; the strength of the top lift ledgers; and the strength of the standards.

(i) The couplers

The safe load on a coupler is 6.23 kN (635 kg), so the coupler joining the ledgers to the standards can support 6.23 m² of platform. If the uprights are arranged in a square pattern, the spacing should be 2.5 m in each direction. This is a frequently used spacing for the supports of a birdcage or a slung mattress.

(ii) The transoms

These are continuous over their supports and are uniformly loaded but with joints. They will have to sustain a bending moment of $\frac{wL}{10}$, which is equal to a resisting moment of fz, where w is the allowable load, L is the span, f is the allowable stress in bending, and z is the modulus of the section of tube. The allowable stress for used tubes will be taken to be 116 N/mm².

The safe load for the 2.5 m span of the transoms is derived from the formula $w = \frac{10\,fz}{L}$.

Putting figures into this formula,

$$w = \frac{10 \times 116 \text{ N/mm}^2 \times 5.7 \times 10^3 \text{ mm}^3}{2500 \text{ mm} \times 1000 \text{ N/kN}}$$

$$= 2.65 \text{ kN (270 kg, 595 lb)}.$$

In Table 31.18 for used tubes with partial end fixings, the corresponding figure is 275 kg. For new tubes, the safe load is 323 kg. Consider the case of a transom placed mid-span along the ledgers. It has to support 2.5 m × 1.25 m × 1 kN/m² = 3.13 kN (319 kg, 703 lb). New tube transoms are safe for this load.

The transoms may not be spaced at 1.25 m centres, i.e. one for each line of uprights and one in between. They may be spaced on site at the full allowable spacing of 1.5 m. If this is the case, they will be loaded with 2.5 m × 1.5 m × 1 kN = 3.75 kN (382 kg, 843 lb). Used tube will be overstressed by 39% and new tube by 18% on the tabulated values.

(iii) The ledgers

The transoms lying under a wide fleet of boards are uniformly loaded, but the ledgers are not. At worst, there may be a transom at the centre of each 2.5 m ledger applying a centre point load. This will apply a bending moment of $\dfrac{wL}{6}$ for both continuous and jointed conditions.

For used tube with this amount of fixing, the safe load is

$$\frac{6 \times 116 \text{ N/mm}^2 \times 5.7 \times 10^3 \text{ mm}^3}{2500 \text{ mm} \times 1000 \text{ N/kN}}$$

$$= 1.59 \text{ kN (162 kg, 357 lb).}$$

(Table 31.18 gives 165 kg for used tube and 194 kg for new tube.) The applied load is 1 kN/mm² × 2.5 mm × 2.5 mm = 6.25 kN (637 kg, 1405 lb). Used tube will be nearly four times overloaded, and new tube will be 3.25 times overloaded. This situation is unacceptable. To overcome the deficiency, two more ledgers, one either side of the one just fixed, must be slung below the transoms.

(iv) The standards

The load on the standards will be 6.25 kN (637 kg, 1405 lb), which is acceptable for lift heights of less than 4 m.

These calculations show that it is impossible to construct a 2.5 × 2.5 m birdcage on single ledgers without them being overstressed. In practice, many birdcages are built with standards at this spacing because the strength of the boards is expected to come into play to assist the ledgers. If the boards are butted on transoms at the centre of the ledgers, however, it is possible that they will not be able to assist the ledgers, and a collapse may occur.

Therefore, one of four methods of overcoming the ledger overstress must be adopted:

- Reduce the grid dimensions of the uprights and particularly the spacing governing the length of the ledgers.
- Double the number of ledgers.
- Arrange the platform boards so that they cannot fail to assist the ledgers.
- Prop up the centre point of the ledgers by A frames or reduce their spans by knee bracing.

(i) Reducing the grid dimensions

The spacing on site might be the full allowable board span of 1.5 m. Starting from the relation,

$$\frac{wL}{10} = fz,$$

$w = 1.5 \text{ mm} \times L \text{ m} \times 1 \text{ kN/m}^2 = 1.5 \, L \text{ kN}$, $fz = 661200 \text{ N/mm}$. For used tube,

$$L2 = \frac{661200 \times 10}{1.5} = 4408000,$$

$$L = 2100 \text{ mm (2.1 m).}$$

The load transferred from the transom to the centre of the ledger is 1.5 m × 2.1 m × 1 kN/m² = 3.15 kN (321 kg, 708 lb). Referring to Table 31.18, with used tube the ledger length can be 1270 mm. With new tube, the ledger length can be 1.5 m.

A birdcage or a slung mattress with supports at a 2.1 m by 1.5 m spacing is a costly structure, but this is what the stresses demand if the boards are to play no part and the top lift has single ledgers.

(ii) Doubling the number of ledgers

The ledgers cannot be doubled one below another with each one attached to the upright. The method of doubling is to couple the second ledger underneath the transoms beside the first ledger and close to it. This relies on the shear strength of the transoms near the uprights and the shear strengths of the first ledgers to transfer the load to the uprights. There is ample shear strength available. Trebling the ledgers for extra-heavy loads is achieved by under-slinging one on each side of the first ledger. The underslung ledgers are coupled to every transom.

When this artifice is adopted, it is as well to reduce the transom spacing from 1.5 m to 1.2 m, which is no hardship with 3.96 m boards. Replacing 1.5 by 1.2 in the calculation in (i) above, the transom length becomes approximately the traditional value of 2.4 m (8 ft). The transom load is 2.4 m × 1.2 m × 1 kN/m² = 2.88 kN (293 kg, 647 lb).

Using doubled ledgers, the ledger load is 14.7 kg and, from Table 31.21 for used tube, the required ledger length is 2.75 m. This system produces a more economical spacing of the uprights of 2.4 m by 2.75 m, or 2.5 m by 2.5 m in practice.

(iii) Arranging the platform boards to bring their strength into play to assist the ledgers

In order to ensure that the boards play their part, the birdcage must be dimensioned so that the joints in the boards are always at the location of the uprights. Then, if only

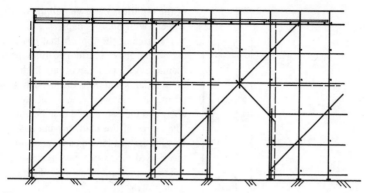

Figure 12.1 A multi-lift birdcage.

one bay is loaded and the loading passes straight downwards through the ends of the boards on to the doubled transom at the joint, it does not apply a centre point load on the ledger. This arrangement, which is shown in Figure 12.1, has the second advantage that the intermediate transoms along the length of the board are not at the centre of the ledgers in any bay but at one-third of the way along the ledgers, which relieves the stresses in them. By this technique, single ledgers can be used for the Code of Practice loadings, and economical layouts can be bought that will support much heavier loads than the Code values. A circumstance where this is required is for the materials storage bays in a birdcage, e.g. when fire precaution sprinkler mains have to be raised and stored on the birdcage platform.

An analysis of the arrangement shown in Figure 12.1 can start by considering the ledger with a span of 1.98 m and loaded not at its centre but about 660 mm from one end. When loaded here, its supporting ability for a point load is about 1.125 greater than its centre point load strength. Using Table 31.18 for used tube partially restrained, it will carry 2.28 kN (233 kg, 514 lb). The load actually applied to the ledger is 6.98 m × 2.4 m × 1 kN/m² = 4.752 kN (484 kg, 1068 lb). The difference of 2.47 kN has to be taken across the span by the boards, of which there are ten.

Table 33.3 gives an allowable bending moment for 38 mm boards of 468 N. With a span of the full bay width of 1.98 m, each board will sustain $\dfrac{0.468 \times 8}{1.98} = 1.89$ kN.

Two boards adjacent to the uprights will relieve the ledgers of 3.78 kN and so make the single ledger safe.

(iv) Propping the ledgers

This method is by far the most satisfactory. It avoids the hazards of errors of construction on site and it avoids any consequence of different flexibilities between boards and tubes. Section 12.4 deals with this method.

12.3 Hazards with the top lift of birdcages

- Failure to realise the degree of overstress in the top ledger of a birdcage with the standards set at 2.4 m × 2.4 m spacings can result in ledger failure.
- When the tube joints and the board joints are both at the centre of the span, failure is almost certain.
- Deviation from the layout correctly set down on the drawing can be very serious.
- Inadequate bracing can allow distortions, which increase the stresses on the already highly stressed ledgers.
- Where it is necessary to avoid the top of the standards projecting through the decking causing 50 mm gaps, the use of a joint such as that shown in Figure 12.2(a) results in a rotational weakness, even if there is adequate vertical strength. Overstressing of the ledgers can sometimes be overcome by knee bracing from the standards to the mid-point of the ledgers or by knee bracing to shorten the span. See Section 12.9.

12.4 Propping the ledgers

Where the ledgers in a birdcage are overstressed, or there is a temporary local loading that causes this for a short time, intermediate standards can be introduced. A typical application is where there is a materials storage bay.

Figure 12.2(a) shows how eccentric loads on the uprights result from the use of a single coupler on the supplementary standard. Figure 12.2(b) shows that if an additional ledger is introduced so that the uprights are coupled with two couplers, the end rigidity of the standards is greatly increased. In some cases, rotation may be prevented completely, resulting in nearly axial loads on the standards.

12.5 Construction details

If the birdcage is within a building, protection of the floor surface may be important, and the spreading of the load

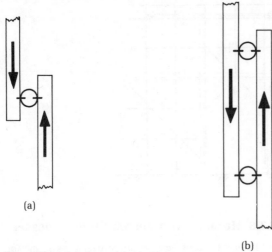

Figure 12.2 Splicing uprights in birdcages.

may require timbers to be laid at right angles to the joists in the case of wooden floors. Some of the standards may have to be placed on stairs and, if these are of timber construction, their strength must be carefully considered.

Construction of the tubular framework differs from ordinary access scaffolding in that the transoms are not fixed to the ledgers but to the uprights, using right-angle couplers. This results in the birdcage having right-angle couplers on the uprights facing in both directions, which improves the stability of the whole structure.

In the top lift, however, this technique is not possible because there must be board-bearing transoms between those attached to the standards. These have to be attached to the ledgers with putlog couplers, which would result in board supports at different levels. Accordingly, the whole

of the top lift is built with all the transoms attached to the ledgers with putlog couplers.

The area of the top lift that has to be used is boarded out with ordinary scaffold boards supported in the orthodox manner. A 50 mm gap occurs in the platform in line with the standards that protrude through it, and this is generally permissible. However, in some cases it may not be acceptable, and the gap should then be covered with hardboard. The edges of the platform and any ladder holes or hoist ways in it must have guard rails and toe boards similar to those used on an ordinary access scaffold.

The user should be notified of the allowable working load in any area.

If free standing, the structure requires bracing in both directions. The bracing should be fixed at the foot of one standard and connected to all the lifts to the top. For interior birdcages, alternate lines of uprights should be braced, starting at the foot of every sixth standard, or every 15 m. The brace can zigzag to the top or be in a continuous diagonal. For outside birdcages subject to normal wind forces, twice this amount of bracing will be required, i.e. a brace every 15 m in every line of uprights. Places of exceptional exposure will need wind force calculations to establish the bracing patterns.

It is preferable to start the bracing on the line of standards one line in from the edge because the edge line has standards with only half the vertical load of the interior ones.

Where practical, the ledgers and transoms should be butted up against the walls of the main structure and box-tied around any columns suitably placed. The butting of the transoms and ledgers may reduce the amount of bracing required, but should not be relied on fully because in some plan shapes a spiral failure may occur, even in a birdcage butted on all sides. This is especially possible with those that are approximately circular.

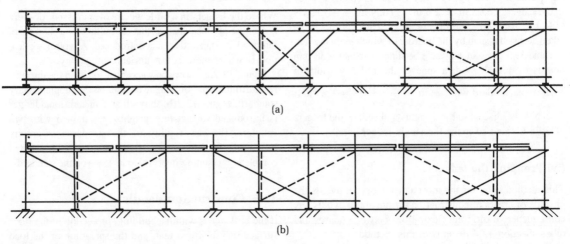

Figure 12.3 Single-lift birdcages.

The joints in the ledgers and transoms should be staggered so that the stabilising effect of the bracing and the butting can be distributed safely throughout the birdcage.

The erector must watch out for cases where opposite sides can be butted on to the walls so that reduced bracing is required in the cross direction, while only one end can be butted in the other direction. Full bracing in that direction will then be needed.

If the walls of the main structure have to be decorated, it will not be possible to butt the tubes on to them. In this case, the perimeter bays of the birdcage can take the form of an ordinary access scaffold and contain the bracing normally used.

In external birdcages, wind force calculations will be needed to establish the stability of the structure. Any birdcage where the height is about equal to the shorter base dimension is liable to overturn, and anchors should be installed where necessary. Chapter 37 gives the relevant data on the overturning of external structures.

12.6 Hazard with spiral failure in birdcages

- In addition to the spiral failure that may occur in a birdcage in a circular tank where bracing has been omitted in the belief that butting up to the circular wall in several directions will suffice, there is the same weakness in a scaffold on the outside of a circular tank. The butted tubes offer no lateral restraint to tangential twisting.

12.7 Single-lift birdcages

These differ from multi-lift birdcages in that there are no intermediate lifts in which the transoms are coupled to the standards with right-angle couplers. In the direction parallel to the transoms, there is very little gain in stability from the stiffness of the couplers being aided only by the rotational grip of the putlog couplers around the tube. Bracing is important. It is desirable to have a foot lift in place so that at least that lift has right-angle couplers in both directions.

Records show that there are more failures in single-lift birdcages than in multi-lift ones. The reason is that the erector is lulled into a false sense of security by the low height of the structure rather than being alert to its dangers.

The upper lifts in a multi-lift birdcage and the bracing in these lifts help to maintain the standards upright. This advantage is not present in the single-lift birdcage.

Suggested forms of bracing, leaving corridors for access, are shown in Figure 12.3. Where practicable, the ledgers and transoms should be butted up against the walls of the main structure and box-tied around any columns suitably placed.

12.8 Hazards in single-lift birdcages

- Single-lift birdcages are frequently built inside buildings, and it may be undesirable to butt on to the walls if these are to be decorated. This leads to instability.
- Small areas frequently have to be isolated because of equipment in the building. These areas of single-lift birdcages are very susceptible to instability.
- Birdcages are frequently progressive jobs. They are continually dismantled at one end and extended at the other to enable the main work on the building to be continued. Such cases are the fitting of suspended

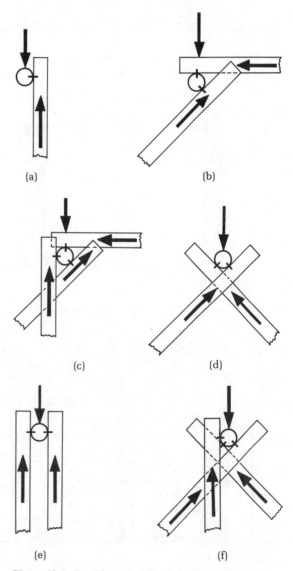

(a)   (b)

(c)   (d)

(e)   (f)

Figure 12.4 Top-lift couplers for birdcages supplementing the top ledger and transom strengths.

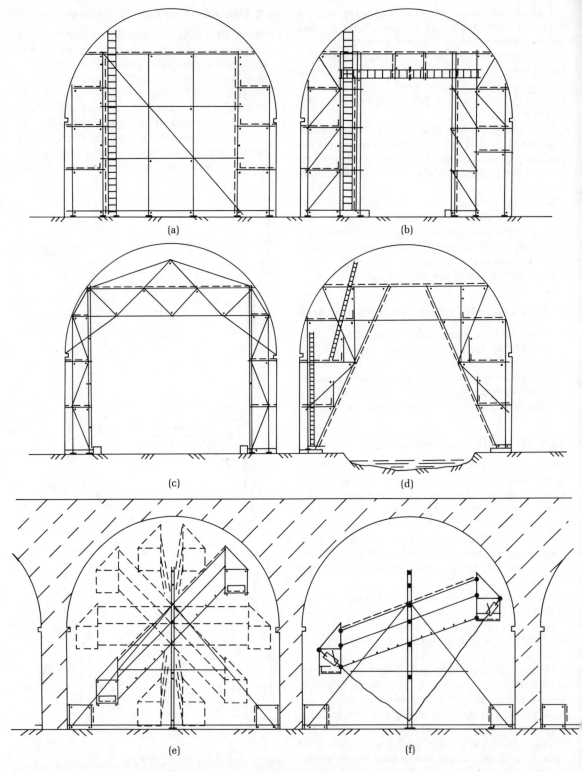

Figure 12.5 Pier-and-beam structures can be used as alternatives to birdcages.

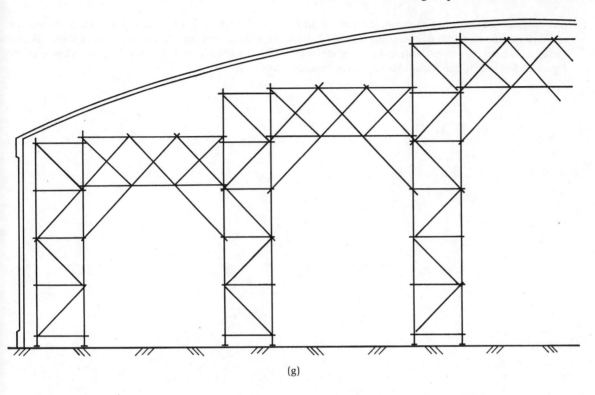

(g)

ceilings, electrical wiring, and the fitting of lights and pipework for sprinklers. The scaffold may be well built on the first occasion, but its accuracy of detail may deteriorate every time each section is reconstructed until it becomes unsafe. It may lose its bracing and the staggering of the joints in its top lift may be lost.
- The user may become overconfident about its load-carrying capacity, forgetting that it is designed for an imposed load of only 0.75 kN/m^2 (15 lb/ft^2).
- The user may also carry on activities beneath it and damage or displace its members.

12.9 Arrangements using one or two couplers to support the top lift

Different arrangements of the supporting tubes for the top lift are worthy of note, especially for a birdcage required to carry up to 2 kN/m^2. In the normal arrangement, only one coupler supports the load at the top of each upright. Figure 12.4(a) and (b) shows this.

By supplementing the support tubes, the load can be supported on two couplers at each node point, as shown in Figure 12.4(c), (d) and (e).

For very high loads, three couplers can be incorporated, with one coupler on the upright and one on each of two inclined rakers. Figure 12.4(f) shows this arrangement.

The following list gives the maximum safe load supported by couplers on rakers at various inclinations and in various combinations:

One right-angle coupler on a vertical leg	635 kg
One 45° knee brace	556 kg
One right-angle coupler on a vertical leg plus one 45° knee brace	1253 kg
Two 45° knee braces	1734 kg
Two right-angle couplers on two vertical legs	1341 kg
One right-angle coupler on a vertical leg plus two 45° knee braces	1948 kg
One right-angle coupler on each part of a staggered vertical leg	635 kg
Two right-angle couplers on each part of a staggered vertical leg	1341 kg

Refer to Table 32.1 for the relaxation of the safety factor for supplementary couplers used in this ledger support system.

12.10 Alternatives to birdcages: pier-and-beam scaffolds

In many cases, economies can be achieved by bridging over areas in birdcages or by designing movable scaffolds on which are rotating platforms.

Figure 12.5(a) shows a standard birdcage that can service

both the sides and soffits of a vault, while Figure 12.5(b) is a variation allowing central access, and (c) and (d) are variations to cater for various loads and floor conditions. Figure 12.5(e) and (f) shows scaffolds that can be rotated around a central point. Similar systems can be mounted on trolleys running on tracks on the ground. Figure 12.5(g) is suitable for dealing with large shells and reservoir roofs.

13 Access towers in tubes and couplers

13.1 Description and dimensions

The term 'tower' is here intended to mean a square or rectangular construction not longer than three or four times its width. It is assumed to be made of either 8 gauge steel scaffold tube or 7 gauge aluminium scaffold tube. Steel couplers are assumed. A full deck area of scaffold board is assumed, with guard rails and toe boards all around. The calculations given for stability against wind forces on aluminium towers do not apply to thin-walled prefabricated tower tubes with an external diameter of 50 mm, which are much lighter than standard aluminium scaffold tubes.

The Code recommends specific height to base width ratios for static and mobile steel towers used inside and outside:

Internal steel towers: static   maximum $\dfrac{\text{height}}{\text{base}}$ ratio 4

                    mobile maximum $\dfrac{\text{height}}{\text{base}}$ ratio 3.5

External steel towers: static   maximum $\dfrac{\text{height}}{\text{base}}$ ratio 3.5

                    mobile maximum $\dfrac{\text{height}}{\text{base}}$ ratio 3

These are empirical rules based on experience with steel towers. The internal ratios take into account that towers are frequently used on imperfect floors. The external ratios must not be assumed to give a tower that will be stable in all wind conditions. The ratios are intended for factory maintenance jobs of short duration. A tower with a height to base ratio of 3 will blow down in a 51 mph wind. This is not an unexceptional wind speed. It quite frequently occurs at the corner of a factory on an industrial estate.

A tower made of aluminium scaffold tubes will blow down in a 40 mph wind. A thin-walled tube prefabricated tower of the same proportions is very much lighter. For such a tower, the supplier should be consulted about the limitations of its use.

13.2 Stability

Figure 13.1 shows towers of six different heights and weights with different face areas to accumulate wind force. As the height of the centre of gravity increases, the slope of the wind force and weight resultant must steepen if the

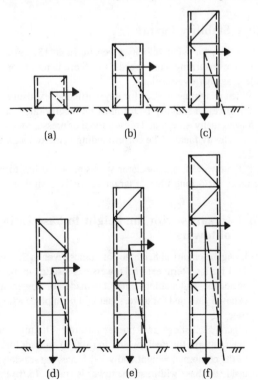

Figure 13.1 Safe wind speeds for external towers without anchors.

resultant is not to pass outside the base area, so the critical wind force expressed as a fraction of the weight must decrease. The weight of a six-lift tower can only resist a total wind force of about 7/12ths of its weight. The face area of the taller tower is greater than that of the smaller tower, so the wind pressure that it can sustain is very much less than can be resisted by the smaller tower. Hence the safe wind speed is very small, being proportional to the square root of the pressure.

In practice, the centre of the wind force is always higher than half height because the wind speed increases at higher levels, and the top deck and its toe board add face area near the top. The centre of the wind force can usually be taken to be 5/8ths of the height of the tower.

The above discussion demonstrates how essential it is to calculate the stability of towers. A stability safety factor must also be applied.

Aluminium towers are only about one-third of the weight of steel towers, so stability is very much reduced. Many accidents occur every year with aluminium access towers. Because aluminium is so light, the centre of gravity of the tower can be as high as the underside of the top decking. Thus, smaller horizontal loads displace the line of the resultant force towards the edge of the tower much more quickly than with a steel tower.

13.3 Stability factors

The values for wind velocity given in Table 13.1(a) and (b) are overturning wind speeds, i.e. there is no factor of safety against overturning.

When designing towers for any particular exposure to wind forces, a stability factor must be incorporated, and 1.5 is normally accepted. This is a force or pressure factor, not a velocity factor. The corresponding velocity factor is 1.22.

If the tower is in a spectator sports area and has to be in place for a long time, a larger factor is desirable.

13.4 Hazards with the height to base ratio of towers

- Using the empirical height to base ratios given in Section 13.1 for long-term external towers has resulted in many accidents. Calculations must be made for towers in exposed sites and for towers that will be in place a long time.
- Figure 13.2 shows two further hazards. The top four diagrams are for steel towers. In (a) the resultant force of the combined weight and wind forces passes down inside the base width, so the tower is stable. In (b) the resultant force just passes through one side pair of legs, and the tower is about to fall over. In (c) the resultant

Table 13.1(a). Wind velocities that will overturn square steel towers of various widths and heights, including the platform, toe boards, guard rails and one man on the top

Number of lifts (each of 2 m high)		Wind velocity		
			Base width	
		2 m	3 m	4 m
1	m/s	35	46	57
	mph	78	102	128
2	m/s	26	35	42
	mph	58	78	94
3	m/s	23	30	36
	mph	51	67	81
4	m/s	20	26	32
	mph	45	58	72
5	m/s	18	23	28
	mph	40	51	63
6	m/s	17	21	26
	mph	38	47	58

Note: There is no safety factor against overturning in these figures.

Table 13.1(b). Wind velocities that will overturn square aluminium towers of various widths and heights, including the platform, toe boards, guard rails and one man on the top

Number of lifts (each of 2 m high)		Wind velocity		
			Base width	
		2 m	3 m	4 m
1	m/s	30	40	50
	mph	67	89	112
2	m/s	22	29	36
	mph	49	65	81
3	m/s	18	24	29
	mph	40	54	65
4	m/s	16	22	26
	mph	36	49	58
5	m/s	14	20	24
	mph	31	45	54
6	m/s	13	17	22
	mph	29	38	49

Note: There is no safety factor against overturning in these figures.

force passes outside the base width, and the tower will fall over. The horizontal force is the overturning force of the wind.

When the tower begins to fall over, as in (d), the righting moment lever arm decreases rapidly. l_4 is less than l_3, so the tower rapidly becomes less stable and the overturning proceeds with accelerating speed. Recovery is impossible once the wheels on the windward side lift from the ground.

The lower four diagrams are for aluminium towers. The vertical force from the weight is only about half of that for

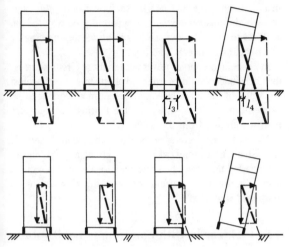

Figure 13.2 Hazards with steel and aluminium towers.

the steel towers because aluminium parts are only about one-third of the weight of steel parts. Hence the corresponding horizontal wind force for conditions (e), (f) and (g) are only about half of those for (a), (b) and (c). The condition represented by (d) also applies in (h). As soon as the upwind wheels lift off the ground, the lever arm for the weight righting moment reduces rapidly and the tower will continue to fall with a wind force that is half of that for the steel tower.

13.5 Construction details

Foot ties

Towers should have foot ties set as low as is practicable.

Castors

Castors should be attached to the tower legs and the wheels should have brakes.

Plan bracing

In static towers, the ground is a good plan brace at the bottom, and the working platform is usually a good enough plan brace at the top. If the tower is higher than four lifts, a plan brace is advisable.

In mobile towers, the castors prevent the ground fixing the location of the legs, so a castored tower can be pushed out of square into a diamond shape. A plan brace must always be used at the bottom of a mobile tower and preferably at every alternate lift above.

Contrary to popular belief, plan bracing will not prevent

torsional rotation in a tower. The lifts containing the plan bracing will retain their shape but will twist in the same way that they might if the plan bracing were not present. Twisting can only be prevented by the provision of adequate face bracing.

Couplers

For maximum stability, the ledgers and transoms should be attached to the uprights with right-angle couplers, except for the top lift, where putlog couplers may be necessary for the transoms.

Braces

All four sides on all lifts should be braced, preferably using right-angle couplers on the horizontal tubes rather than swivels on the uprights.

Guard rails and toe boards

The platform must have guard rails and toe boards in compliance with the Regulations. There is a problem at the side of the access ladder. If a gap is left in the guard rails and toe boards, a person may fall through it but, if there is no gap, he has to climb from the ladder over the toe board and under the guard rail.

The Code proposes a hinged guard rail and a loose piece of toe board. This may comply with the Regulations, but it is a hazardous solution of the difficulty. A person is not safe on a vertical ladder when holding on with only one hand, and even less safe if the other hand is manipulating a guard rail that moves. A loose toe board is not acceptable because it may be used as a step by the person entering the platform.

The best solution of the difficulty is to fix a normal full-length toe board over which one treads, and to set the guard rail about 600 mm back on to the platform so that there is a small landing inside the toe board but outside the guard rail.

13.6 Vertical loads

Unless otherwise stated, a tower should be designed for a vertical load of 1.5 kN/m² (30 lb/ft²). Any special duty requiring a larger load or a point load should be given very careful consideration, especially in the case of a mobile tower.

If loads such as television cameras are to be raised on to the tower, their weight must be ascertained, and the force resulting from the pulley system calculated and inserted into stability calculations. Account must be taken of the lever arm of the force outside the tower.

13.7 Hazard with overloading towers

- Because work involving towers is frequently in connection with buildings that are completed and in use, there is often a tendency to place all operational materials and plant on the tower, e.g. compressors, drills, lifting gear, cleaning water tanks, paint storage, ventilation plant, etc. The design load may be considerably exceeded in the case of a prefabricated lightweight mobile tower designed only for 0.75 kN/m^2 (15 lb/ft^2). Also, the imposed weight may be stored on one side of the platform. If the users obediently come to the ground when the tower has to be moved but leave the weight on the platform when they push the tower along, there is an overturning potential.

13.8 Ladders on towers and external vertical loads

If a ladder is attached to the side of a tower, it will contribute to the overturning moment in one direction. If the ladder is offset a considerable distance, and castors on the tower are turned inwards, the ladder weight may be very significant.

A person climbing a ladder exerts a further overturning moment, which will require consideration in the case of small towers.

The raising of loads using a rope or more particularly a rope over a gin wheel will apply an overturning moment that has to be considered.

Sloping ladders may be dangerous against a mobile tower. Stepping off a ladder resting on a wall on to a mobile tower may also be dangerous.

13.9 Horizontal forces

Horizontal forces of any sort should be avoided unless they are specially catered for. Pulling mobile towers along from the top is a recipe for disaster. They must be moved only by forces at the bottom. No person or load should be on the top of a tower when it is moved.

A workman drilling into a wall with a power tool can exert a horizontal force at guard rail level of 20−25 kg, and with greater effort he may push with a force of 30−40 kg. When the tower is used specifically for horizontal drilling, a good factor of safety against overturning must be used − at least 2. To calculate stability and ensure safety, the user must ascertain the technical data of the tower that he intends to use and the rules for its use, as set down by the manufacturer.

13.10 Towers in confined spaces

Attention must be paid to stabilising a tower, even if it is in a confined space such as a lift shaft or inside a chimney. In such circumstances, there is apparently no way for the scaffold to fall sideways, so it is built without bracing or other means of lateral stability.

In a tube-and-coupler tower used in these circumstances, the horizontals should be extended alternately to butt against the walls of the main building. Sometimes work is to be carried out on these walls, and the butting transoms are then withdrawn from the walls.

The legs may collapse because the strut length is excessive, or they may twist around, giving a spiral failure.

Outside a building, the same situation can occur with an access scaffold or scaffold tower in the recesses of an H-shaped building.

13.11 Hazards with continuous horizontal force on towers

- When using a tower that is not tied or guyed, its stability is dependent on the product of its weight and half its width, i.e. on the righting moment of the self weight. If a continuous horizontal force is applied to the tower, giving a continuing overturning moment, and this is increased to the critical value, the tower will begin to overturn and the legs on one side will lift off the ground. The centre of gravity will drift quite quickly towards the pivot point and the restoring moment will decrease quickly. Hence the tendency for the tower to turn over will increase as soon as the legs on one face of the tower come off the ground and if the force is maintained, the tower is doomed to overturn. Many stability problems have the failsafe characteristic that the further the structure is deflected the more it resists deformation, but the opposite is true with unanchored scaffold towers, and this must never be ignored.
- If a tower is being pushed along a factory floor, and the leading castors trip over debris or a joint in the floor, the rear castors may come off the ground. The system described above then takes over and, if the force is maintained, the tower will fall.
- Special attention must be given to the manner in which a tower is anchored or tied. If it is tied to a wall to prevent it falling away from the wall, the tie may not stop it blowing over laterally.
- The manufacturer's literature giving the weights of the prefabricated elements and of the assembled tower must be obtained and used in stability calculations.
- In the case of small-bore steel tube and aluminium towers, there is very little weight available to provide the righting moment. The designer must therefore assess the horizontal forces very carefully and ensure that only one man applies them at any one time.
- A workman can exert a much greater horizontal force than might be expected. The decking will probably be

of timber, and the user may have composition soles on his footwear. The coefficient of friction between his feet and the tower platform may rise to 1. The designer can test the friction in any given circumstance by standing on the platform on the ground and leaning forward on to a wall while preventing himself falling by his hands. The flattest angle that he can achieve will indicate the friction grip of his shoes on the platform material. He will find that he can lean progressively against the wall until he is only about 30° above the horizontal. If he weighs 80 kg and is 1.8 m tall, he will exert a 'falling over' moment of 72 kg/m. The righting friction moment of the force couple formed by the floor friction and the wall reaction on his hands will also be 72 kg/m, with a lever arm of 0.9 m, giving a friction force of 72 kg. Thus, a workman drilling into a wall from the platform of a tower may exert a horizontal force that is much greater than the customarily assumed value.

- Workmen ordered to use towers not tied to the structure must be specifically warned of the hazards.
- Several accidents have been caused by progressive deterioration in the accuracy of assembling a tower during a job requiring repeated use of the tower. The tower is correctly built, tied and used for the first few moves, but carelessness and familiarity result in pieces being left out or badly fixed later in the job. Getting behind with the progress of the job and having to assemble the tower in a hurry results in bad assembly. The inclusion of these errors then becomes the normal habit of the job.
- There have been accidents caused by leaving the couplers on the uprights of a tube-and-coupler tower to save the effort of refixing them during re-erection. In several cases, the couplers left on the uprights have caught in a workman's clothing as he lowered the elements to the ground and he was pulled off the tower.
- A hazard frequently arises from the use of a tower built around the columns of a structural steel framework during its construction. The structural steel beams are held by the crane and bolted into place by the workman on the platform. The tower is easily box-tied around the column and is safe to use, but it has to be dismantled to be re-erected around the next column. The steelwork rigger then decides to build the tower outside the column so that it can be wheeled to the next column without being dismantled. The tower is used untied and blows over or is knocked over.
- At the start of a factory extension job, the towers may be in a sheltered location. As open steelwork erection progresses, the towers are moved to locations more exposed to the wind and may blow over if changes in the tying method are not made.
- In the reverse situation, towers at the start of a job may be correctly anchored, guyed or held but then moved

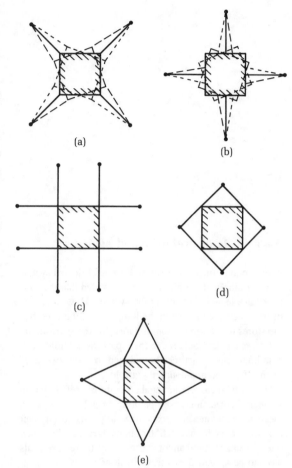

(a)

(b)

(c)

(d)

(e)

Figure 13.3 Methods of buttressing or guying towers to resist torsion.

into what appears to be a sheltered vicinity. The anchoring system is omitted and overturning results.

13.12 Guying towers for stability

Figure 13.3 shows five methods of guying a static tower. Methods (a) and (b) will provide resistance against the tower overturning but, as demonstrated by the dotted lines, these methods will not resist torsions in the structure.

For television cameras, it is necessary to prevent torsion as well as overturning. Methods (c), (d) and (e) provide restraint against torsion. Method (c) has eight guys on eight anchors, and such an arrangement may be necessary in exposed places on poor ground. Methods (d) and (e) have fewer anchors.

Section 36.21 gives details of attaching guy ropes to the anchors and to the scaffold. Methods (c), (d) and (e) in Figure 13.3 are feasible with only four ropes using the methods of attachment shown in Figure 36.1.

14 Tower-and-bridge scaffolds, frame scaffolds and system scaffolds

14.1 Costs of prefabricated systems

The erection time on site is less for prefabricated systems than for tube-and-coupler towers, provided that the job is of a suitable shape for using the system. However, some of the time saved is lost in the longer handling time from the stores to the lorry, from the lorry to the scaffold area and from the scaffold area to the top of the scaffold. Total man hours can sometimes be reduced to about two-thirds of that for tube-and-coupler scaffolds.

However, the overall annual costs do not reflect the site savings. Fabrication costs, painting or galvanising costs, mechanical maintenance costs, handling costs during maintenance and the high rejection of units during maintenance, due to damage that cannot successfully be repaired, also have to be spread over the work done.

Then there are job design charges and technical supervision. On the job, there is the high cost of the necessary adjustable legs.

When all this is added up, there is no saving in prefabrication over tubes and couplers. In fact, there is an adverse costing. The pundits of prefabricated systems do not agree with this, but the proof of the pudding is that in the last 20 years more than a dozen 'systems' have been marketed and none has survived to claim a decent share of the market.

14.2 Design principles for tower-and-bridge scaffolds

Obviously, the towers must be able to carry the loads imposed, and the beams must be able to support the loads that they collect and transfer to the towers. The whole must be laterally rigid in both directions, along and at right angles to the building. Stability must be maintained as the scaffold structure is progressively built and dismantled.

The towers are not loaded down their centre lines. The loads from the beams pass to the ground through the uprights at one side of each tower. With alternately loaded bays, the towers may be loaded on only two legs.

It is difficult to assess the supporting ability of a tower loaded only down two adjacent legs. In addition to the vertical load stress, there is the bending stress from the attachment of the beams to one side of the uprights and from the deflection of the beams. There is the rotational stress due to the beams trying to turn over sideways under their load. Another bending stress is caused by the addition of knee braces or V braces below and above the beams. All these factors must be assessed for the irregular loading conditions found on a construction site.

Ties to the building must be more frequent than for conventional tube-and-coupler scaffolds because a tie can only affect the lateral stability of the tower to which it is attached, whereas in tube-and-coupler structures it can supplement others quite distant from it.

If the frames making up the tower are not joined one on top of another with joints capable of tension, the designer must investigate the wind uplift possibility. There are several cases on record of the top part of a prefabricated scaffold being lifted off the bottom part.

The problems associated with the joints in the vertical members of prefabricated tower-and-bridge scaffolds are not present in tower-and-bridge scaffolds constructed with tubes and couplers. The economies of the tower-and-bridge scaffold can be achieved in tubes and couplers.

14.3 Hazards with tower-and-bridge scaffolds

- There are difficulties handling ledger beams safely and in holding them still when in place to enable them to be coupled into the assembly.
- If one of the towers is hit by a vehicle or if its foundations are made inadequate, there will be a major collapse.
- There is a danger that with repeated use and abuse, the prefabricated elements may develop cracked welds, which pass unseen in the scaffold depot.

- There can be local overloads in tube-and-coupler scaffolds, giving local zones of overstress, but the rest of the scaffold will come to the rescue of the overstressed zone following its deflections. This is unlikely to be possible in tower-and-bridge scaffolds, so misuse may produce a collapse condition more readily.

14.4 Frame scaffolds

Several systems have been tried using frames 1250 mm or 1375 mm high for brickwork and frames 2 m high for other duties. Ledgers, transoms and longitudinal bracing can be prefabricated or made of straight tubes.

The system requires that all the frames are set in a horizontal plane, so the lowest lift has to be mounted on adjustable legs. This is costly.

With all prefabricated systems, if the profile of the building suits the frame system, erection is rapid but if the building is unsuitable for scaffolding by frames, too much site time is lost manipulating the system to match the building.

14.5 Design principles for frame scaffolds

Suppliers of frames and associated elements must provide the safe working loads for each unit and the rules that must be followed to ensure that the assembly can carry the loads claimed.

A frame system with some of the elements missing can be a structural disaster, so designers must avoid setting out a deficient structure on the drawing board.

Tying must follow the same rules as for tubes and couplers. Bracing must also follow the same rules unless provided for by welded elements in the system.

The platform supports must be spaced according to the thickness of the boards and not to the frame system to comply with the Regulations.

14.6 Hazards with frame scaffolds

- A frame scaffold appears very robust, but prefabrication does not enhance the ability of its uprights to resist loads better than normal scaffold tubes laced at the same vertical intervals.
- The lateral resistance for small deflections may be worse than for tubes and couplers. After a considerable deflection, the frames may begin to contribute to stability by their welded corners, but the deformation of the whole scaffold when this is achieved may be unacceptable.

14.7 System scaffolds

This term is currently applied to prefabricated systems in which the uprights are not welded into frames but have lugs welded on to them into which fit the prefabricated ledgers and transoms.

Some systems have prefabricated or fixed-length boards, while others have movable transoms so that orthodox scaffold boards can be used.

Bracing is prefabricated, and corner adaptors, ladder fixings, guard rails, toe boards and other accessories are usually available for each system.

14.8 Design principles for system scaffolds

Suppliers of system scaffolds must provide the safe working loads and the rules that must be followed to ensure that the assembly can carry the loads claimed.

The designer should ascertain the structural section properties of all the elements in the system. This is more important in system scaffolds than in frame scaffolds because the elements are simple structural tubes and not welded frames.

It is important that the designer investigates the rotational stiffness of the joints because the rules for bracing depend on this. In tube-and-coupler scaffolds, designers do not customarily attribute any stiffness to the coupler joints as far as angular stability is concerned. They rely wholly on the bracing for lateral stability. In system scaffolds, suppliers sometimes claim that the structure can be built with less bracing than in conventional tubes and couplers.

When enquiries are made of the suppliers about joint stiffness test results, i.e. the moment per radian of deformation of the joints and its variation from the initial value at the commencement of deformation to the ultimate, they do not have the data available. Until such time as they provide this information, the only advice that can be given is to provide bracing to the full extent necessary to transfer lateral forces to the ground.

14.9 Hazards with system scaffolds

- Most system scaffolds have a lift containing four joints in the uprights. This is not usually in an isolated bay but is repeated throughout the structure, despite most systems having two lengths of uprights. The same difficulty occurs in frame scaffolds. This weakness can be made good by tying to the building at the appropriate height and frequency, but this is often not done.
- The longitudinal continuity of the whole scaffold may be small compared with a tube-and-coupler scaffold due to the short ledgers. It is important to get the tying pattern right in a system scaffold, or there may be local problems of alignment and strength.
- The use of prefabricated elements may result in an inadequate supply of ordinary tubes and couplers to make the ties.

15 Cantilever scaffolds

15.1 General description

A cantilever scaffold or a truss-out scaffold is used when it is undesirable or impracticable to found the scaffold on the ground. Cantilevers, usually of steel beams but sometimes of prefabricated scaffold beams, are projected outside the building resting on the edge of a floor slab and held down at the inside end by bolts through the floor slab or by a scaffold frame above the beam and below the soffit of the slab higher up.

One projecting joist or beam is placed for each pair of uprights in the scaffold to be built, and the scaffold is then built on the steel beams or directly coupled to the prefabricated beam. It is not necessary to space the uprights in the scaffold at regular intervals, and advantage can be taken of this to place the cantilever joints at the most suitable places on the slabs rather than at even spacings.

Figure 15.1 shows cross-sections of the scaffolds at the location of the cantilevers.

15.2 Calculations

Calculations are essential to find the downward loads that will be applied to the edge of the building slab and the upward loads that will be applied on the interior of the slab. The downward loads on the edge of the slab are inevitably greater than the weight of the scaffold. The agreement of the building owner or the architect must be obtained before a scaffold is built on the cantilever principle. He will need to know the weight of the scaffold and the cantilever arm projection and tailing lengths before he can agree to the layout of the beams and the loadings on the main structure. If the scaffolding contractor has decided on the layout, he should make the load calculations for the architect and obtain his approval for both the downward and upward loads, the places at which these are to be applied, and the means of padding the fulcrum point of the beam and of the attachment of its tailing end.

15.3 Hazards related to the loads on the building

Many failures have occurred because:

- The load calculations have not been done.
- The slab is loaded with more than the weight of the scaffold.
- The slab is loaded at its edge.
- The slab is point loaded and not uniformly.
- Spreaders meant to distribute point loads have not worked.
- The uplift at the tailing end of the cantilever beam loads the slab in an upward direction, for which it is not reinforced.

When the load calculations have been done for one or two spacings of the cantilevers, it will be necessary to study the strength of the supporting slab edge and find the places where it can be loaded to cause the least stress.

15.4 Fulcrum points and tailing down points

Figure 15.1(a) shows a ladder beam cantilever. This can be padded off the slab with a spreader timber, which should be approved by the architect. The ladder beam should be prevented from rolling over by attaching the scaffold to both beam chords. The tailing end of the ladder beam should also be prevented from rolling over by attaching both chords to the tie-down bolts.

Figure 15.1(b) shows the same arrangement using a lattice beam.

Figure 15.1(c) shows a rolled steel joist as a cantilever, with the scaffold founded on forkheads upside down resting on the top flange. The joist may require web stiffeners at the fulcrum point.

Figure 15.1(d) shows a double-fulcrum bearing pad with supplementary support for the slab by a scaffold pier in the storey below the cantilever. This diagram also shows

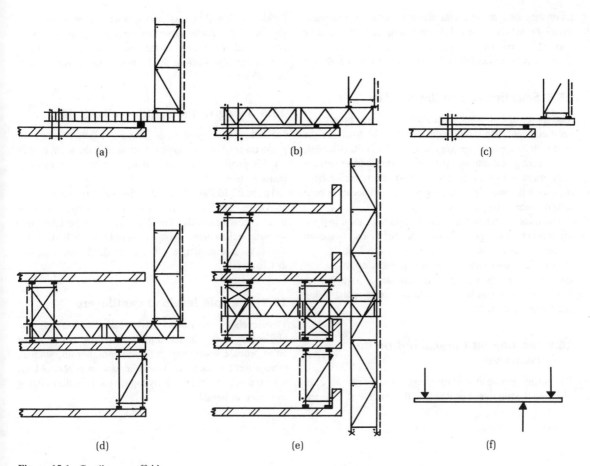

Figure 15.1 Cantilever scaffolds.

the tailing down force catered for by a scaffold pier bearing upwards on the soffit of the slab above the cantilever.

Figure 15.1(e) shows two slabs being brought into play under the fulcrum and two slabs resisting the uplift.

15.5 Calculations

Figure 15.1(f) shows the forces on the cantilever beam, whatever its type. It shows the total weight of the scaffold acting at the centre of its width. This is adequate if both the standards of the scaffold are equally loaded, but it will be necessary to work out the overturning moment of each leg at its appropriate lever arm separately if the scaffold standards are very differently loaded.

The fulcrum point is taken as the outside support, if there are more than one, because the beam will arch upwards away from the inner support.

The tailing down load is shown centred on the centre of the tailing down bolts or scaffold tubes.

The scaffold load multiplied by its lever arm from the fulcrum point is the overturning moment, and the tailing down force multiplied by its lever arm is the stabilising moment. By equating the two moments to represent the balancing condition, the tailing down force can be calculated. The attachment or propping must be at least twice the balancing force to give a safety factor of 2 for the whole system. If the balancing force is not provided by a structural attachment of the tailing end of the beam to the structure but only by Kentledge, then this Kentledge should be three times the amount calculated for the balancing condition.

The downward force applied to the slab at the fulcrum point is the sum of the actual weight of the scaffold plus the tailing down force calculated for the balancing condition.

The slab edge must be capable of supporting this downward force, which is often as much as 1.5 times the scaffold weight, at the fulcrum point.

In these calculations, the scaffold weight must include the imposed load on the scaffold as well as the dead weight.

The slab at the tailing end must be capable of resisting the uplift applied to it, always bearing in mind that this is an upward force for which the slab may not have been intentionally designed. The slab at the tailing end must also be capable of supporting the Kentledge downward dead weight if the cantilever scaffold relies on this for its stability.

15.6 Selection of cantilever joists

The choice of cantilever joists should be made by a structural engineer. He should consider that the loads on the joist may be applied over very short lengths of the flanges, which may necessitate web stiffeners. He should be particularly attentive to the low torsional restraint resulting from the fact that none of the loading or reaction points is built into or cleated on to the structure.

The rules for the choice of cantilevers to be used in these circumstances are given in the relevant British Standard Codes of Practice and will not be repeated here.

Consideration should be given to preventing rotation of the joist on to its weaker side and lateral sway of its projected end by tight box tying it at the loading points using scaffolding materials.

15.7 Starting off the scaffold on the cantilevers

It is not acceptable to start off the scaffold by placing base plates on the top flange of RSJ or UB cantilevers. Inverted forkheads should be used and spot welded or bolted on to the end of the cantilever so that the scaffold cannot vibrate off the end of the joist. In the case of scaffold tube beam cantilevers, the scaffold should be coupled to both chords of each beam.

15.8 Tying in the cantilever scaffold

Inevitably, there will be a downward deflection of the projecting end of the cantilever beam, and the scaffold built on this projecting length will tend to lean outwards in consequence.

The scaffold should be tied to the main structure 2–3 m above the cantilever, irrespective of the normal rules for tying access scaffolds. These ties can then be taken into account in the normal tying system of the scaffold, which should be in accordance with the standard rules set down in Chapter 6.

15.9 Multiple levels of cantilevers

Two levels of cantilevers are sometimes installed. The higher one, if it is attached to the scaffold, can be relied on to provide some support for loads higher up. Such an arrangement is necessary when the cantilever scaffold has to be replaced at progressively higher levels as the building increases in height.

16 Truss-out scaffolds and drop scaffolds

16.1 General description

In cases where an access scaffold cannot be founded on the ground or when it is inconvenient to do so, the scaffold can be supported on a framework fixed to the building and projecting outside it. The framework attached to the building or fixed within it is called the truss-out. The access scaffold fixed to the truss-out is an orthodox scaffold in every respect, except that it is founded on and attached to the truss-out. The access scaffold can also be built downwards below the truss-out, a method of construction sometimes called a drop scaffold.

The truss-out can be contained within the height of one storey of the building or spread over two or three. It can be on the roof and support only a drop scaffold.

There are two types of truss-out. One is a box-type structure constructed in the form of a pair of 2.5 m deep scaffold tube beams projecting outwards as far as is necessary. The other is a simple triangular bracket, in which the stability is dependent on the ties at the top of the triangle. Sometimes, a combination of both can be built, in which case the designer must be certain that both of the methods of support will work independently and that each structure will be strong enough to support the loads resulting from a failure of the other.

Figure 16.1(a) shows the normal form of truss-out scaffold with one frame inside the building for every pair of uprights outside. The holding down force is provided by screw jacks, which can be at the top or bottom of the inside frame.

Figure 16.1(b) shows a drop scaffold constructed on a similar inside frame.

Figure 16.1(c) shows a multi-lift truss-out scaffold in which the inside frames incorporate ladder beams. This scaffold should be constructed so that it can be dismantled at the bottom and reconstructed at the top as the building increases in height.

Figure 16.1(d) shows a variety of single-lift truss-outs in tubes and couplers. Note that the guard rails are braced and do not rely on short guard rail posts.

Figure 16.1(e) shows a variety of prefabricated frames on which truss-out scaffolds can be built. One of these is a ladder beam unit with adjustable height and an adjustable platform. This assembly can be raised and lowered by a tower crane.

16.2 The forces to be dealt with

The truss-out is essentially a beam of a special design that is cantilevered out from the building to perform the same functions as the cantilever scaffold described in Chapter 15. The calculations are the same, i.e. the outward moment about the outermost fulcrum point is calculated and balanced by the restraining moment. From this, the tailing down force is calculated and multiplied by a safety factor of 2 if the truss-out is structurally attached to the building, or 3 if it is restrained only by Kentledge.

In the case of the truss-out structure, the restraining force can be provided not only by bolts into the slabs, but also by the truss-out being tucked behind a downstand beam in the slab over it or bolted into this slab as well as the slab below it.

Figure 16.1(a) is a case where only the tailing down bolts can be taken into account for the balancing moment because the top of the truss-out can act only by lateral friction. Figure 16.1(d) shows restraint only at the top by being tucked behind the downstand beam.

Figure 16.1(d) also shows cases where there is no downstand beam, and hence the top friction grip must be supplemented by a bolt anchor in the top slab soffit.

Figure 16.1(e) is a case where there is a tailing down frame and a tuck behind the downstand beam. A ladder beam is incorporated. The lower truss-out can be taken away and rebuilt higher up as the building height increases.

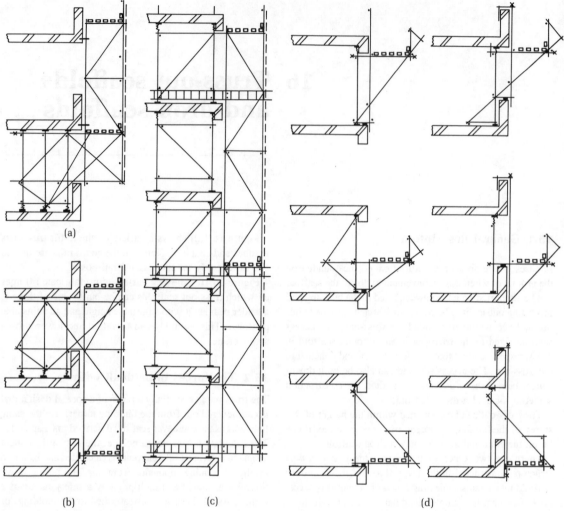

Figure 16.1   Truss-out scaffolds.

The truss-out scaffold should be tied to the building about 3 m above the truss-out slab, as shown in Figure 16.1(b), but these ties must not be taken into account as providing restraining moment because they may be removed at some stage in the building construction.

16.3 Calculations for a beam-type truss-out

The scaffold self weight plus imposed load = W.

This is not necessarily in the middle of the scaffold. It may be near the outside.

Projected length	= p.
Overturning moment	= $W \times p$.
Tailing length	= L.
Tailing down force	= $T = \dfrac{Wp}{L}$.

Bending moment in the beam = $TL = Wp$.
Slab edge reaction $\qquad\quad = R$.

The truss-out should be designed to have a strength in excess of these values to enable it to accommodate impacts due to oscillation and vertical wind surges. This emergency strength should be at least 50%, giving a load factor of 1.5.

The beam structure must be fully analysed structurally. The forces in its chords and braces must be assessed, and the couplers at each end of these elements should not be loaded in excess of 635 kg in any case where the moment is $1.5TL$ or $1.5Wp$.

R and T should be assessed for the applied load of $1.5W$, and the structure must be capable of sustaining the resulting loads. It will be seen that $R = T + W$, so $R > W$.

If $L = p$, $T = W$ and $R = 2W$, the edge of the slab

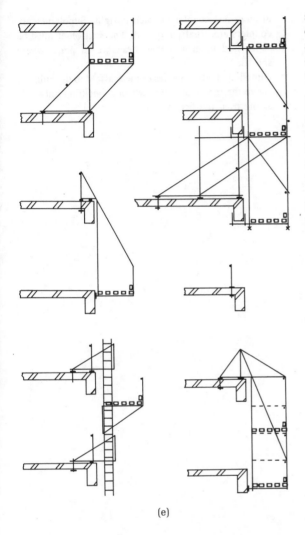

(e)

supporting the beam has to sustain a load of twice the weight of the scaffold and its imposed load. This is such a large amount over the weight supported that many hazards and damage result. Permission of the building owner or his engineer or architect should be obtained before a truss-out is constructed, and the scaffold designer should inform the owner, engineer or architect of the value of R for which he has to give his approval.

The scaffold designer knows the length of the lever arms used in the beam design and so can work out the value of R. The owner, engineer and architect cannot make this assessment because they will not know the dimensions of the truss-out and may not know the value of W without liaison with the building or maintenance contractor.

T may be greater or less than W, depending on the values of L and p. This force is upwards and is applied under the soffit of the building storey slab. The slabs are usually designed to resist downward bending and may be quite weak against upward forces. For this reason, the building owner, engineer and architect must be consulted to approve the value of T and its location.

When the values of R and T exceed the amount that can be sustained safely by the floors, the forces must either be spread over a wider area to achieve the designed reduction of effect, or the floors above and below must be brought into play by support scaffolds, as shown in Figures 15.1(e) and 16.1(e), which spread the load between two floors in each case.

It is inconvenient sometimes to build the shores in the lower and upper floors. In this case, a system of bolting must be devised that will produce the same result.

Generally, it is safer to build the truss-out beams in pairs and lace them together as a box beam so that they have adequate lateral stability. Single beams are laterally weak and need stabilising by being butted on to the cross walls of the building to prevent them buckling sideways. If single beams are used, a system of plan bracing must be installed between the access scaffolding and the building edge to deal with lateral sway and horizontal wind forces.

Attachment of the access scaffold to the truss-out beam should be by the number of couplers necessary to support the load. It may not be sufficient to make the attachment only to the top and bottom chords of the truss-out beam. Supplementary couplers may be needed on one or both chords.

16.4 Calculations for a triangular bracket truss-out

The force diagram for this arrangement is shown in Figure 16.2. Here, the overturning moment is $W \times p$, and the resisting moment is xh or yh. Force y can usually be accommodated by tubes butting on to the wall, but force x has to be a tension attachment to the building.

The values of x and y will be considerable and must be assessed. Contact with the building owner, engineer or architect must be made and approval obtained for the forces to be applied to the building in the places desired.

After x and y have been calculated by equating the moments, a load factor of 1.5 must be applied to them and the anchors at x made adequate for this. The contact points at y must also be capable of resisting 1.5 times the forces expected.

16.5 Hazards with truss-out scaffolds

- The most frequently occurring problem is that the slab edge cannot support the bearing loads on its edge.
- The scaffold is sometimes connected to the truss-out

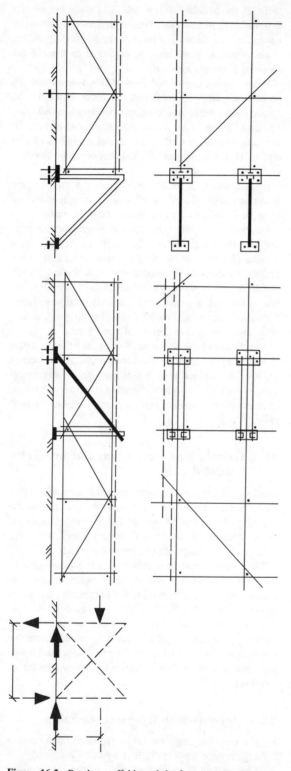

through one coupler on each upright. Supplementary couplers are usually required. The scaffolder must be supervised to ensure that these are in place where specified.

• Omission of adequate ties of the scaffold to the building above the truss-out can result in it buckling outwards as the truss-out beds down.

Figure 16.2   Bracket scaffolds and the forces on the fixings.

17 Slung scaffolds

17.1 Description

A slung scaffold is one that hangs down from the structure that supports it but cannot be raised or lowered by the operators on it. It can be suspended on tubes, on wire ropes or on specially made steel hooks. It can be slung beneath a tubular truss-out projecting from a building face or roof. The term 'drop scaffold' is sometimes used to identify a tubular slung scaffold built downwards from a tubular truss-out.

17.2 Scaffolds slung on tubes

Figure 17.1(a) and (b) is an example of scaffolds slung on tubes. Sometimes only one drop lift is required, and sometimes two or more. The width of the scaffold may not match the spacing of the node points in the roof truss. This is dealt with by arrangements such as that shown in Figure 17.1(b).

The first operation is to fix the horizontal tubes in the roof truss or on the projecting portion of a truss-out from a wall.

Trapezes are then made up consisting of two hanger tubes and a cross tube. Check couplers are fixed at the top and bottom of the hanger tubes, which are then lowered through couplers on the horizontal tubes to the required level and firmly coupled in place. When several trapezes have been fixed, the platform ledgers can be fixed over the cross tubes at the bottom of the trapezes. Finally, the transoms for the platform boards are fixed.

If the roof trusses are not more than 6 m apart, the ledgers can be single lengths of tube. They will need to be supported at intervals between the roof trusses. Supplementary trapezes are fixed as shown in Figure 17.1(a) elevation.

If the roof trusses are more than 6 m apart, a 6.4 m tube may not span the gap, and a difficulty arises that requires considerable skill to overcome. First, short horizontal tubes are fixed in the trusses projecting each way to such a distance that a 6 m tube will span the gap, as shown in Figure 17.1(c). The short horizontal tubes need stabilising by triangulation from the roof truss.

By building out triangles of tube from each side, a position will be reached where an intermediate support or two such supports can be inserted to strengthen the platform ledgers.

This type of slung scaffold, and in fact all drop scaffolds, must be constructed by scaffolders using safety harnesses. The scaffolders should ensure that they have put check couplers on the top and bottom of all hanger tubes at the beginning of the operation for their own benefit and not leave these items until the end of the construction.

Several levels of platform are needed if the length of the drop is large. A structure such as is shown in Figure 17.1(d) or (e) can then be constructed.

Bracing

Despite being built on trapezes that hang vertically, a slung scaffold will need diagonal bracing against side sway in both directions. It is particularly important that it is so braced to avoid any of the couplers being subjected to continuous small rotary movements.

Tying

When possible, the working platforms of a slung scaffold should be tied into the parent structure to maintain them rigidly in their intended position.

Check couplers

Every hanger tube must have its structural coupler tightened and rigorously checked for tightness and, in addition, have a check coupler mounted on the extensions of the hanger tube upwards and downwards beyond the structural coupler.

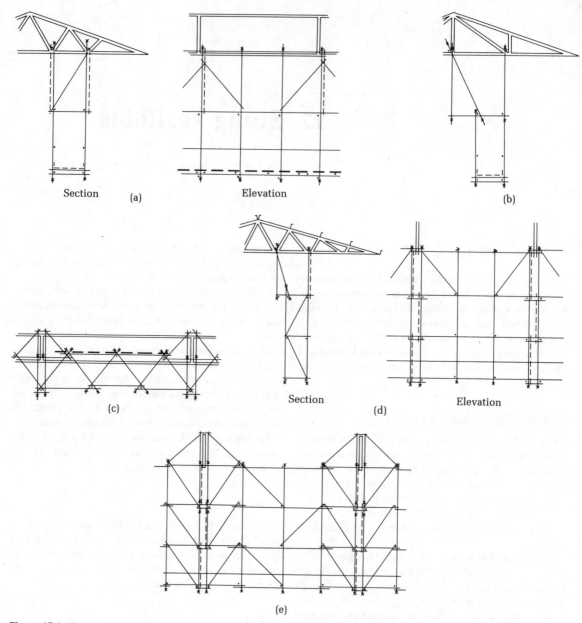

Figure 17.1 Typical slung scaffolds using tubes.

Horizontal tubes should not just rest in the node points of the trusses, but be locked into position by check couplers when the trapezes do not perform this function.

Boards, guard rails, toe boards and access

The working platform of a slung scaffold should be treated as any other working platform and be correctly boarded with board support transoms at the correct centres, and with guard rails and toe boards in the orthodox positions.

Access should be arranged particularly carefully to be firm and safe, and not to cause lateral sway.

17.3 Hazard with scaffolds slung on tubes

- Scaffolders, aware that check couplers are necessary at the top and bottom of the hanger tubes, often construct the hanger tubes at first without the checks and put the check couplers on only when they have finished the assembly. This means that they build the slung scaffold without the safeguard of the check couplers for themselves during construction.

17.4 Scaffolds slung on wire ropes

In many circumstances, it is more convenient to hang the platform on wires instead of tubes, such as where there is a false ceiling below the roof trusses containing small holes only suitable for wires. Another case is where the main structure is of pipework or very large structural tubes, over which wires are more easily rigged. Another is where the platform can be assembled at ground level and subsequently raised on wires into its final position.

The wires should be attached to the node points of the structure, which should first be soft packed to prevent damage to the structure or the rope. The top knot should be a round turn and two half hitches with the loose end moused off with light line. This is to ensure that slight movements of the platform do not allow the knots to work loose. A round turn and three bulldog grips is satisfactory.

The bottom knot should be around a node point of the main platform framework of ledgers and transoms, and the knots should be as described above.

The suspension wires should be as vertical as possible. The Regulations state that they must be vertical, but this is sometimes impracticable. When this is the case, the H and SE inspector must be asked to agree to the arrangement used.

Suspension wires that are continuous over the top support should have a round turn at the top if possible and should be anchored with a short separate length of wire with a round turn and two bulldog grips at each side. Suspension wires must not be jointed or spliced.

Spans can be divided by intermediate support points on sloping wires suitably attached to the main wires or separately to the roof trusses.

Bracing should be incorporated using diagonal wires with crossing slopes in both directions.

Ties should be achieved by horizontal lashings where possible. Other orthodox tubular methods at columns and wall surfaces can also be used.

The platform should be constructed in an orthodox manner.

The simplest platform slung on wires is one that has its board bearers spaced as for a normal access scaffold at intervals of 300 mm, 1200 mm, 1200 mm, 1200 mm and 300 mm, and continuously at these intervals so that the boards are correctly supported (see Figure 17.2).

The ledgers are then slung below the board bearers and spaced so that neither they nor the board bearers are over-stressed. The general practice is to space the ledgers at intervals of 2400 mm, providing for a bay ten boards wide.

The suspension connections are made at the ends of the boards and halfway along their length, and every ten boards

in the direction at right angles. This gives an area supported by each suspension point of 1.98 m $\times$ 2.4 m = 4.75 m^2 (51.13 ft^2).

If the platform self weight is 0.25 kN/m^2 (25 kg/m^2, 5 lb/ft^2), and the imposed load is 0.75 kN/m^2 (75 kg/m^2, 15 lb/ft^2), totalling 100 kg/m^2, the load on the suspension wire will be 475 kg. An 8 mm wire rope, which has a breaking load of 3800 kg as delivered and 3040 kg after being bent around the support points, coupled or knotted, provides a safety factor of 6.6 on the above load at the spacing given.

The board bearers in the middle of the boards support an area of 1.2 m $\times$ 2.4 m = 2.88 m^2 (31 ft^2) and carry a load of 288 kg. The safe uniformly distributed load for continuous tubes with partial end fixing over a 2.4 m span, assuming used tubes, is also 288 kg (see Table 31.18).

The ledgers span 1.98 m and in the worst case would receive the board bearer loads as a near centre point load. They are not simply supported beams over this span but have partial fixity and have a safe centre point load of 207 kg with used tubes (see Table 31.18). New tubes can carry 243 kg under these circumstances.

The ledgers supported at 1.98 m centres will be over-stressed for an imposed load of 0.75 kN/m^2. A reduction in the imposed load to 0.47 kN/m^2, together with the self weight of 0.25 kN/m^2, will reduce the ledger load to 207 kg. This rate of imposed load 0.47 kN/m^2 (47.91 kg/m^2, 9.8 lb/ft^2), is below the British Standard for slung scaffolds. Only the lightest of inspection work can be done using the platform.

To ensure an imposed load of 0.75 kN/m^2, either the spacing of the ledgers, i.e. the transom span, must be reduced to 1.75 m (seven boards wide), or the supports for the ledgers must be fixed closer together. This last method is mostly used, and Figure 17.2(a) shows on the left-hand side how this can be done either by vertical or inclined and vertical suspensions. In this case, with a transom span of 2.4 m and a ledger span of 1.2 m, the total platform load for an imposed load of 0.75 kN/m^2 becomes 288 kg at each transom to ledger function, but there are no intermediate loads causing bending in the ledgers because the suspension points are located to match the board end supports.

Now consider the right-hand side of Figure 17.2(a). The suspension at the middle location of the board length has been achieved by slinging a transom below the ledgers. This results in a ledger span of 1.98 m, but the transom load is applied about 660 mm from its end. For this condition with used tube partially end-fixed, the supporting value of the ledger is 2.28 kN (233 kg, 514 lb) for old tube and 2.68 kN (273 kg, 603 lb) for new tube. The transom span to give the 2.28 kN load is about 2 m (eight boards wide).

Derived from the above calculations, the best practice for wide-area or mattress scaffolds slung on wires (or

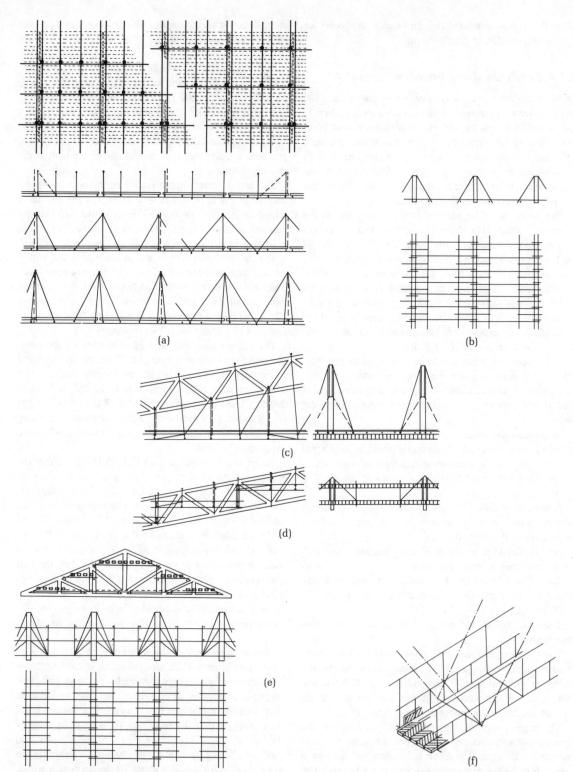

Figure 17.2 Typical slung scaffolds using wire ropes.

suspended on tubes) is that shown on the left of Figure 17.2(a), with the width between the ledgers, i.e. the transom span, calculated to deal adequately with the imposed load.

It will be noted that if the load in two suspension wires comes together on to one scaffold coupler, the coupler will not be overloaded, but check fittings should be used at the location.

If an imposed load of 1.5 kN/m^2 (154 kg/m^2, 31.4 lb/ft^2) is required, giving a total platform load of 1.75 kN/m^2 (178 kg/m^2, 36.5 lb/ft^2), the system shown on the left of Figure 17.2(a) is most suitable, with the ledgers spaced at 1.65 m (seven boards wide).

It is unlikely that the structure being served by the slung scaffold will have suitable locations for the top anchoring of the suspension wires in such a convenient pattern. The alternatives shown in the diagrams show how sloping wires can be brought up to one suspension point. Such suspensions need not be symmetrical. The configuration of the roof beam governs the layout of the suspension wires. The diagrams show single or double wire suspensions at the board ends.

Figure 17.2(b) is an example of a common case where the ledgers can be placed between the trusses resting in the structural nodes of the bottom chords of the trusses. This can be done provided that the trusses are not spaced more than 6.1 m (20 ft) apart.

With such a span, the ledgers will be too weak. The figure shows how the span is reduced by fixing sloping wires from the trusses to the quarter points of the ledger spans.

This technique is most useful when the trusses are lattice beams of constant depth. If they are ridge-type trusses, they are shallow towards the eaves, and the inclined suspension wires are at a very flat angle. This problem can be overcome by slinging a second ledger below the platform level and propping the platform off this with tubes.

The short tubes at right angles to the trusses are frequently necessary to transfer some of the large bay load directly on to the truss and so relieve the load on the wires.

Figure 17.2(c) shows a large roof beam. When slinging a platform below it, both the top and bottom structural nodes in the beam can be brought into use as support points. In this case, the large spans require ladder beams. Alternative suspensions do not hang vertically to equalise the spacing of the ladder beams.

In Figure 17.2(d), the slung scaffold is required within the depth of the roof beam, so only the top chord nodes of the beam are available as suspension points. In this case, the bottom chord node points are used to support scaffold tube columns inserted to support the top chord between the top chord nodes, so this point can be used as a suspension. The vertical connection between the ladder beams is in tubes and couplers.

Figure 17.2(e) is a frequently occurring case where a protection deck is required over a factory floor during maintenance of the roof. The protection deck is constructed on tubes that span between the trusses. Their strength is supplemented by wire ropes from various locations up the slopes of the trusses. The truss is then infilled with tubes and couplers so that the working platforms can be constructed. These are supported on the truss and on tubular puncheons from the protection deck frame.

Figure 17.2(f) is a walkway slung on wire ropes. The beams are ladder beams. The suspension ropes can be vertical or inclined. They may sometimes be inclined when viewed in two directions. This arrangement prevents sway along and across the walkway.

17.5 Correcting the hanging wire tension

Because it is not possible to tie all the knots in a wire rope with identical tightness, when the platform takes up its own weight and its imposed weight, some wires will be less taut. This reduced tension should be taken up by drawing the suspension wires together in pairs by wire loops until by feel they take up their loads. If the platform is to be used for any length of time, this process should be repeated at weekly intervals until further adjustment is unnecessary.

17.6 Hazards with scaffolds slung on wire ropes

- The most frequent cause of problems in scaffolds slung on wires is that the suspension wires that have been spaced and arranged by the designer to carry approximately equal loads do not do so in practice because they cannot be knotted tightly enough or with uniform tension. The tighter wires carry more load than intended, causing damage to the main structure and deformation of the working platform.
- Bracing is left off, and instability results.
- The end suspension points may not be attached to a node point in the platform, so failure bending stresses occur in the platform ledgers.
- Because the suspensions are wire ropes, guard rails and toe boards are omitted from the edge of the platform.

17.7 Scaffolds slung on tubes and wire ropes

This arrangement is sometimes necessary for practical reasons, but should be avoided if possible. While tubes are rigid, wire ropes extend slightly under load, and the knots at the ends extend before taking up their share of the load. This means that the tubes will take more than their fair share of the load, and the ropes will be slack. Local overloads will occur which will be of unpredictable magnitude and perhaps have serious consequences.

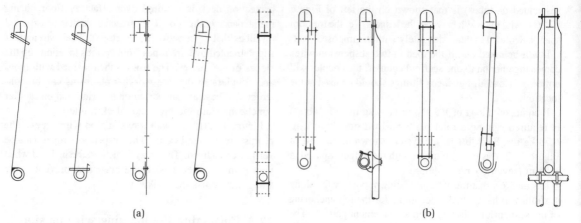

(a)   (b)

Figure 17.3   Prefabricated hangers for slung scaffolds.

When this system has to be used, the wire suspensions must continually be retightened until it is certain that they are accepting the design load.

Crossed wire bracing can satisfactorily be used on a tubular slung system, but tubular bracing should only be used with the greatest care on wire rope slung systems because it may take up suspension loads for which its end fixings are not designed.

When a mixture of tube and wire suspension is necessary for some structural reason connected with the building, it is better to use the two systems independently.

17.8   Scaffolds slung on special steel bar hooks or special components for suspension

In the case of an extensive roof with many trusses of a similar design, it is worthwhile considering prefabricating the hanger elements with a view to cutting down the time involved in installing the tube-and-coupler and check coupler system. It is also worthwhile considering welding safety stops on to tubes or drilling holes through the tubes and inserting safety bolts.

Figure 17.3(a) gives some forms of hanger fabricated from concrete reinforcement bars or steel strip.

Figure 17.3(b) gives some forms of prefabricated tube hanger.

17.9   Hazards with specially designed hangers

- In the case of Figure 17.3(a), the reinforcing bar hook should be carefully designed and also tested because it has a tendency to straighten. The locking loops are usually necessary for strength and as safety devices.
- In the case of Figure 17.3(b), the flattened tube with the drilled hole should not subsequently be used as a

putlog because it will become locked in with mortar in the hole.

17.10   Scaffolds slung on fixed girder couplers

Section 32.18 gives data concerning girder couplers. The essence of the system is that scaffold tubes are coupled to the underside of rolled steel beams or to the side of rolled steel columns. When these starting tubes have been fixed in place, the remainder of the slung scaffold is orthodox and should follow the standard rules for slung scaffolds, particularly the requirement for check couplers.

The ledgers coupled to the rolled steel section must not be loaded to the point where deflections can deform, over-stress or destroy the girder couplers. Because no check couplers can be fixed to the girder coupler, twice the calculated number should be used.

Figure 17.4(a) to (c) gives some examples of the use of fixed girder couplers. In Figure 17.4(a), the scaffold is permanently slung from the bottom flange of the beam. In Figure 17.4(b), it is built up above the beam. In Figure 17.4(c), the scaffold is trussed out from the steel framework and coupled to the stanchions at every lift with girder couplers. Figure 17.4(d) is an example of a roller girder coupler suspension that allows the suspended scaffold to be moved along, either to facilitate its construction or for use as a mobile platform.

17.11   Scaffolds slung on girder roller couplers

The girder roller coupler is a very useful coupler for forming a slung scaffold that can be rolled along under a bridge or roof. The roller couplers should be mounted on the lower flanges of rolled steel sections and support ledger tubes that are linked together with transoms into a movable

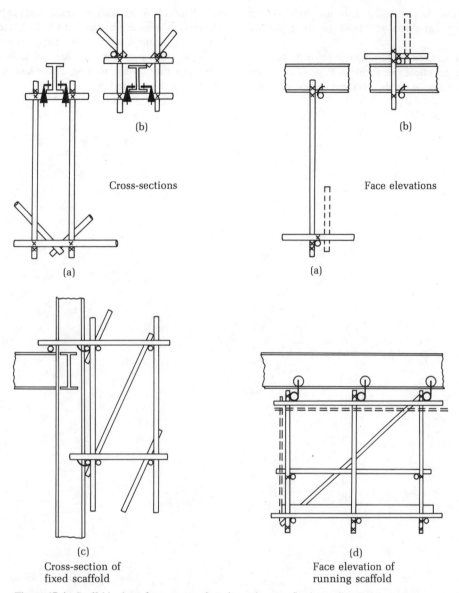

(b)

Cross-sections

(a)

(b)

Face elevations

(a)

(c)
Cross-section of
fixed scaffold

(d)
Face elevation of
running scaffold

Figure 17.4 Scaffolds slung from structural steel members on fixed or roller girder couplers.

carriage. The scaffold is then slung from this carriage in the orthodox manner.

Both the rolling carriage and the scaffold should be firmly plan braced to avoid skew deformation, which will cause the carriage to jam on its supports.

It is usually not possible to provide check couplers to the roller couplers because these are attached directly to the ledgers of the carriage. Therefore, twice the calculated number of roller couplers should be used.

Figure 17.4(d) shows an example of a rolling carriage and a slung scaffold beneath it.

17.12 Special precautions for slung scaffolds

A slung scaffold should not be constructed while other persons are working below. This precaution applies not only to the precise area of the platform being built, but also to the access route along the slung platform to the place of work and to the area where materials are being raised.

Slung scaffolds are often used in places where the need for the scaffold repeats every few years. Even with intervals such as this, an effort should be made to employ the same chargehand and the same gang. It is also very desirable that the same scaffolders who built the slung platform dismantle it.

The supervisor has a special responsibility in the construction of a slung scaffold. He should note particularly how the scaffolders carried out the work and the order in which each element is fixed. This information not only leads to safety when the same job is repeated in the future, but also helps during the dismantling. Another duty of the supervisor is to ensure that only one scaffolder works in one place at the same time so that one man cannot loosen a coupler on which another is relying.

18 Suspended scaffolds

18.1 General description

A suspended scaffold differs from a slung scaffold in that the platform can be raised and lowered on its suspension ropes. The winch or climbing device can be situated at the top or the lower end of the suspension rope and can be manual or power operated.

A suspended scaffold achieves the same objective as a truss-out or cantilever scaffold in that it avoids the foundations on the ground and the cost of the full height of a ground-based scaffold. The suspension points can be on a scaffold projecting from the building at high level. This high-level truss-out is cheaper than the low-level one required for a truss-out scaffold because it does not have to carry the weight of an orthodox scaffold. The suspension points can sometimes be on the main structure, in which case there is no cost involved in a high-level scaffold framework. Because there is no cost from foundations and no cost from idle intermediate lifts, the suspended scaffold is often the most economical of the various types considered for any job.

A suitable part of the main structure is selected as the suspension point. If there is no such place, a truss-out frame of scaffolding or a series of joist outriggers is assembled to form temporary suspension points. Wires are then fixed to these and the platform to the wires. A winch or climbing device is fixed at either the top or the bottom end of the suspension to enable the platform to be raised and lowered.

The four elements of the system − the top frame, the suspension ropes, the lifting appliance and the platform − can each be assembled from scaffolding components, prefabricated or from a purposely designed kit.

Even though a suspended scaffold may not descend to the ground, it may interfere with the use of the pavement in a High Street job, and it may need the same local authority approvals and have to comply with the same requirements as orthodox scaffolding. Public protection may also be needed.

Figure 18.1(a) to (c) is an example of simple roof rigs using pine poles, with the thick end outboard and the thin, tapered end inboard. The scheme in Figure 18.1(c) incorporates a zigzag diagonal bracing of poles to ensure lateral stability.

Figure 18.1(d) shows a propped jib. In this case, lateral stability is achieved by cross bracing with wire ropes.

Figure 18.1(e) incorporates a 'horse' frame on the roof and a method of overcoming a large overhang.

On pitched roofs, arrangements such as Figure 18.1(f) and (g) can be used. For mansard roofs, type (h) is a rig using poles. Type (i) is a system using jib poles held back by wire ropes to a safe anchorage. Lateral stability is achieved by a zigzag of poles behind the jib poles.

When there is no firm edge to the roof, the jib poles can be supported on corbelled brickwork or sills, to which they should be very firmly attached. The jib poles in Figure 18.1(j) are held in place by wire ropes to a firm anchorage, and lateral stability is achieved by cross wire ropes.

Most of these rigs can be constructed in scaffold tubes and couplers. They can be fitted with fibre rope suspensions, travelling joists and hand or power winches on wire ropes.

Figure 18.1(k) to (m) shows the plans of the outriggers and various arrangements of the counterweights. The case shown in Figure 18.1(m) needs careful consideration. The total counterweight is not reduced by this arrangement.

When large projections are necessary to reach over balconies, rigs such as those in Figure 18.1(n) to (q) are useful. In cases (n) and (o), the fulcrum point is on or behind the outer parapet, while in cases (p) and (q) it is on or just outside the inner parapet.

In cases such as (n), the bending moment on the outrigger tube or beam must be calculated. It may be necessary to use one tube directly above another joined by parallel couplers.

In all roof rigs where scaffold tubes are used and the tailing end is held down by weights placed in the roof

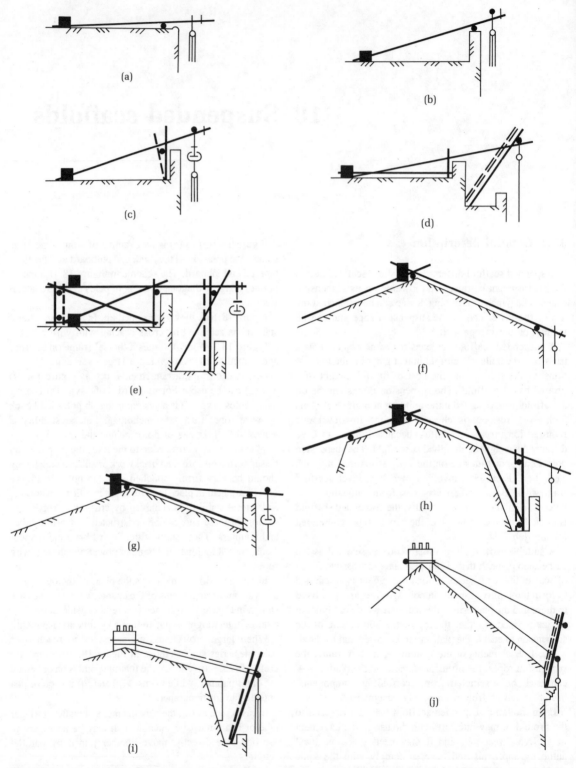

Figure 18.1 Suspended scaffold roof rigs.

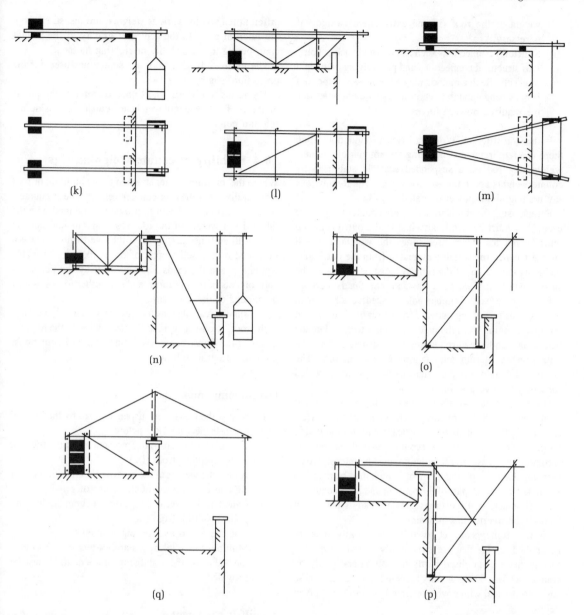

(k) (l) (m)

(n) (o)

(q) (p)

'horse', the outrigger pole should be coupled to the 'horse', assisted with supplementary couplers or check couplers after calculations have been done. The supplementary or check couplers should be above the tailing end of the outrigger and below the outrigger support at the outside edge.

18.2 Codes and Regulations

The relevant standards and codes are BS 2830 relating to suspended safety chairs and cradles, BS CP 5974 for temporarily installed suspended scaffolds and access equipment, and the many standards relating to lifting equipment.

If the scaffold is for building or engineering construction, the platform and roof rig must comply with the Construction (Working Places) Regulations 1966, and the ropes, lifting gear and lifting appliances must comply with the Construction (Lifting Operations) Regulations 1961. Other statutory requirements may apply in other locations, e.g. in mines, shipyards and factory stock yards.

There are three systems for fixing the roof trolley or rig to the roof:

(i) No physical connections but ensuring a greater righting moment than overturning moment by using the self weight of the roof rig and extra counterweights if necessary.

(ii) Structural connections such as bolts and anchor bolts.

(iii) Part structural connections and part relying on weight moments. Such a method may be necessary if the roof is not strong enough to carry a large weight or sustain the required upward forces.

In the Construction (Working Places) Regulations 1966 there is a requirement concerning the amount of counterweight required on a suspended scaffold stabilised by counterweights and not power-operated. The Regulations say nothing about power-operated scaffolds.

Paraphrasing the Regulations: where counterweights are used, they must be firmly attached to the outriggers and must be not less than three times the weight that will counterbalance the weight suspended from the outrigger, including the weight of the runway, joist or rail track, the suspended scaffold, and persons and other loads thereon.

For a single pair of timber poles placed on a roof with the butt ends projecting outwards and a scaffold suspended on ropes hanging from the poles, it is easy to calculate the balancing weight required at any fixed distance along the pole from the fulcrum point near the edge of the roof. The calculated weight that just balances the system is trebled for safety to give a stability factor of 3.

However, if the roof rig is not a simple pole system but a substantial scaffold structure, as shown in Figure 18.1(e), (l) and (n), that contributes significantly to the stabilising moment, it can be taken into account as part of the counterweight. In this case, the Regulation must be interpreted as meaning that the righting moment, which may be made up of the self weight of part of the roof rig and added counterweights, must be three times greater than is needed to balance the overturning moment.

No requirements are set down for the stability factor to be applied to rigs that are structurally bolted down, nor to rigs that are stabilised partly by bolting and partly by counterweights. The generally accepted rules to meet these cases and to deal with power-operated platforms are given below.

18.3 The fulcrum point and the moment lever arms

Figure 18.1 shows 17 types of roof rig for a suspended scaffold. In each case the fulcrum point is the point about which rotation would take place if the cradle were overloaded. The projection length and the tailing length are measured from this fulcrum point and not from the front face of the building. The projection length is the distance from the fulcrum point to the centre line of the suspension system. The tailing lengths, and there may be several

different lengths for various weights, are measured from the fulcrum point to the centre of gravity of the weight contributing to the stability or righting moment.

The terms 'nib' and 'nip' are sometimes used for the projection length.

The terms 'outboard' and 'inboard' refer to the points of the roof rig or cantilever beam outside or inside the fulcrum point.

18.4 Stability of counterweighted rigs

Where the top frame is not attached to the building or roof structurally but relies on counterweights, these counterweights must be sufficient to provide, combined with the inboard self weight of the roof rig, a stabilising moment at least three times greater than the overturning moment produced by the self weight of the outboard part of the equipment and the loads imposed on it by the personnel and contents of the platform, and such other forces as may be applied to the platform.

The stability calculations for rigs not structurally attached to the roof, to ensure that the rig complies with the requirements of the Construction (Working Places) Regulations, are carried out as follows:

Overturning moment

(i) Assess the maximum suspended load by the method given in Section 18.9 below.

(ii) Calculate the overturning moment about the fulcrum point resulting from (i).

(iii) Assess the weight of the outboard section of the roof rig and the position of its centre of gravity.

(iv) Calculate the overturning moment about the fulcrum point resulting from (iii).

(v) Add these two overturning moments.

(vi) Multiply this total by a stability factor of 3 to give the value of the stabilising moment that must be provided.

Stabilising moment

(vii) Assess the weight of the inboard section of the roof rig and the position of its centre of gravity.

(viii) Calculate the stabilising moment about the fulcrum point resulting from (vii).

(ix) Subtract the stabilising moment of (viii) from the stabilising moment calculated in (vi). This gives the extra stabilising moment that has to be provided by extra counterweights.

(x) Decide where the counterweights are to be attached to the inboard section of the roof rig and measure their lever arm about the fulcrum point.

(xi) Divide the extra stabilising moment calculated in (ix) by the lever arm determined in (x) to give the value of the extra counterweights to be added at the point chosen in (x).

The counterweights must be firmly attached to the outrigger, and any other weight that is taken into account, e.g. the stand in Figure 18.1(l), must be firmly attached to the outrigger so that it can contribute to the balancing moment.

18.5 Hazards with suspended scaffolds

- Not attaching the outrigger to the roof frame in the case of Figure 18.1(e) and (l).
- Not realising where the fulcrum point is in cases such as Figure 18.1(e) and (p).
- There must never be any confusion between the projection length measured from the fulcrum point and the distance that the outrigger extends beyond the face of the building. Such confusion can result in inadequate counterweighting.
- The cantilevered length of the outrigger may be less than the projection length, as in Figure 18.1(l) and (n), but it must not be used in stability calculations against overturning.

18.6 Stability of structurally attached rigs

The stability calculations for rigs that are structurally attached to the roof, to ensure that the rig complies with the Construction (Working Places) Regulations, are carried out as follows:

If the stabilising moment is provided by a structural fixing such as a pair of bolts through a flat concrete roof or on to a structural steel member, the stability factor of 3 does not apply. All the elements of the assembly except the bolt must be designed in accordance with the appropriate safety factor for the material. This is usually 1.7.

Anchor bolts can be of two types:

(i) Those that pass through the roof and have a washer and nut at the lower end.
(ii) Those that terminate in the thickness of the roof with an expanding or chemical anchor.

Through bolts should be selected from the loading tables of the manufacturer and have a safety factor of not less than 2.

The holes for expanding anchors should be drilled to the size and depth required by the anchor manufacturer and preferably using drills supplied by him. The anchor should be installed by the method and with the tools recommended by the manufacturer.

The bolts should be tested after insertion and should sustain a load that gives a safety factor of 2.

A test bolt not subsequently to be built into the equipment restraint system should be inserted in the roof and tested to failure to ascertain what emergency strength there is. A judgement on the suitability of the fixing system can then be made.

In the case of chemical anchors, the manufacturer's instructions should be followed very carefully. All the bolts should be tested to ensure there is a safety factor of 2. A bolt not to be used in the restraint system should be tested to failure to ascertain the emergency strength available.

18.7 Stability of roof rigs partly reliant on structural connections and partly on counterweights

There are circumstances where the roof is not strong enough to resist the upward forces that would be applied to it by the tailing end of the roof rig, or where holding-down bolts cannot be set in it to have a safety factor of 2. Such roofs may also be insufficiently strong to support the weight of the inboard section of the roof rig and the counterweights attached to it. In these cases, the stability of the rig may have to be achieved by a combination of the available structural strength of the roof and added counterweights.

The strength of the roof or the allowable strength of bolts in it must first be assessed. This should be done using the working stresses of the materials in the roof and the bolts so that the calculated limit of the strength will have a safety factor of 2. If these calculations cannot be done, the roof should be tested and its reliability assessed with the same factor of safety. Anchor bolts in the roof should be tested to failure and their allowable loads assessed with a safety factor of 2.

The weakest link in this structurally attached part of the anchorage governs the contribution that the roof can make to the stability of the rig.

The remainder of the required balancing force must be provided by the self weight of the inboard portion of the roof rig and extra counterweights added to it. Sufficient counterweights must be added to ensure that this part of the balancing system not attached to the roof has a safety factor of 3.

18.8 Hazards with partially counterweighted system

- There is the obvious hazard in this system that if the counterweights are inadvertently changed because the system is apparently a bolted one, the strength of the roof may be inadequate.
- The roof bolts may be taken out in the belief that the

rig is fully counterweighted, leaving an unsafe system.

- A further hazard is that in moving the rig from one place to another, one of the stabilising systems has to be removed before the other. It must be ensured that there is no overturning moment on the system when the changeover is made.

The user must take appropriate measures to overcome these hazards and before use ensure that the whole balancing system has not been made defective.

18.9 Calculations of the maximum suspended load

The paraphrase of the Regulations given in Section 18.2 stated that some weights were required to be counter-weighted, and that the expression 'and other load thereon' was added.

This puts the onus on the designer to incorporate in his calculations all the forces that may lead to overturning of the roof rig.

In addition to the self weight of the platform and the static imposed load, there is the variation in the imposed load when this is assembled under one suspension wire. There is the weight of the outboard part of the roof rig, and the weights of the suspension ropes, safety ropes and lifting appliances. There are the impact forces that the lifting devices will apply to the suspension ropes.

Accordingly, the Code uses the term 'maximum sus-pended load'. This is intended to incorporate all the forces that are applied to the roof rig that tend to overturn it or cause stress in any element of it.

The imposed load is usually from the persons on the platform, and they may congregate under or near one suspension rope. Therefore, the maximum suspended load on any one suspension must take this into account. All the suspensions must be assumed to have this accumulated load on them at some time. The other suspensions will be relieved of some load, which may also be taken into account in the case where the counterweights serve more than one suspension point.

18.10 Calculations of the suspended load on the ropes other than that containing the maximum suspended load

When the imposed loads, i.e. the personnel and their tools, have congregated under or near one suspension rope, the other suspension rope or ropes will not be as heavily loaded.

The other rope will carry that part of the self weight of the platform and its lifting gear that remains on it, together with a small portion of the imposed load, usually about one-eighth, resulting from the fact that all the personnel cannot congregate exactly under one rope.

18.11 Impact forces

The impact forces are due to the operation of the winches under the combined self weight and imposed loads. For manually operated winches, the impact load should be an additional 10%. For power-operated winches, the impact allowance should be 25%.

Where there are traversing rails, jockeys, suspension rings and safety wires, the weight of these must be taken into account after the impact force has been calculated because they are not suspended below the winches.

18.12 The outboard part of the roof rig

The suspended load is not the only load that tends to tip the roof rig over. The outboard portion of the rig, i.e. that portion of it outside the fulcrum point, also tends to tip the rig over.

The weight of the outboard part of the roof rig must be assessed and its centre of gravity located. The overturning moment of this portion of the roof rig must be added to the overturning moment of the suspended loads to find the overturning moment on the roof assembly and enable the holding down system to be calculated.

18.13 The inboard part of the roof rig

The weight of the inboard portion of the roof rig must also be assessed and its centre of gravity located. The righting moment from the inboard section of the roof rig can be taken into account.

Where this is insufficient to balance the system ade-quately, counterweights or bolts have to be added to supplement the righting moment, or a combination of the two.

18.14 Symbols used in suspended scaffold design

Sample calculations for each of the three anchorage systems given in Section 18.2 have been worked out for each of the three types of roof rig shown in Figure 18.1(k), (l) and (m). Only independent two-point suspension platforms have been considered.

The symbols used are as follows:

M The most heavily loaded line in the system, incor-porating all the weights, congregated imposed loads, impacts and other forces (corrected for pulley systems), and the lifting gear, appliances and tracks, etc. attri-butable only to that line that apply an overturning effect on the roof rig, i.e. the maximum suspended load on one suspension.

L The suspended load on the second suspension, calcu-lated the same way as M but assuming the imposed load to be congregated at the first suspension.

U   The weight of the outboard part of the roof rig.
V   The weight of the inboard part of the roof rig.
C_1 The counterweight without a factor of safety.
C   The counterweight to be added to the rig to provide the specified stability factors.
T   The safe holding-down bolt tension required to provide the specified stability factors.
A   The strength of the roof or the anchor bolts, if these are only partly adequate and require supplementing.
m   The projection length, i.e. the lever arm of M from the fulcrum.
u   The lever arm of U.
v   The lever arm of V.
c   The lever arm of C.
t   The lever arm of T.
O   The overturning moment.
R   The righting moment.

18.15 Calculations for system (i), i.e. rigs relying only on counterweights for stability and having no structural connections to the roof

This system is used when there is no strength in the roof against uplift or when it is undesirable or impracticable to attach anything to it.

The righting moment must be greater than the overturning moment by a stability factor of 3.

Case (a)

Figure 18.1(k). Each suspension has its own jib and its own counterweight.

M is calculated for each suspension and assumed to act on either jib.
U and V are calculated for each jib.
$O = Mm + Uu$.
Multiply this by 3 to obtain the righting moment that must be provided, giving $3Mm + 3Uu$.
$R = Vv + Cc$.
Equate and calculate C.

$$C = \frac{3Mm + 3Uu - Vv}{c},$$

which is added to the inboard end of each jib.

Each jib must be capable of resisting the bending moment from M acting at its end, i.e. Mm.

Case (b)

Figure 18.1(l). Both jibs are part of one roof rig with a single common counterweight.

M is calculated for one suspension jib.

0.5U is half the weight of the whole outboard part of the roof rig.

L and 0.5U are calculated for the other suspension and jib.

V is calculated for the whole of the inboard part of the roof rig.

$O = Mm + 0.5Uu + Lm + 0.5Uu$.

Multiply this by 3 to obtain the righting moment that must be provided, giving $3Mm + 3Lm + 3Uu$.

$R = Vv + Cc$.

Equate and calculate C.

$$C = \frac{3Mm + 3Lm + 3Uu - Vv}{c}.$$

Counterweight C is then added to the centre of the combined roof rig, or the same total weight is evenly distributed along the back edge.

The roof rig as a whole must be capable of resisting the bending moments applied to it by the sum of $M + L$ acting at the end of the jibs, i.e. $Mm + Lm$. Each jib must be capable of resisting the bending moments from M acting at its end, i.e. Mm.

Case (c)

Figure 18.1(m). Each suspension has its own jib, but these are angled together to one common counterweight.

M and U are calculated for one suspension and its jib.
L and U are calculated for the other suspension and its jib.
V is calculated for each jib.
m, u and v are measured at right angles to the building face and not along the centre lines of the jibs.
$O = Mm + Uu + Lm + Uu$.
Multiply this by 3 to obtain the righting moment that must be provided, giving $3Mm + 3Lm + 3Uu + 3Uu$.
$R = Vv + Vv + Cc$ for two jibs and one counterweight.
Equate and calculate C.

$$C = \frac{3Mm + 3Lm + 6Uu - 2Vv}{c}.$$

The counterweight C is then fixed firmly to both jibs at the apex.

This is approximately twice that attached to each of the separate jibs in case (a).

Each jib should be capable of resisting the bending moment from M acting at its end, with the lever arm that is measured for this strength calculation along the centre lines of the jib.

18.16 Calculations for system (ii), i.e. rigs that are structurally attached to a sound roof on a building and have no counterweights

This system can be used when the roof strength to resist

upward forces is at least twice as much as is required, and the anchor bolts are twice as strong as is required.

The structure of the rig is designed with the normal safety factors appropriate to the materials. Its anchoring down should have a stability factor of 2.

Case (a)

Figure 18.1(k). Each suspension has its own jib and its own counterweight.

M is calculated for each suspension and assumed to act on either jib.

U and V are calculated for each jib.

$O = Mm + Uu$.

Multiply this by 2 to obtain the righting moment that must be provided, giving $2Mm + 2Uu$.

$R = Vv + Tt$.

Equate and calculate T.

$$T = \frac{2Mm + 2Uu - Vv}{t}.$$

T is the roof resistance required and the strength of the anchor bolts to be used at the inboard end of each jib. Each jib must be capable of resisting the bending moment from M acting at its end, i.e. Mm.

Case (b)

Figure 18.1(l). Both jibs are part of one roof rig with a single common counterweight.

M is calculated for one jib.

$0.5U$ is half the weight of the whole outboard part of the roofing.

L and $0.5U$ are calculated for the other jib.

V is calculated for the whole of the inboard part of the roof rig.

$O = Mm + 0.5Uu + Lm + 0.5Uu$.

Multiply this by 2 to obtain the righting moment that must be provided, giving $2Mm + 2Lm + 2Uu$.

$R = Vv + Tt$.

Equate and calculate T.

$$T = \frac{2Mm + 2Lm + 2Uu - Vv}{t}.$$

The roof rig as a whole must be capable of resisting the bending moments applied to it by the sum of $M + L$ acting at the end of the jibs, i.e. $Mm + Lm$. Each jib must be capable of resisting the bending moment from M acting at its end, i.e. Mm.

Case (c)

Figure 18.1(m). Each suspension has its own jib, but these are angled to one common bolting system.

M and U are calculated for one jib.

L and U are calculated for the other jib.

V is calculated for each jib.

m, u and v are measured at right angles to the building face and not along the centre line of the jibs.

$O = Mm + Uu + Lm + Uu$.

Multiply this by 2 to obtain the righting moment that must be providing, giving $2Mm + 2Lm + 4Uu$.

$R = Vv + Vv + Tt$.

Equate and calculate T.

$$T = \frac{2Mm + 2Lm + 4Uu - 2Vv}{t}.$$

The anchor bolts are then fixed firmly to both jibs at the apex and into the roof to sustain load T safely. The value of T is approximately twice that attached to each of the separate jibs in case (a).

Each jib must be capable of resisting the bending moment from M acting at its end, with the lever arm that is measured for this strength calculation along the centre lines of the jib.

18.17 Calculations for system (iii), i.e. the roof rig relying on a combination of anchor bolts in a weak roof and on counterweights

This system is not to be preferred, but it is sometimes necessary when the roof is not strong enough in uplift to satisfy system (ii) or insufficiently strong against the downward load of the counterweights required in system (i).

The method of calculating the anchorages required is to take the overturning moment and apply a factor of 2 to this, assuming that the system will be anchored first by bolts. The strength of the roof with anchorage bolts therein is then subtracted from the total required, leaving the remainder to be provided by counterweights, which will have a safety factor of 2 from the calculations carried out. The counterweight so calculated is then increased by 1.5 to give that portion of the anchorage the safety factor of 3 required by the Regulations.

The weakness in the roof may be in its structural design or in its inability to hold anchors firmly in its material. The advice of the owner or his architect or structural engineer should be obtained with regard to the structural design of the roof and its bolt-holding ability. In the absence of reliable information, tests can be carried out from which its reliability can be assessed and its safe loading upwards and downwards chosen to give a safety factor of not less than 2.

System (iii) comes into play when this assessed safe loading will not anchor the suspended platform of system (ii).

Case (a)

Figure 18.1(k). Each suspension has its own jib and its own counterweight.

M is calculated for each suspension and assumed to act on either jib.

U and V are calculated for each jib.

$O = Mm + Uu$.

Multiply this by 2 to achieve the righting moment that must be provided, giving $2Mm + 2Uu$.

$R = Vv + At + Cpt$, where A is the safe strength of the roof at the anchor point and Cpu is the parked counterweight to be applied.

Equate and calculate $A + Cp$.

$$A + Cp = \frac{2Mm + 2Uu - Vv}{t}.$$

From this subtract A, which is the fraction of $(A + Cp)$ that can safely be accepted by the roof and its anchors, leaving Cp to be provided by counterweights that already have a safety factor of 2 incorporated. Multiply Cp by 1.5 to provide the counterweights with a safety factor of 3.

Each jib must be capable of resisting the bending moment from M acting at its end, i.e. Mm.

Case (b)

Figure 18.1(l). Both jibs are part of one roof rig with a single common counterweight.

M is calculated for one jib.

$0.5U$ is half the weight of the whole outboard part of the roof rig.

L and $0.5U$ are calculated for the other jib.

V is calculated for the whole of the inboard part of the roof rig.

$O = Mm + 0.5U + Lm + 0.5Uu$.

Multiply this by 2 to obtain the righting moment that must be provided, giving $2Mm + 2Lm + 2Uu$.

$R = Vv + At + Cpt$.

Equate and calculate $A + Cp$.

$$A + Cp = \frac{2Mm + 2Lm + 2Uu - Vv}{t}.$$

From this subtract A, leaving Cp to be provided by counterweights, which already have a safety factor of 2 incorporated. Multiply Cp by 1.5 to provide counterweights with a safety factor of 3.

The roof rig as a whole must be capable of resisting the bending moments applied to it by the sum of $M + L$ acting at the end of the jibs, i.e. $Mm + Lm$. Each jib must be capable of resisting the bending moment from M at its end, i.e. Mm.

Case (c)

Figure 18.1(m). Each suspension has its own jib, but these are angled to one common bolting system.

M and U are calculated for one jib.

L and U are calculated for the other jib.

V is calculated for each jib.

m, u and v are measured at right angles to the building face and not along the centre line of the jibs.

$O = Mu + Uu + Lm + Uu$.

Multiply this by 2 to obtain the righting moment that must be provided, giving $2Mm + 2Lm + 4Uu$.

$R = Vv + At + Cpt$.

Equate and calculate $A + Cp$.

$$A + Cp = \frac{2Mm + 2Lm + 4Uu - Vv}{t}.$$

From this subtract A, leaving Cp to be provided by counterweights, which already have a safety factor of 2 incorporated. Multiply Cp by 1.5 to provide counterweights with a safety factor of 3.

Each jib must be capable of resisting the bending moment from M acting at its end, with the lever arm that is measured for this strength calculation along the centre lines of the jib.

18.18 Hazard in calculations for counterweights

- There is a dangerous hazard if the calculations demonstrated in the previous three sections are not performed in the manner given.

    Referring back to Section 18.15, the overturning moment was given as $Mm + Uu$. This was multiplied by 3 to give the righting moment that must be provided. From this the counterweight was calculated as:

$$C = \frac{3Mm + 3Uu - Vv}{c}.$$

If the overturning moment of the rig has not been calculated on the outboard moments only, because the righting moment of the inboard section of the rig has been subtracted first, the residual overturning moment becomes $Mm + Uu - Vv$.

If this is multiplied by 3 to give the righting moment that has to be provided, we obtain:

$$C_2 = \frac{3Mm + 3Uu - 3Vv}{c}.$$

This latter value C_2 is less than the correct value C. If Vv is small, as in the case of a single-pole outrigger, it will not make much difference, but if Vv is large, as in the case of a raised roof rig of tubes and fittings to surmount a parapet, it will make a very great difference.

For instance, if Vv is nearly equal to $Mm + Uu$, C_2 turns out to be very small, and the rig will be on the balancing point instead of having a stability factor of 3. This is such a dangerous hazard that it is worth giving a numerical example.

Consider a tubular roof rig 1.5 m wide, 2 m from front to back and 1 m high to enable the top jib arms to go over a parapet, as in Figure 18.1(l).

Assume the two suspension loads add up to 360 kg at a projection length of 0.75 m.

The outboard section of the roof rig will weigh about 54 kg and have a lever arm of 0.33 m.

The inboard section of the roof rig will weigh about 240 kg and have a lever arm of 1 m.

Counterweight by the correct method

$O = (360 \times 0.75) + (54 \times 0.33) = 288$ kg.
$3O = 864$ kg to be provided.
$R = (240 \times 1) + (C_1 \times 2)$.
Equate and calculate C_1.

$$C_1 = \frac{864 - 240}{2} = 312 \text{ kg.}$$

$$\text{Stability factor} = \frac{(312 \times 2) + (240 \times 1)}{(360 \times 0.75) + (54 \times 0.33)}$$

$$= 3.$$

Counterweight by the incorrect method

$O = (360 \times 0.75) + (54 \times 0.33) - (240 \times 1) = 48$ kg.
$3O = 144$ kg to be provided.
$R = C_2 \times 2$.
Equate and calculate C_2.

$$C_2 = \frac{144 = 72 \text{ kg}}{2}.$$

$$\text{Stability factor} = \frac{(72 \times 2) + (240 \times 1)}{(360 \times 0.75) + (54 \times 0.33)}$$

$$= 1.33.$$

If the same roof rig were to be used for a painter's cradle weighing only 300 kg:

Counterweight by the correct method

$O = (300 \times 0.75) + (54 \times 0.33) = 243$ kg.
$3O = 729$ kg to be provided.
$R = (240 \times 1) + (C_1 \times 2)$.
Equate and calculate C_1.

$$C_1 = \frac{729 - 240}{2} = 244.5 \text{ kg.}$$

Counterweight by the incorrect method

$O = (300 \times 0.75) + (54 \times 0.33) - (240 \times 1) = 3$ kg.
$3O = 9$ kg to be provided.
$R = C_2 \times 2$.
Equate and calculate C_2.

$$C_2 = \frac{9}{2} = 4.5 \text{ kg.}$$

The necessity for counterweights nearly vanishes and the painter in the cradle will be on the balancing point, with no stability factor worth talking about.

18.19 Externally applied lifting force

There are infrequent cases where the lifting force is external to the system, e.g. from a winch anchored to the ground with its rope over a pulley on the roof rig.

Another case is that of a winch mounted on a part of a building such as a balcony or side walkway around a mansard roof with the rope over a pulley in a roof rig at a higher level.

In these cases, the forces exerted by the top pulley may be hazardously greater than the lifted load, sometimes two or three times this value. This must be taken into account in assessing the maximum suspended load.

18.20 Types of suspended scaffold

(i) Bosun's chairs (see Figure 18.2(a)).
(ii) One-man work cages of various sizes and shapes, rigged on fibre ropes as a bosun's chair, fitted with a manual or power-driven winch or hung from a crane jib.
(iii) Manually operated fibre-rope painter's cradles, usually 450 mm wide (see Figure 18.2(b)).
(iv) Painter's cradles rigged with wire ropes and manually operated reeling winches or climbing devices.
(v) Painter's cradles rigged with power-operated winches.
(vi) Independent platforms, usually 640 mm wide, rigged with manual winches or climbing devices, or fitted with power-operated winches. These can be up to 20 m long (see Figure 18.2(c)).
(vii) Double and treble-deck independent platforms, usually used for sheeting work, fitted with manual winches or climbing devices, or power-operated winches (see Figure 18.2(d)).
(viii) Hinged continuous platforms, usually 600–640 mm wide, fitted with manual winches or climbing devices, or power-operated winches (see Figure 18.2(e)).

Types (vi), (vii) and (viii) may have the lifting appliance mounted on the platform or on the building roof.

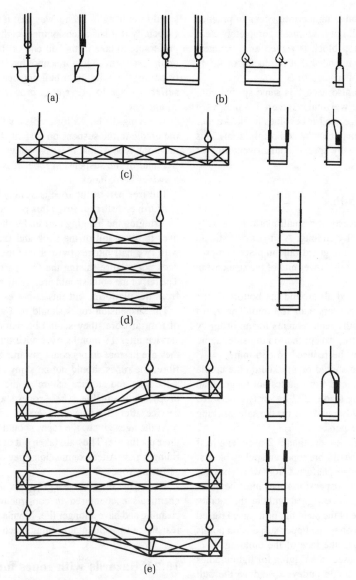

Figure 18.2 Types of suspended platform.

18.21 Bosun's chairs

The bosun's chair is a suspended scaffold. It consists of a chair suspended on a rope rigged through pulley blocks and operated by the person in the chair. The top block is tied off on the building or on a jib projecting from the building. The lower block is attached to the chair. The fall rope or downfall rope is tied off on the supports of the chair.

Traditionally, the chair was a simple plank of wood or a round leg in a shipyard slung from the lower pulley block by loops of rope. Modern chairs are of plastic, with a steel hanging bar.

When the chair is rigged through two single pulleys, the operator pulls with about half his weight. The output from the top of the upper block is the weight of the tackle, the chair and the person when the downhaul rope is tied off on the chair. If, as the operator pulls himself up, the chair hooks under an obstruction and he raises his weight from the seat, he may apply a tension in the rope equal to his weight, in which case the tension in the link supporting the top pulley may rise to three times his weight.

If the chair is rigged on a double pulley and a single pulley, the operator will have to pull only one-third of his weight. In this case, if the chair hooks under an obstruction, the tension in the link supporting the top block may rise to four times the operator's weight.

This extreme condition is unlikely to occur in practice because the operator will not lift himself completely clear of the seat, but if the top block is rigged on a counter-weighted jib, it is recommended that the factor of safety against overturning be 4 instead of 3.

One-man work cages also need this same special consideration of their counterweighting. If there is a motorised means of raising the cage, and it becomes hooked under a window, the roof rig may be winched off the roof. This must be prevented either by special counterweighting or by load-limiting devices.

18.22 Painter's cradles

The painter's cradle consists of a small platform, usually 432−450 mm wide and 1.8 m long. A suspension stirrup supports each end, holds the guard rails in place and provides the anchor point for the lower end of the suspension rope.

Toe boards are provided all around the bottom of the platform to a height of not less than 153 mm. The guard rails must comply with the requirements for an ordinary scaffold working platform, except that on the side facing the work, the height can be reduced to 686 mm.

The safe working load should be marked on the cradle and on the lifting equipment. For cradles not longer than 1.98 m, the safe working load is 227 kg, or two persons. For cradles between 1.98 m and 3.2 m, the safe working load is 295 kg, or three persons.

The suspension for cradles less than 3.2 m in length, if used for light work, can be of fibre ropes and pulley blocks. For longer cradles and other platforms from which heavier work is carried out, wire suspension ropes must be used.

The painter's cradle can be rigged without the facility to traverse across the face of the building, or it can be rigged on a rail fixed at the top on to the upper suspension points so that it can travel along the face of the building.

When fibre ropes are used, the rigging for light cradles should be through two single pulleys to reduce the pull necessary to raise the platform to half of the lifted weight at each end of the platform. For heavy cradles, the rigging should be through a single pulley and a double pulley to reduce the pull to one-third of the lifted weight at each end of the platform.

When wire ropes and mechanical lifting equipment are used, the mechanical advantage of the winch, lifting appliance or climbing device should be such that the manual effort to work the equipment is not greater than 20 kg.

If winches, other lifting appliances or climbing devices are used, they should be provided with a brake or similar device to sustain the load when the operating handle or lever is released.

When climbing devices are used, the tail end of the rope should not hang in such a way that it can cause a hazard, especially if it has a tensioning weight attached to it. It is preferable to take in the tail end of the rope into a reeler on the platform so that it cannot interfere with any person or part of the building or building operation, and it cannot suffer damage to prevent its proper use with the lifting equipment.

The tying-off hooks for cradles with fibre ropes to join the cradle to the suspension system should be of the long tail-hook type so that there is plenty of room to tie off the free end of the suspension rope. It is not essential to use a swivel hook block.

The free part of the rope is usually hanging outside the platform end stirrup. From this position, tying off is done by bringing the hanging part under the stirrup, hooking it over the long tail of the hook and fixing it on to the tail with two half hitches twisted into the rope. This method does not require raising the free part or finding its end. The tail of the hook should preferably point outwards away from the platform, but this is not essential.

The suspension ropes should be 18 mm in diameter. If of natural fibre, they should be manila and taken out of service after 18 months of service on site. Periods when they are in store do not count towards this. If of man-made fibre, the ropes should not be shiny and should be of the same size and surface nature as the manila rope. Polypropylene is suitable. These ropes should be taken out of service after 200 weeks of service on site.

At the ages given, the ropes should not be at the end of their useful life. They are taken out of service only as man riding ropes. After examination, they can continue usefully as lifting ropes.

Man-made fibres have a greater resistance to the sort of chemicals encountered in construction work and stone cleaning and have a longer life. Manila ropes have a greater resistance to accidental burning, such as by a blow lamp.

18.23 Hazards with ropes for suspension

- A loaded polypropylene rope may not last a minute if accidentally subjected to a blow lamp or gas torch flame.
- Tying off the loose end of the rope directly on to the hook tail without passing it under the top crossbar of the stirrup, so that when it takes the tension it can pull off the tail and allow the suspension rope to run free.

18.24 Power-operated independent platforms

A variety of these platforms are available in lengths up to 20 m. They can be dismantled for easy handling and transportation. These platforms are usually suspended on one wire rope at each end and are also fitted with an emergency safety rope at each end.

Operation can be by manual winches or climbing devices, or by power-operated winches mounted either on the platform or on the roof.

These platforms cannot have a fixed safe working load because this depends on the number of units put together to form the required platform. The manufacturer's instructions and rules for use must be strictly followed. In no case can the platform be overloaded or the approved distribution of the load be changed. The safe working load for the job must be plainly marked on the platform and on the lifting appliances.

If the platform is dismantled and re-rigged elsewhere on the construction site, the greatest care must be taken to retain all the features that were incorporated in the first location.

The electrical controls and circuits must comply with the requirements of BS 5974.

18.25 Hinged continuous platforms

These systems are frequently made up from the same elements used for independent suspended scaffolds. The width should be 640 mm if used for personnel only, but 870 mm if materials are to be deposited on the platform.

At each hinge there will be a suspension point with one or two suspension wires and one or two safety wires. The lifting gear can be manual or power operated.

The guard rails and toe boards should permit a good angle of slope to be adopted at any hinge so that the continuous platform can be used as a ramp from one storey of a building to another.

Continuous suspended platforms are frequently in use for a long time during the construction of a building and so should receive adequate servicing on site.

The rules for loading should be strictly followed and a limit put on the number of men who can congregate at any one suspension point.

Provision should be made for tying the platforms into the building.

18.26 Interchanging the equipment

By the nature of the equipment, the various components are taken from a depot and assembled on site in an arrangement suitable for that site. When the job has been completed, it is returned to the depot and each element is stored with others of the same type awaiting the next job.

This process results in different components being assembled repeatedly. Sometimes, the elements may not be ideally compatible. For instance, power cables for cradles 5–6 m long may be used on shorter cradles with dangerous loops in the cable. Stirrups from one system may be used with another system. Hinges and decking units may be unsuitable for use in any system but the one for which they were designed.

The contracting company must incorporate controls that prevent dangerous situations arising, and these controls must operate at all levels in the contracting team.

19 Roof scaffolds, house chimney scaffolds and roof edge protection

19.1 Hazards in work on roofs

Work people who have to go on to roofs are exposed to many hazards unless proper protection is afforded by carefully planned scaffolding. These hazards include:

- Falling off the edge of the roof.
- Falling through the roof sheeting.
- Slipping down the slope.
- Attempting to raise material over the eaves or the gable end without proper lifting tackle.
- Carrying materials that are too heavy.
- Carrying sheets, which may be caught by the wind.
- Gaining access to the roof from the top of a ladder.
- Stepping off the end of a roof ladder or crawling board that is too short.
- Taking short cuts from one crawling board to another across fragile materials.
- Falling through translucent sheets.
- Roof ladders or crawling boards slipping off their hooks over the ridge.
- Roof ladders resting on or against the scaffold around the edge and not firmly attached to the roof.
- Roof ladders attached to the scaffold but the scaffold not attached to the building.
- Walking along gutters that are not sufficiently strong or that have deteriorated.
- Falling over the edge of the sheeting fixed when extending the roof.

19.2 Definitions of a sloping roof

Figure 19.1 shows four slopes that are referred to in the Regulations and Codes.

(i) 1 in 1.5 is the maximum slope of any gangway (Regulation 24(2)).
(ii) 30° is the slope above which only 'suitable workmen' can be employed on good surfaces (Regulation 35(2)).

(iii) 1 in 4 is the maximum slope of gangways without stepping laths (Regulation 24(3)).
(iv) A sloping roof is defined as one that has a pitch of more than 10° (Regulation 35(1)).

19.3 The scaffolding contract specification

The above list of hazards is both extensive and formidable. It should alert the main contractor, the roofing subcontractor, the users and the scaffolding contractor to the necessity to provide the exact access and protection scaffolds required to make the operation that is being undertaken safe. In other words, the scaffolding contractor should agree with the employer of the roof operatives precisely what hazards need to be overcome and what safety measures need to be provided.

Roofers are frequently subcontractors, who rely on the main contractor to provide the equipment to enable them to carry out the subcontract works in a safe manner. Therefore, the way in which the workers are to carry out their objectives has to be set down and the essential protection scaffolds agreed on and provided. Access ways, working platforms and protection decks frequently have to be moved along the area of a roof as the roof is clad or repaired. The equipment provided must be capable of being moved, and the responsibility for moving it must be properly allocated to one or other party. The scaffolding contractor should ensure that his scaffold provides all the safety measures

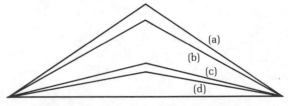

Figure 19.1 Angles of roof slope referred to in various UK Regulations.

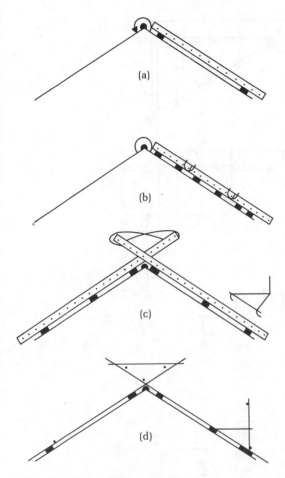

Figure 19.2 Roof crawling boards and ladders.

ladder is to be used skew to the ridge or eaves, as may be necessary near a hip or valley. Figure 19.2 shows some ridge ladder fixings.

Stagings placed horizontally should be supported on specially prepared angular brackets firmly fixed above the rafters. Access to the horizontal stagings should be by crawling ladders fixed firmly by separate means.

In the absence of roof ladders with ridge hooks, or when these or the ridge are unsuitable, extending ladders can be split and each half placed down each of the two slopes of the roof. They should be interlaced with three or four rungs projecting upwards. The intersection point at the ridge should be firmly lashed, and the projecting tops of the ladders should also be lashed together to prevent the angle opening up and allowing the unused section to slide over the ridge. Weights should be attached where necessary.

When old sheeting has to be modified or new ventilators or new translucent sheets fixed, there are the combined dangers from the old fragile materials, the problems of getting the old roof bolts undone, the difficulty of working at the edge and the short duration of the work at each location. It is essential to avoid carelessness and to ensure that all the elements necessary for safe access are reinstated at every change of location. This is the sort of circumstance where the use of safety nets slung on the trusses below the working area using an internal mobile tower should be considered.

Figure 19.2(a) shows a fixed-length ridge-hook roof ladder, which must be long enough to come down to the eaves if access to it is at the bottom from a ladder leaning on the wall. The ridge hook can be attached four rungs below the top of the ladder to afford a handhold if it is necessary to cross over the ridge and down the other side on a similar ladder.

Figure 19.2(b) shows an extendable ridge-hook roof ladder.

In Figure 19.2(c), the two parts of an extending ladder are interlaced over the ridge and tied across the top as well as at the intersection point. The top lashing prevents the angle of the intersection altering. The figure also shows a ladder bracket to support a walking board or boards between two such assemblies.

Figure 19.2(d) shows a scaffold tube rig to form a roof frame, with an edge walking platform and edge protection.

In all the above cases, the ladders should not lie flat on the roof. Timber battens must be lashed into place below them.

asked for, and he would be well advised to obtain the approval of his client, the main contractor, for the scaffold provided.

Many roof accidents would not occur if the subcontract enquiry, the tender and the order stated precisely what was wanted and supplied, and how many of each item was to be supplied. This would ensure that both parties to the subcontract had thought out the access problems in advance, and this would in turn help towards its provision and use on the job. The subcontract condition 'main contractor to provide scaffolding' frequently leads to wrong or inadequate scaffolds being provided.

19.4 Simple roof maintenance equipment

Crawling boards up the slope with ridge hooks should be long enough to reach from the ridge to the eaves and not leave a gap at the bottom without safe access. Roof ladders can be extended either with special extension pieces or by careful lashing. Very careful tying must be made if a roof

19.5 Scaffolds around roof chimneys

These fall into two categories, for eaves chimneys and for ridge chimneys. Although they are small jobs, there is always some difficulty in fixing a platform completely around the chimney, especially in the case of a chimney

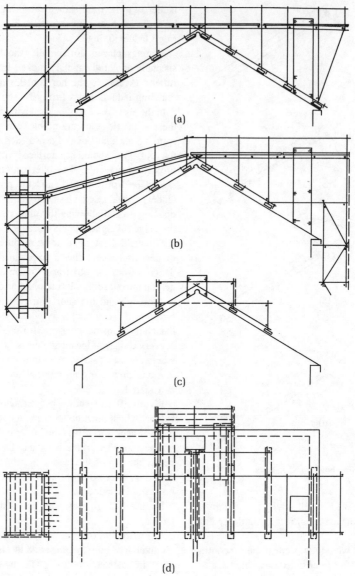

Figure 19.3 Domestic chimney scaffolds.

at the end of a ridge, which requires a platform over the gable-end wall. The access scaffold to reach the chimney scaffold frequently causes some problems.

Figure 19.3(a) and (b) is an example of eaves chimney scaffolds with access on the opposite side. Figure 19.3(d) is a plan showing an extension to reach the ridge-end scaffold shown in elevation in Figure 19.3(c).

These diagrams are drawn square to the eaves and ridge lines, but the whole access sometimes has to be on the skew. Careful attention to detail is required when this is the case.

19.6 Bearing boards on roofs

The object of placing boards on the roof before a roof scaffold is built is to spread the load over the slates or tiles on to safe members existing in the roof and to minimise damage to the slates or tiles.

Neither of these objectives can be achieved unless the space below the roof is inspected to ascertain where the rafters are and whether there are any purlins. If the rafters are 100 by 50 mm (4 by 2 in) timbers spaced at 400 mm

(16 in), and tiles are fixed on tile battens every 100 mm (4 in) up the roof, it is generally immaterial which way the scaffold roof boards are placed. However, an old slated roof may have slate battens spaced at 150—200 mm intervals, and these may be rotted and span between rafters at 600 mm (24 in) intervals. In this case, if scaffold roof boards are placed up the slope, they may push through the roof. There is no hard and fast preference for the manner in which scaffold roof boards should be laid.

Which ever way they are placed, they must be prevented from slipping down and over the eaves.

19.7 Roof edge protection

Falls from the edges of roofs, of both workmen and materials, are so numerous that special consideration must be given to the problem. BS 5973 1981 does not deal with the subject.

The Construction (Working Places) Regulations 1966, Section 35(4), require that where work is done on or from any sloping roof, suitable crawling ladders or crawling boards must be provided on that roof. Except where the work is not extensive, either a barrier must be provided at the lower edge and so constructed as to prevent any person falling from that edge, or the work must be done from a securely supported working platform not less than 432 mm (17 in) wide fitted with guard rails and toe boards.

For the purpose of Regulation 35, a sloping roof is defined as one having a pitch of more than 10°.

For roofs that are not sloping, Section 33(2)(a) of the Regulations requires guard rails and toe boards along the edges.

There is some difficulty in interpreting these requirements for work at the edge of a roof as opposed to work in the centre.

The scaffold contractor should ascertain from his client the precise duty of the scaffold requested.

(i) Is it to prevent materials slipping off, in the case where the workmen are made safe by other means?

(ii) Is it to prevent workmen falling off, in the case where there is no substantial amount of materials, e.g. replacing a few tiles?

(iii) Is there any work to be performed on the edge of the roof or within arm's length of the edge?

(iv) Are there any materials to be stacked near the edge ready for use?

(v) Is the edge protection to include a walkway, a barrow run or a materials delivery gangway?

(vi) What trades will be engaged, and what weights will their work involve?

Each of these functional specifications will require a different edge protection scaffold, and sometimes several of the functional requirements may have to be met.

(i) To prevent materials slipping off a sloping roof when the workmen are made safe by other means

All that is needed for this duty is to fix a scaffold baord on edge along the bottom edge of the roof. This can be done by arranging a vertical or near vertical tube hard up against the roof edge or the gutter and maintaining it in this position in any convenient way.

Figure 19.4(a) and (b) is an example. In case (a), when the windows cannot be opened, a bridle tube is fixed on the outside of the window using scaffold anchor rings. The rakers are tucked behind this and rested in pairs on the sills. Alternatively, the rakers can be lashed back to the rafters if these project outside the wall and are accessible.

(ii) To prevent workmen slipping off a sloping roof

This can be achieved similarly to (i) but with the vertical tube extended 1 m above the tile level and fitted with two guard rails.

Figure 19.4(c) shows this arrangement. To prevent the fall of a person, it must be much more robust than case (a). Through ties should be used as shown.

Figure 19.4(d), (e) and (f) is a variant of (c).

In cases (d) and (f), the tie tubes pass through the windows horizontally. This can be achieved by placing them in the corners of the frames and setting a 100 by 50 mm timber between them. The windows are then closed on these battens. This can be done only with sash windows. Casements present a problem that can probably be dealt with only by using an external scaffold.

(iii) To provide a platform at the edge of a sloping roof to enable work to be done on or near the edge or on the fascia boards or gutter from the platform

If the work does not entail the workmen climbing on to the roof or reaching up the slope more than arm's length, such work can be done from the top deck of an ordinary external access scaffold, which can be of the independent tied or putlog type. The deck should have toe boards and guard rails in the orthodox manner and be placed at a convenient level below the eaves.

Figure 19.4(g), (h), (i) and (j) is an example of edge protection by a platform. In some circumstances, particularly between close houses in city areas, a single-lift horizontal diaphragm can be used butted up against both buildings and supported on a single line of uprights down

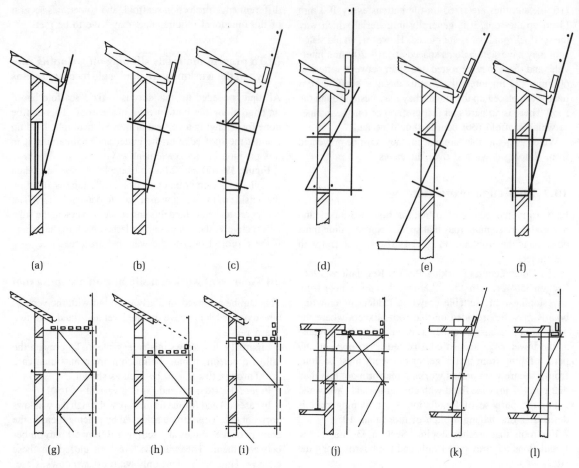

Figure 19.4   Roof edge protection.

the centre. Case (g) is for working from the sloping surface. Case (h) is for working from the platform. Case (i) is a combination of (g) and (h).

(iv) To provide an edge working platform to a flat slab or roof

Figure 19.4(j) provides adequate protection if an inside board is fitted.

(v) To provide edge protection only to a flat slab or roof without placing a rail on trestles on its top surface

Figure 19.4(k) and (l) is an example of edge protection that does not interfere with roofing work such as asphalting or sheeting.

(vi) Platforms to act as working platforms, edge protection and for materials storage

In this case, the platform should be five boards wide and should be fixed higher than in case (h) so that material sliding down the roof cannot knock off material stored. A double-height toe board should be considered, which will prevent materials being knocked over by a man slipping off the edge of the roof. A second lower storage level can be installed if necessary.

(vii) Edge protection and a walkway or barrow run

A three-board deck is adequate as a walkway or barrow run, but it must have guard rails and toe boards. It is good practice to place this at gutter level and to put guard rails and toe boards on both sides of it. Those on the inside form the edge protection and those on the outside are the normal working platform guards. Materials from such a gangway should not be stored on it but transferred directly to the roof.

19.8 Loads

Gutter painting work can be carried out from a scaffold designed as a very light-duty scaffold, 0.75 kN/m² (75 kg/m², 15 lb/ft²). See Table 2.8.

Gutter replacement requires 1.5 kN/m² (150 kg/m², 30 lb/ft²).

Slating and tiling require careful consideration. The work is fast. Barrow loads of old material have to be taken off the roof and new materials delivered. For this, the general-purpose loading of 20 kN/m² (200 kg/m², 40 lb/ft²) is required.

If material is to be stored, it will accumulate very quickly and be stored in local areas. A definite agreement on the disposition of this load and its distribution must be made between the roofing contractor and the scaffolding contractor, who should jointly consider a heavy-duty rating of 2.5 kN/m² (250 kg/m², 50 lb/ft²).

19.9 Access equipment for use below roofs

The remarks made previously about both the subcontractor and the main contractor, stating clearly what elements they want and what they can supply, apply also to internal access under roofs.

A slung scaffold is probably the best solution, but this may be disproportionately expensive for a painting or cleaning job where the rate of the work is high.

Lightweight stagings are the most frequently used equipment for this circumstance. Access to these stagings must be safe and this may have to be as fast-moving as the stagings themselves.

Mobile towers are frequently built within factories for roof painting, reglazing and cleaning. These are not without their hazards, even when used internally. There is a risk of overturning when the castors are obstructed by articles on the floor, and from the repeated changing of the guard rails and repeated dismantling and re-erection of the top few sections to enable the tower to pass beneath the roof trusses.

Factories frequently have internal travelling crane systems. The travelling beams of these are usually wide enough to mount a scaffold gantry along the top of them. This can be done successfully using scaffold tubes resting on the crane gantry beams and anchored thereon by girder clamps. If the scaffold gantry has to be continually dismantled and reconstructed, care must be taken that its construction design does not deteriorate as the job progresses until an unsafe or detached structure results. Use of the crane may not be possible when the scaffold gantry is on it.

The use of aluminium tubes should be considered for this type of work.

19.10 Lightweight stagings

These are platforms designed to span up to 7.25 m (24 ft) without any intermediate support. They are 457 mm (18 in) wide. They are constructed from two side battens, which can be of timber or aluminium joined by timber cross members and tie bolts. The space between the side battens is decked with five longitudinal timber slats.

The side battens are strengthened by a pre-tensioned wire running along the lower edge. Because of this arrangement, they must not be used upside down or as a loaded cantilever projecting beyond a support.

Maximum slope

Because they span such a distance, lightweight stagings are flexible and are subjected to some vertical oscillation, which lowers the friction between the support and the staging. This friction may be relatively small because:

- The support is a steel member.
- The lower side of the staging battens has the stressing wire not buried fully in the wood but acting as a skid.

Accidents have resulted when the slope has been about 1 in 4 and the staging has worked its way downwards off its top support.

Tying

Manufacturers can supply specially made ties to lash the stagings to their supports. Even if the staging is horizontal, it is desirable to tie it down.

Safe working loads

Lightweight stagings are manufactured in various lengths, but their construction is different for different lengths, so their safe working loads are more or less the same. Stagings manufactured to BS 1129 are intended to support three persons spaced reasonably apart, which is equivalent to about 2.5 kN UDL (275 kg UDL, 600 lb UDL), all of which are rounded-off values.

Hazard with the loads

- Stagings are frequently used side by side to give a platform about 510 mm (36 in) wide. It cannot be assumed that two platforms side by side will safely support a load of six men reasonably spaced apart, because the six men are moving loads, and all six may stand on one of the pair of stagings. Platforms made with two or more lightweight stagings should be rated for three men spaced apart on the whole platform between the supports.

Guard rails and toe boards

Manufacturers make special components for attaching toe boards and guard rails to lightweight stagings. These elements are fastenings that fix to the stagings and pass over or under them so that toe boards and guard rails can be attached to one or both sides of a single- or double-width stage. Guard rails and toe boards to attach to the fastenings may be supplied by the manufacturer.

End guard rails

If the staging overshoots its end support, workmen must be prevented from walking on the overhang by an end guard rail, rope or chain.

19.11 Hazard with guard rails on stagings

- Stagings used in roof maintenance work frequently span distances of up to 7.25 m (24 ft) between trusses. Because there is no intermediate support to which a guard rail can be attached, frequent use is made of the special guard rail fastenings referred to in the previous section.

If a lightweight stage 460 mm (18 in) wide is fitted with guard rails at a height of 910 mm (3 ft), and a person or the three persons allowed on the staging lean on the guard rail, a lateral force of about 30 kg per person (90 kg for three persons) is applied to the guard rail, giving an overturning moment of 27.3 kg m per person (82 kg m for three persons).

The righting moment for a stage 7.5 m long weighing 7 kg/m is 12.12 kg m from the self weight and about 49 kg m if three men stand in the middle of the staging, giving a total righting moment of 61 kg m. This means that the staging will roll over. If the platform is shorter than 7.5 m, it will lose some self weight. If it has aluminium styles, it will lose about a quarter of its self weight, and overturning will be very likely.

If the workmen use drills or other tools that exert horizontal forces, there may be further overturning effects due to impacts and the weight of the hanging cables.

To prevent overturning, it is imperative to lash the guard rails to the roof truss or to some other firm points.

19.12 Regulations governing guard rails and toe boards on gangways in roofs

Wherever possible, the ordinary rules concerning toe boards and guard rails on working platforms should be applied to gangways and to stagings in roofs, but there are difficulties, which the Regulations acknowledge.

The normal rules do not apply to a temporary platform passing between two glazing bars of a roof with a sloping surface if those bars or the roof framework afford a secure handhold along the full length of the platform.

The normal rules do not apply to the means of access or egress for operations or work on a roof or part of a sloping roof.

The normal rules do not apply to a temporary gangway that is used only in the course of erecting any framework forming part of a building or other permanent structure for work of such short duration as to make the provision of a gangway with guard rails and toe boards or other barriers unreasonable.

Guard rails, toe boards and other barriers normally required can be removed or remain unerected for the time and to the extent necessary for the access of persons, the movement of materials or other purposes of the work. However, the guard rails, toe boards and barriers must be replaced or erected as soon as is practical.

19.13 Hazards in roof scaffolds

- Using the edge protection to support a roof ladder causes many accidents. Instead of anchoring the roof ladders or crawling boards over the ridge or otherwise to the roof, they are rested on the edge protection platform. They slide down and push through the toe board, thus displacing the edge protection.
- If the roof is of such a width and nature that the roof ladders cannot be anchored on to it and have to be supported by the edge protection, the forces involved must be assessed and catered for by special ties. These need to be at fairly close centres because the roof ladders are moved along in 0.5 m stages and may have old or new materials stored on them.
- There is a problem when using equipment such as that shown in Figure 19.2 where the access is on the skew, as at a roof valley or hip. Ladders or crawling boards placed up the maximum fall line of a slope can be hooked over the ridge quite safely but, when placed on the skew, they tend to slip down sideways. This puts the scaffolder at risk, especially during dismantling from the top down. Roof ladders are unsatisfactory when used on the skew without any other tying.
- Experience has revealed another hazard when repairing all the chimneys on a housing estate. Sound scaffolding is erected for perhaps the first half dozen cases, and the job progresses with the scaffolds being taken down one at a time and re-erected for the next repair. As the work progresses, the scaffolds are built more carelessly until either damage to the property occurs or there is an accident.
- If the job progresses along a curved road, the details of each scaffold have to be modified for each re-erection. The safety precautions for the earlier ones are sometimes lost in the later ones.

20 Gangways, pedestrian bridges and vehicle ramps

20.1 Access ways

The subjects dealt with in this chapter are not places from which work on a building is carried out but the means of access to these places. This does not mean that they escape the scope of the Construction (Working Places) Regulations 1966, because these have specific requirements concerning access.

The definition of a scaffold in the Regulations contains the clause 'any temporarily provided structure that enables persons to obtain access to or that enables materials to be taken to any place at which such work is performed'.

Regulation 6(1) states that 'there shall, so far as is reasonably practicable, be suitable and sufficient safe access to and egress from every place at which any person at any time works, which access and egress shall be properly maintained'.

It is thus clear that all the rules for scaffolding also apply to the means of access. This is as it should be, because there are more falls from scaffolds by persons moving along them and carrying materials along them than there are by persons actually working from the platform. The same requirement appears in the Factories Act 1961 and applies not only to permanent access but to any scaffolding temporarily erected in a factory for work that does not come under the Construction (Working Places) Regulations.

The Occupier's Liability Act 1957 imposes upon an occupier of premises the obligation to take such care as is reasonable to see that his visitors are reasonably safe, and this applies to a temporary gangway or stairway built as a means of access.

The Offices, Shops and Railway Premises Act 1963 requires the provision of sound floors, stairs, steps, passages and gangways, and this includes any temporary structures built within the premises.

Returning to the Construction Regulations, the scaffold gangways, ramps, stairway towers and bridges must be as closely scrutinised by the scaffold inspector as the rest of the temporary structure. They must be closely boarded and fitted with toe boards and guard rails. A ramp must not have a slope of more than 1 in 1.5. If the slope is more than 1 in 4, the ramp must be fitted with stepping laths, with a gap of 100 mm (4 in) for barrow wheels.

20.2 Horizontal forces

The forces to be considered fall into two classes:

(i) The horizontal reaction of the feet of pedestrians walking along an access way. This can be taken as 15% of the weight of the person, i.e. his body is assumed to incline by about 300 mm (12 in) in a height of 1.8 m (6 ft).

This occurs on an access way loaded with moving persons, for which condition an area of 2 m^2 (20 ft^2) per person is reasonable. This results in a uniformly distributed horizontal force of about 0.07 kN/m^2 (7.5 kg/m^2, 1.5 lb/ft^2).

(ii) In circumstances where many persons are using a gangway and have to pass down steps at more or less the same speed, the area per person can be taken as 0.25 m^2. The vertical load and impact forces will be 5 kN/m^2 (500 kg/m^2, 102 lb/ft^2) from an inclination of the body of about 150 mm per 1.8 m, or 1.5%. This will result in a uniformly distributed horizontal force of 0.375 kN/m^2 (37.5 kg/m^2, 7.7 lb/ft^2).

This value is five times that for the horizontal part of the gangway.

Both the vertical and horizontal loads on access ways should be discussed by the scaffolder and the purchaser, and the loads should be confirmed as acceptable by the specifying authority.

20.3 General construction of gangways

Width

Gangways for persons should be not less than 432 mm

(17 in) wide, and gangways where there is passage of materials should be not less than 635 mm (25 in) wide.

Slope

Gangways should be horizontal in cross-section. In long section, a gangway should not be used if the slope exceeds 1 in 1.5 and, if it exceeds 1 in 4, stepping laths should be provided that have a 100 mm (4 in) gap in them to enable barrow wheels to pass.

20.4 Timber and lightweight stagings

The smallest gangway normally encountered on a building site is a pair of planks across a ditch or from the general site level to the building floor level.

It would be an unanswered plea for scaffold boards not to be used for this purpose. A person and a barrow of building materials crossing a scaffold board spanning 4 m stresses it to 20 N/mm², which is twice its allowable stress and not far off the point where it will be irreparably damaged, if it does not fail.

Timber 50–60 mm thick should be used for this purpose.

Spans of 4–7 m can be dealt with by lightweight stagings with a clear span, but they can support only three persons on each board. Thus, they are no use for an access way intended for the public to use or for the whole construction gang entering the site.

20.5 Scaffolding for access

Gangways intended for use by many persons should be designed for a distributed load of 5 kN/m² (510 kg/m², 104 lb/ft²), which allows about six persons per square metre. This can be dealt with by ordinary scaffold boards supported at 1 m centres, i.e. five transoms in each board length instead of the usual four.

Such a gangway is therefore constructed as an ordinary access scaffold with a line of uprights along each side. The uprights will be loaded through a single coupler supporting the ledgers and so are limited to 635 kg each, which will result in reasonable foundation loads. A pair of uprights will carry 1270 kg, corresponding to a deck area of 2.5 m². Assuming that the platform is five boards wide, i.e. 1.2 m, there will be a pair of uprights every 2 m along the gangway, which results in a reasonable design.

If for any reason the pairs of uprights have to be spaced wider than 2 m, or the gangway is wider, it will be necessary to supplement the ledger right-angle couplers or to take some of the load from the centre of the bay to the uprights at a lower level. Figure 20.1 shows how this can be done. For heavy pedestrian traffic, the spans of the boards and the top ledgers may need to be calculated.

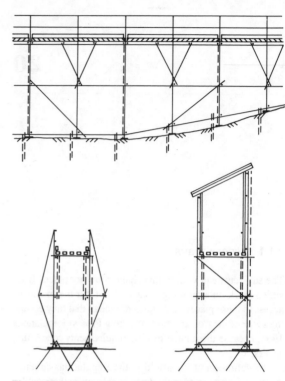

Figure 20.1 Open and covered gangways.

20.6 The decking of gangways

When scaffold boards are used, experience shows that five transoms per board give a very satisfactory platform.

When plywood is used, it must be of adequate strength and capable of maintaining its strength in the particular environment. Special attention is needed when plywood is used exposed to the atmosphere, to chemical vapours or to spray in river and marine situations.

Open-texture decks can be used, provided that the apertures are not greater than 20 cm² (3 in²) in area. The space between planks can be 25 mm (1 in). In these two circumstances, there must be no risk of persons below the platforms being injured by falling debris or materials.

20.7 Covered access

Figure 20.1 gives the general form of an open and covered gangway. It is essential to calculate wind forces for sheeted gangways to evaluate the adequacy of the standards and to check the amount of transverse bracing required.

20.8 Pedestrian bridges

Bridges of up to 30 m have been successfully designed and built in scaffolding materials, but they must be designed and supervised by an engineer. It is not within the scope of this book to set down the details of a tube-and-coupler design. It follows the basic principle of any lattice girder design, with added safeguards to overcome the deficiencies caused by the use of in-line friction joints in the chords and friction couplers in the bracing. In an orthodox access scaffold or a birdcage, if there is a local overstress or a local failure, there is usually an adjacent element that can come to the rescue of the affected zone. In a bridge this is not so. If a joint gives way, the bridge gives way if its construction follows the theoretical design closely.

In practice, in pedestrian and traffic bridges and ramps, every element of scaffolding should be duplicated in a special way. Where there is a joint in a chord tube, even if this is well lapped, another chord tube must be present without a joint. Where a brace is attached to a transom on a chord, there must be another brace on another transom.

The bridge should be designed with a very deep section, twice that for a structural steel girder bridge of the same duty. If this is not done, each chord will be found to need ten tubes for a 30 m span and quadruple bracing will be needed in the end bay. This creates a very cumbersome design, which is partly avoided if the span to depth ratio is kept at 5 or 6.

The modern practice is to use prefabricated scaffold beams or proprietary demountable structural beams. This simplifies the details of design and construction. The scaffolding work is then confined to battening, bracing and anchoring the prefabricated beams, the decking and guard rails, and the erection of a canopy or complete cover if required.

20.9 Traffic ramps

These are also structures to be engineer designed and supervised. The most common requirements are for a ramp down into an excavation for muck-shifting lorries or an access ramp crossing a retaining wall excavation.

The Ministry of Transport standard bridge loading may not be applicable. Sometimes the traffic may be light, requiring only a 4–5 tonne loading. On another site, 40 tonne excavation lorries and equipment may use the bridge.

As in the case of pedestrian bridges, it is not intended to go into detailed structural design. The same special requirements for the scaffold materials and joints must be incorporated into a scaffold ramp design. One difficult problem to be dealt with is a consequence of the deck being of separate elements, e.g. railway sleepers. These are excellent for distributing a load that occurs in the middle of the sleeper but, where the sleepers join end to end, there is no continuity and the applied load passes straight through the deck to the support below without any distribution.

Thus on the drawing board the ramp may look like a well-designed birdcage in which four groups of wheels, each group carrying 8 tonnes, appear to be well distributed over 24 standards, giving 1.33 tonnes per standard.

In practice, if one wheel group passes over the end of a sleeper, the whole 8 tonnes may go through on to two standards, with disastrous consequences. The wheel groups also have to pass from one sleeper to the next, even if not at the ends, and this puts a point load on the primary beam below the deck.

Dealing with matters of this sort requires the experience of the scaffolding companies and the cooperation of the user in placing the fendering in a critical position.

Surge loads on an excavation site are difficult to predict in quantity or location, and arranging foundations to cater for horizontal surges is difficult in a site being excavated.

20.10 Hazard with load spreading

• In the case of vehicle ramps, the wheel loads of construction vehicles may be very large and have high impact values. It is sometimes necessary to use the decking material of the ramp or bridge to spread the load between the supporting uprights. In wide ramps, it is sometimes necessary to extend the decking planks end to end, or to extend the primary or secondary beams end to end. These joints may be butt joints over a beam or over a forkhead or an upright. If a vehicle wheel passes over this spot, its load passes straight through the decking planks or beam ends without any distribution, i.e. the column has to accept the full wheel load, even though it may have neighbours within 0.5 m. Failure to recognise this has led to many failures.

The decking elements must be designed to match the supporting elements, and the whole has then to be supervised carefully on site to ensure compliance with the design.

20.11 Horizontal loads on guard rails and fenders

In orthodox access scaffolding, these have always been dealt with successfully by scaffold tubes spanning up to 3 m between the uprights of the scaffold or between the corner posts of a tower.

In gangways on sites, the span should be not more than 2.5 m. In mainstream gangways, this can be reduced to 2 m.

In traffic ramps, the loads should be negotiated with the site owner or contractor.

Special rules apply in places of public entertainment, and the site owner must be consulted.

A table of the forces to be used in design is given in Part IV.

Table 20.1. Access loading

Vertical loads	kN/m²	kg/m²	lb/ft²
Pedestrian standing places such as observation platforms	4	408	83.5
Entertainment stands for standing	5	510	104
Entertainment stands fitted with seats	4	408	83.5
Pedestrian gangways giving access to places of work	4	408	83.5
Pedestrian gangways giving access to places of entertainment	5	510	104
Staircases at places of work	4	408	83.5
Staircases at places of entertainment, workmen's gangways on building sites	5	510	104
Workmen's gangways under or on roofs : moving concentrated load (three men)	3 kN	300 kg	661 lb
	Moving to any place		
Pedestrian footbridges in public places	4	408	83.5
Traffic bridges on sites	As directed by the main contractor		
Traffic bridges in public places	As directed by the local authority		
Traffic ramps to various levels on construction sites	As directed by the main contractor (especially with regard to muck-shifting lorries)		

The deficiency of puncheon guard rail posts on one coupler must be borne in mind.

20.12 Suspended gangways using the combined strength of the scaffold tubes and scaffold boards

It is sometimes not practical to arrange supports either from the ground or from a structure above at centres closer than about 3.66 m (12 ft). Such cases occur in boiler cleaning operations, in chemical works, and in buildings with false ceilings and predetermined holes for suspension wires. In some cases, the structure may not be able to sustain the horizontal forces that raking intermediate supports for the gangway apply to it.

In these various circumstances, the timber of the decking can be used as a structural member, i.e. as a load spreader in conjunction with the scaffold framework that supports it. The hazard referred to in Section 20.10 applies here.

20.13 Hazards with gangway guard rails

- One hazard that escapes attention is for the scaffolder. During construction of the gangway, it cannot be in its final and safe form. The scaffolder builds it with the usual care and attention to his own safety. When construction moves ahead to another part of the work, the gangway built for future use by other workers becomes the access way for the scaffolders. The scaffolder is not building the passage that he is using as a normal access to his own work. For this purpose, it should be completed and fully guarded. If it is not, the scaffolder is at risk and he has created this risk for himself by not completing the platform as necessary.
- When the walkway or ramp has been completed, there are many accidents because the guard rails are not strong enough for the loads they were intended to resist.
- The loads are sometimes underestimated.

21 Tall chimneys, churches, monuments, cooling towers and water towers

21.1 Tall structures

The scaffolding for tall structures should be engineer designed and supervised. This is mainly on account of the high wind forces, but also because of the large self weight on the standards.

Most tall structures are relatively narrow, so there is a stability problem that has to be dealt with by ties. They are frequently circular in section, which causes problems in constructing the platforms to be near the work all around and to ensure the correct spacing of supports for the working platforms.

It is frequently desirable to put protection sheeting in place at large heights, and this increases the wind force problem.

Finally, there is the difficulty of access to the high platforms for both men and materials, especially if these are required in large quantities.

It has generally been found most satisfactory to build scaffolds that are square in plan and to modify the inner line of standards and ledgers as necessary to deal with various changes in the shape of the main structure.

The designer must pay close attention to thermal expansion and contraction in the long vertical tubes. A 150 m (492 ft) standard can vary in length by as much as 35 mm with changes from day to night temperatures.

21.2 High chimney scaffolds

Several different jobs may be necessary on a high chimney:

- Cleaning and reposition or reconstruction of the top section of 5–15 m either inside or outside, or both.
- Removal of the refractory lining and reconstruction either of the top section or of the whole depth.
- Removal of both the lining and external structure of the top section.
- Demolition of the whole chimney by means other than blasting or toppling it.

For all these processes, three special factors need careful consideration and calculations:

(i) The width of the platform and the nature of the access.
(ii) The imposed loading of the old and new materials that will be stored on the platform.
(iii) The large wind force that may occur at the top level of the chimney, especially in the third case above.

Special attention must be paid to the means of attaching the scaffold to the outer skin of the chimney in a manner that can resist the vertical loads, which may include the forces resulting from the hoisting gear, and the horizontal wind loads.

The choice between a cylindrical scaffold and a square one has to be made. There are disadvantages with both.

A cylindrical scaffold has discontinuous ledgers, and in consequence the uprights, when viewed along the tangent of the chimney, can go out of plumb away from the chimney between tie points. There is a further disadvantage with the cylindrical form in that it is impossible to fix longitudinal bracing over any width greater than one bay. Figure 21.1(a) shows that both the bracing and the ledgers have to be discontinuous. The bracing in one bay does not have any effect on the next. If the bracing could pass over two or more adjacent bays, it would help to stiffen the scaffold in plan, as would continuous ledgers.

Consideration should be given to rolling the tubes into preformed curves for the ledgers and face bracing. The quantity of tubes and couplers is reduced, so the weight and wind forces are also reduced.

A square scaffold for a round chimney, Figure 21.1(b), has the disadvantage that there is more material in it, so it is heavier and collects more horizontal wind force. On both these counts, the tying of the scaffold to the chimney is more difficult and the tie tubes are not always at right angles to the scaffold.

There are problems boarding out the deck in either type because of the curvature. Even when boards are lapped,

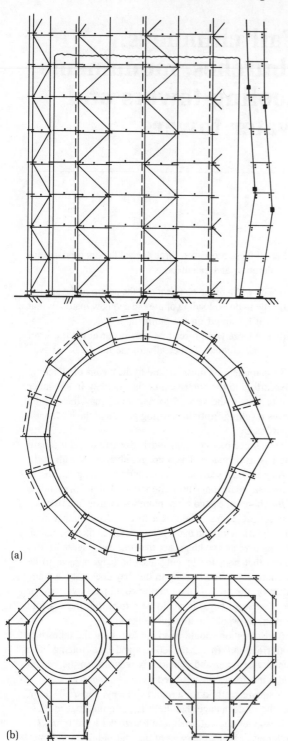

(a)

(b)

Figure 21.1 Tall structure and chimney scaffolds showing a weakness due to discontinuous ledgers in curved structures overcome in rectangular shapes.

which is permissible when the structure is curved, there are problems maintaining a reasonably small space between the decking and the structure.

A cylindrical scaffold is easier to tie to the structure, and it is easier to use prefabricated brackets if the scaffold has to be started above ground level.

21.3 Foundations

The foundations are usually heavily loaded because of the height of the scaffold. In addition to this, they suffer wide variations of load due to the influence of thermal expansion of the long vertical lengths of tubing. The foundation loads of the hoist tower or mast alter significantly due to wind forces on the hoist structure and the cage.

There is a variation in the foundation load due to the changes in the imposed load from the materials stored. This is a more severe effect in a cylindrical scaffold than in a square one because, when the ledgers are discontinuous, all the load being taken up by any standard at its upper loaded level has to pass directly to its foundation. In the case of a rectangular scaffold, with its continuous ledgers and bracing, the load on any standard is partially shared with its neighbours through the continuous ledgers and the bracing.

In consequence of the large dead load and its variations due to the causes described above, the foundations 'pump' in the ground. Because the scaffold has to be used for a considerable time, this can have serious consequences. After the scaffold has been designed and the locations of the standards have been fixed, it is preferable to install concrete foundations. Concrete strips 400–500 mm wide are superior to 228 mm wide scaffold boards or 75 mm thick timbers.

The foundation contact level should be 200–300 mm below the surface, at which level the soil has less freedom to be forced out sideways beneath the loaded area.

21.4 Load-shedding brackets

An attempt is often made to relieve the loads on the standards and, at the same time, to reduce the load on the foundations by taking some of the load on to the chimney through brackets joining the scaffold to the outer skin of the brickwork or concrete.

This is a difficult procedure because it is never certain which way the load will go. It usually goes down the stiffest path because the less stiff support settles, compresses or deflects under its share of the load and leaves more than the anticipated share on the stiff load path. Because the brackets are likely to be stiffer than the ground, they are unlikely to take only a portion of the load. They will probably take all of it. Therefore, they must be designed to do so, and the concept that they will only relieve some of the load is not borne out in practice.

21.5 Support brackets

These are intended to transfer the full vertical and horizontal loads from the scaffold to the chimney structure. They may be fixed at any level on the chimney and the scaffold constructed upwards over them. With some types of bracket, a portion of the structure can be hung below the brackets.

When a multiplicity of brackets is used, their rigidity of fixing to the chimney structure must be reasonably equal so that they will share the load fairly. Above a certain level, the outer skin of the chimney may have deteriorated badly, especially in the top 5–7 m, and brackets in this zone should not be relied on to take their fair share of the load. If possible, it is better to ignore them entirely and to carry the load on brackets lower down.

Figure 21.2 gives some examples. When the chimney surface is inclined, an allowance must be made for this in the design of the bracket or by pads under its fixing. In the case of a cooling tower scaffold, the angle of the structure from the vertical may vary considerably and an adjustable bracket may be essential. Figure 21.2(a) and (b) are hinged brackets. In (b), the specially designed transom is supported by doubled ties.

A greater inclination of the diagonal member of the bracket gives a greater direct pull on the anchor bolts. Thus if the outer skin of the chimney is suspect, the brackets should be tall with a slope of not greater than 30° from the vertical. If the material is very poor, it may be necessary to cut needle holes in the brickwork to obtain the necessary bearing and to insert tubes or rolled sections.

If all the ledger bracing tubes on a chimney scaffold are inserted inclined at their tops towards the chimney, some of them can be extended upwards until they meet the chimney surface. At this point, the tube can be pressed flat to one side, drilled and bolted to the chimney. Figure 21.2(e) shows this.

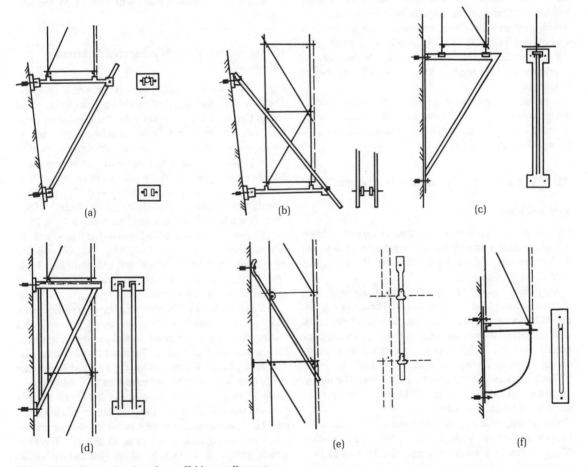

(a)   (b)   (c)

(d)   (e)   (f)

Figure 21.2 Support brackets for scaffolds on tall structures.

The transom opposite the bottom of these 'tension tube' anchors must butt up against the wall. The number to insert depends on the loads and the test results of some typical assemblies in the brickwork of the particular job.

Lifting tackle is sometimes needed inside the chimney, and this necessitates beams being placed over the top of the chimney from one side of the scaffold to the other. Such an assembly will exert a considerable local load on the scaffold, its supporting brackets and its foundations.

Figure 21.2(d) is a prefabricated double bracket to permit the scaffold to be built through it.

Figure 2.12(f) is a tall chimney bracket in large-diameter tube. The sockets above the transom can be designed for a constant or near verticality, or for a small variation in the inclination of the wall.

21.6 Ties

The load-bearing brackets described above are the best form of chimney ties, but they are only required at intervals up the scaffold. Ordinary ties may be necessary between these brackets. Drilled-in anchors are the only satisfactory type. A bolt through a hole in a structural angle welded to a scaffold tube gives a rigid fixing. A 50 mm ring bolt gives some room for movement. The choice depends on the requirements of the scaffold design and on the number of bracket ties.

A problem may occur in an old brick chimney due to deteriorating brickwork, which can be dealt with by using a multiplicity of the extended ledger brace ties shown in Figures 21.2(e) and 21.6.

21.7 Typical data for a chimney scaffold

Vertical loads

The vertical loads during dismantling and reconstruction need accurate assessment. If an outrigger is attached to one bay on one side of the chimney, this bay should receive special consideration.

A typical imposed load is two men in any bay or six men distributed around the platform, plus 100 bricks in any bay or 300 bricks distributed around the platform (these bricks may be refractory bricks weighing 4.5 kg each), plus special units such as liners, cast iron capping segments or coping stones at 250 kg per element per bay.

These loads must be assessed for each job. The nature of the brackets may be such that all the weight comes down the inside or outside standard.

The hoisting gear may apply a force of twice the lifted load to the structure, perhaps 100 kg vertical and another 50 kg for the dead weight of the tackle. There may be a horizontal force of one-fifth of this due to the inclination of the hoisting rope.

These vertical loads typically add up to something like 900 kg per bay for the processes of dismantling and reconstruction.

To this has to be added the self weight of the working platforms. Two working platforms should be allowed at 50 kg per bay each for the lapped timber decking. The imposed load is thus 1000 kg per bay, or 500 kg per standard.

On the bays where the hoisting gear is attached, the outer standard load may be increased to 650 kg at the top of the scaffold under the platforms.

Below the working level, the self weight of the scaffold has to be added.

If the chimney is 100 m high and there are beams across the top equivalent to 4 m of height, the total self weight of the scaffold will be from 52 lifts with 1.2 m bay lengths = 52×68 kg. The total weight per standard is thus about $1768 + 500$ kg = 2268 kg, which is satisfactory for 2 m lift heights in a properly tied and braced scaffold.

The extra local loads from the hoisting tackle, the wind loads and the horizontal loads will have to be specially catered for.

Horizontal wind loads on scaffolds around chimneys

Figure 21.3 shows a typical flow net around a chimney. The flow along the 'sides' of the chimney has an increased speed. Careful analysis indicates that the increase in speed is about 1.5 times the unobstructed velocity, and in consequence, the pressure will be 2.25 times greater.

At a height of 100 m on a chimney stack in an area where the basic wind speed is 46 m/s, the wind velocity coefficients are $S_1 = 1.1$, $S_2 = 1.2$ and $S_3 = 0.77$. This gives a design wind speed of 46.75 m/s. The 'sides' of the chimney will be subjected to a wind speed of $1.5 \times 46.75 = 70.125$ m/s. The upwind and downwind faces will be subjected to $0.7 \times 46.75 = 32.725$ m/s. The corresponding pressures will be 3004 N/m² and 658 N/m² (see Table 38.2). These pressures are very much greater than will be encountered in a city street.

As an example, consider a circular chimney scaffold with six lifts 1.8 m high containing 16 bays 1.5 m long and 1.2 m wide. These are fixed around the top of a 100 m high, 6 m diameter chimney. There are two working lifts.

In each lift there will be four bays upwind and four downwind, each with an effective tube area of 0.439 m² (see Table 38.4), giving a total tube area of 3.512 m². Six lifts give an area of 21.072 m². Two lifts are boarded, i.e. 16 bays have platforms facing the wind, each with an area of 0.848 m², giving an added area of 13.568 m². The total area of scaffolding tube and boards upwind and downwind is 34.64 m². The pressure on this area is 658 N/m², giving a wind load of 22793 N (2323 kg, 5122 lb).

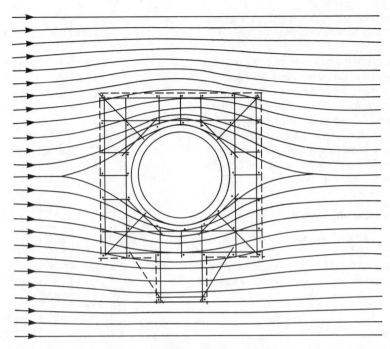

Figure 21.3 Wind modification through a tall chimney scaffold showing the acceleration of the wind along the sides of the scaffold relative to the wind flow.

In each lift there will be four bays on each 'side' with an effective area of 0.466 m². There will be 48 such bays in all, giving a tube area of 22.368 m². Two lifts are boarded on each side, giving the equivalent of four end areas of platform boards, each of 0.565 m² (see Table 38.7), giving an added area of 2.26 m². The total area of scaffold tubes and boards in the side stream is 24.628 m². The pressure on this area is 3004 N/m², giving a wind load of 73983 N (7542 kg, 16626 lb).

The total wind force on the scaffold is therefore 96776 N (9865 kg, 21748 lb), i.e. approximately 10 tonnes.

If each anchor is safe for about 1 tonne, ten anchors will be required, i.e. two per lift staggered around the chimney surface.

The total surface of the chimney scaffolded is 203.5 m², in which 12 ties have to support 17 m² each. This is about twice the number required in a normal building job. It demonstrates the necessity for the sentence in the British Standard Code: 'For scaffolds greater than 50 m high, the tie spacing should be chosen by design.'

Horizontal wind loads on scaffolds projecting above chimneys

When the scaffold has to project above the chimney and to be capped by cross beams to enable the inside to be serviced by a slung scaffold, there is no chimney structure inside the scaffold to alter the basic wind speed. In the examples given above, the unobstructed design wind speed was 46.75 m/s. For this, the design pressure was 1336 N/m². The area of scaffold tubes broadside on to the wind and any platforms is assessed, and the effective area is obtained using a shape factor of 1.2. This multiplied by the design pressure gives the wind load on the part of the scaffold projecting above the chimney. Extra ties are needed to deal with this, having regard to the large lever arm of the wind force on the projecting scaffold.

Horizontal wind loads on scaffolds after the demolition of the chimney within the scaffold

As the chimney is demolished, the deflection of the wind ceases and the whole of the area of the scaffold tubing without the platforms, which will have been lowered, is subjected to the design wind pressure for the site.

Continuing the example cited above, where the basic design pressure is 1336 N/m², consider the case where the top four lifts of the scaffold have become exposed and are not cantilevered up on the ties and extra ties placed in the bottom two lifts.

The top four lifts have a tube area broadside to the wind of about 40 m² which, with a shape factor of 1.2, gives an effective area of 48 m². The wind load is 48 × 1336 = 64128 N (6537 kg, 14411 lb).

This wind force will act as a lever arm of 6 m above the tie points, giving a turning moment of 39.2 tonnes/m.

Two rings of ties providing horizontal resistance spaced 2 m apart and each capable of sustaining 20 tonnes will be required.

Allowing one quarter of the ring to be in direct bearing on the undemolished part of the chimney and three quarters relying on ties, the number of ties required at 1 tonne each will be 16 per ring, i.e. in the bottom two lifts of the scaffold during the demolition of the chimney, each upright in the example given above will require one tie per lift per bay.

Similar calculations and construction on very high chimneys, i.e. above 200 m, show that six, seven or eight rings of ties per lift per bay may be necessary.

Horizontal forces other than wind

Material hoists used by steeplejacks may have inclined winch ropes passing around a single pulley at the top. Such a pulley exerts double the lifted load at an angle on to the scaffold. The horizontal and vertical forces from this source must be assessed and taken into account.

Marine structures

Tall structures may not necessarily project high into the air. They may be underwater. Advice should be taken from hydraulic engineers about the basic pressure exerted by the stream flow, which in some channels may rise to 6 knots. The same acceleration of the rate of flow is caused by an existing structure in water as that described in the preceding sections for a structure in the air.

Scaffolding work off the ends of jetties is particularly affected by modified stream flow and needs careful attention.

21.8 Church spire and monument scaffolds

It is not possible to give rules for church spire scaffolds because the buildings are very different. In the main, the scaffold has to be trussed off the masonry of the spire because it cannot be founded on the ground. Frequently, there are unglazed windows through which cantilevered tubes or joists can be placed. There may be corbels to support rakers, but these must be checked for strength.

Figure 21.4 shows a typical design. A rectangular system has been adopted in preference to a cylindrical one. The functional specification may require access to all surfaces and an imposed load of 10 tonnes spread over three boarded platforms, each of which must be capable of carrying 5 kN/m^2 (510 kg/m^2, 105 lb/ft^2) in any bay.

The method of design is to make drawings of a scaffold that will meet the requirements and assess its weight. The scaffold shown in Figure 21.4 will have a self weight of 31 tonnes in steel tube and 16 tonnes in aluminium tube.

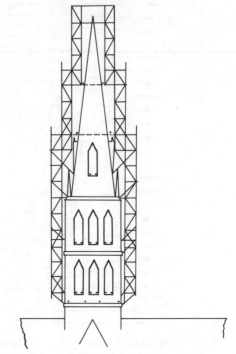

Figure 21.4 A typical stepped scaffold for a spire.

If the masonry is of suspect quality, a church spire is obviously a job for aluminium tubing.

In the case shown in Figure 21.4, there are locations for 56 horizontal cantilever tubes through windows and 96 rakers bearing on the masonry. The total load to be supported is 41 tonnes in steel. If the horizontal tubes are relied on in bending with a lever arm of 900 mm to the centre line of the scaffold, each one will be able to support only 90 kg, and 56 will support 5.04 tonnes, which is obviously inadequate. The cantilevers need to act in shear rather than in bending, when each one will safely support 2.8 tonnes, and 56 will support 156 tonnes.

The ledger bracing should be arranged in the form of rakers projecting downwards and bearing on the building in jobs such as this.

The scaffold must be tied closely to the spire to maintain the shear connection.

Now consider the rakers bearing on masonry corbels as the sole means of support. These 96 are carrying 41 tonnes, giving an average load of 0.43 tonnes per raker. This is a substantial load to apply to a corbel of masonry 50 mm from its root.

The above job was built with both systems of support, thus guaranteeing safety if one or the other failed in any place.

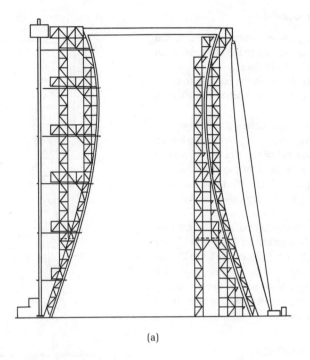

(a)

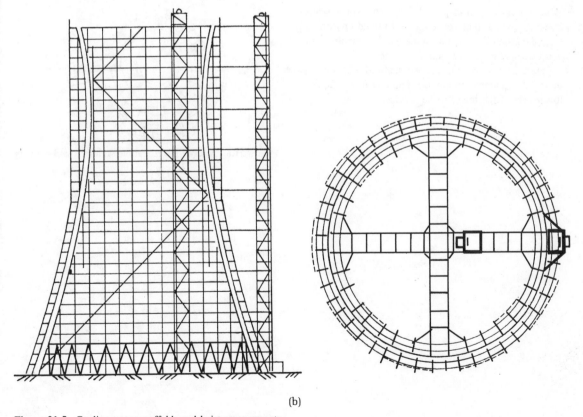

(b)

Figure 21.5 Cooling tower scaffolds and hoist arrangements.

21.9 Cooling tower scaffolds

The diameter of a cooling tower is usually so large that a scaffold on its surface can be built to the same rules that apply to an ordinary building scaffold.

The attachment of the scaffold to the cooling tower is usually through a system of brackets, such as those shown in Figure 21.2.

The outward inclination of the surface at the top is dealt with by raising the scaffold straight up vertically from a point on the shell vertically below the rim and increasing its width inwards to reach the neck of the tower. Such a structure is shown in Figure 21.5(a). Uprights may have to be doubled at the lower levels, depending on the loads involved.

Internally, a normal birdcage or a birdcage with a hollow centre is built. A materials hoist can be fixed either inside or outside. A scaffold such as that shown in Figure 21.5(a) is for maintenance and repair. The cooling tower will usually have been built from platforms on the shuttering.

If the shell needs support, a system such as that shown in Figure 29.5(b) can be used.

21.10 Water tower scaffolds

Water towers vary in shape and size, so no typical scaffold can be drawn. It requires less skill to build up vertically from the ground to encompass the whole structure than to construct a truss-out section at the top.

Figure 21.6 shows an inclined scaffold with rakers to increase its diameter at the top. This scaffold enables the lower part of the tower to be serviced.

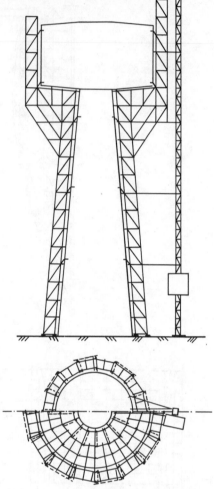

Figure 21.6 A water tower scaffold with a trussed-out upper section.

22 Demolition scaffolds

22.1 Scaffolds with external working platforms

An orthodox independent tied scaffold performs this duty satisfactorily, but it must be designed for two working platforms. One should be assumed to be loaded with debris to 3 kN/m^2 (300 kg/m^2, 60 lb/ft^2), while the other should be assumed to be loaded to 1.5 kN/m^2 (150 kg/m^2, 30 lb/ft^2).

The structure should be braced in an orthodox manner, and it should be tied into the face of the building according to the standard rules for ties. Box ties and lip ties through the windows are usually adequate.

Some precautions must be taken. As the height of the wall is reduced by the demolition, the scaffold will be left standing high above the top of the wall without any ties in it. This portion of the scaffold will catch the wind, either along the building or from the front or back of the building area recently dismantled. It will sway in the wind and exert its effect with very considerable leverage on the uppermost level of ties left in place. The forces exerted by this swaying have been known to be more than the wall can resist, resulting in a collapse of both building and scaffold into the street.

The demolition contractor and the scaffolding contractor must agree a functional specification for the scaffold before the demolition starts if the scaffold is to act as a shore as well as a structure to provide working platforms. An ordinary independent tied scaffold will not act as a shore to provide resistance to the wall against horizontal movement and collapse. If this is required, a special shore will be required built into the access scaffold.

Decisions must be taken on the following points before the design of the shore is started:

- Is the wall to be shored and saved, or demolished with the rest of the building?
- Is the wall to be propped only from falling outwards, with the floors remaining in place to stop it falling inwards as it is progressively demolished?
- Is the wall to be prevented from falling both inwards and outwards as each floor is removed?
- What process of demolition is to be used, and will the shore interfere with it or itself be destroyed by it?
- Is there to be temporary storage of old materials on the platforms during demolition and, if so, what weight per bay, and how are they to be transferred to the ground?
- Do the public have access below the shore, and if so, what means of protection for them is to be built into the shore?
- Which method of protection is to be provided?

  (i) A boarded platform that is reboarded on each lower lift during demolition.
  (ii) A fan at the top that is progressively moved down.
  (iii) A boarded lift and a fan permanently near the bottom.
  (iv) Vertical sheeting over the building face.

- Is protection for the public against water sprays to be provided?
- Will the windows be removed so that the lintels can be jacked against to transfer weight to the heel of the shore?
- Does the access part of the scaffold have to remain in place after the demolition for use during reconstruction?
- Will rubbish chutes be attached to the scaffold?
- Is a lorry gantry required to facilitate the disposal of debris?
- Is any special lifting gantry and winch to be built into the scaffold to lower heavy beams to the ground?
- Is the scaffold to be sheeted over with tarpaulins? If so, how are the wind forces — inward and outward — to be carried to the ground as the building is progressively removed behind the sheeting?

22.2 Demolition shore and access scaffold

Figure 22.1 shows a combined access scaffold and raking shore. It has some special characteristics. Because the wall is to be demolished, the scaffold must be dismantled floor by floor, leaving the shoring below each floor in place.

To save the time and expense of cutting needle holes through the walls at the top of each raker, the rakers are arranged to bear up under timber pads below the window lintels. This is possible on a demolition job because the glass can be taken out.

The outside bottom end of the rakers can be mounted on a relatively small foundation block because this can be tied back towards the wall with the extended foot lift of the access scaffold. The block can also serve as part of a traffic fender.

The inside corner of the shore is held down by steep rakers screwed up under the lintels at the top, attached to the inside foot ledger at the bottom and bearing on a timber sole plate. These inside rakers are connected to the access scaffold at intermediate levels. They transfer the weight of the wall to the inside of the shore and prevent it rotating outwards if the wall leans outwards.

22.3 Pedestrian barriers

As a general rule, the area surrounding the building being demolished should be closed to any sort of public access. Fenders should be erected to prevent the use of the pavement and the parking of cars. In this space and to achieve this object, it is sometimes useful to site materials disposal chutes, and a hoist tower and hoist if required. These two structures can usefully be made an integral part of the access scaffold and so act as buttresses for it.

22.4 Scaffolds with fans

If a fan is to be installed (see Section 11.8), the purpose and construction of this fan should be agreed with the demolition contractor before construction begins. The functions that should be considered are:

- The prevention of pieces of debris falling through it to the ground.
- The prevention of falling debris bouncing off the fan outside its edge.
- The storage of reclaimed material awaiting transport off site.

The scaffold behind the fan can be an orthodox tied scaffold, but it must be strongly tied at the fan level and at the higher level where fan wires are incorporated. As the building is demolished, the upper level of fan wire ties will lose their anchors. The means of support of the fan must take this into account.

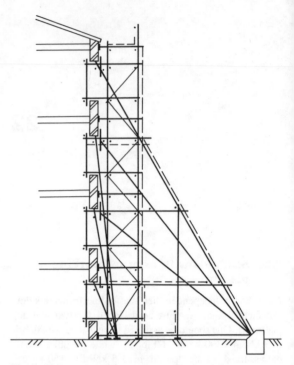

Figure 22.1 A demolition access scaffold and shore designed to be reduced in height in stages.

22.5 Hazard with the strength of fans

- Accidents have happened because no positive decision has been made prior to construction as to the objects to be achieved. Inappropriate design has resulted in fans being not strong enough to protect the street below or so heavy that they have caused dislodgement of the building.

22.6 Scaffolds required to stiffen the building wall

Where the process of demolition leaves a weakened wall, the access scaffold structure can be adapted to contribute to the stiffness of the wall. This is not intended to be a shore to prevent the wall falling over, which is a different type of structure incorporating raking shores and requiring special design.

The first modification of the orthodox independent tied scaffold is to ensure that every transom butts up against the wall, that the transoms are placed at the location of the uprights, and that intermediate transoms are placed opposite the building piers or columns.

Every lift should be converted into a horizontal beam by a continuous arrangement of zigzag plan braces, and every pair of standards should be ledger braced to double the vertical stiffness.

Lift heights should not necessarily be set out at even intervals, but they should be at the sill and lintel level of the windows so that transoms can be projected into the inside of the building through the four corners of every window. The inside of the transoms should be joined by vertical or horizontal bridles and all the tubes firmly wedged on to the structure.

Where extra stiffness is required, ladder beams can be introduced set on the flat when they replace the ledgers or bridles and set at right angles to the building when they replace the inside uprights.

Figure 22.2 is an example. In some instances, the beams can be set horizontally.

22.7 Debris chutes

The chutes can be metal, timber or plastic. The manufacturer's instructions should be obtained with regard to the method of fixing and the height limit.

The downward force exerted by the debris in the chute can be taken as one-third of its weight. Its density will be about 1.75 tonnes/m^3 (109 lb/ft^3). The dead weight in the chute depends on whether it is used as a storage chamber as well as a chute. If material is to be trapped behind a door after falling, the impact should be taken as five times the dead weight of material falling. Large chutes should be referred to the main contractor for a functional specification. The scaffold should be specially strengthened and tied in the locality of the chute.

22.8 Scaffolds with vertical sheeting

Tarpaulins are commonly hung on the outside of demolition scaffolds to protect the public from dust. In this case, they will catch the wind coming through the window openings of the partly demolished building and tend to drag the scaffold away from the building. It is essential that these tarpaulins are lowered as the demolition work proceeds and that the scaffold is box-tied through all the windows in the top exposed storey.

Exposed sites and buildings with large openings should have calculations done for wind forces, as described in Chapter 37.

22.9 Hazard with wind forces

- The acceleration of the wind may be underestimated because the partly demolished building forms a wind corridor between adjacent buildings, and the wind velocity is further accelerated through the window openings, even after the lintels have been removed. The sheeted scaffold, if not tied at sill level, may pull out the brickwork between the windows into the street.

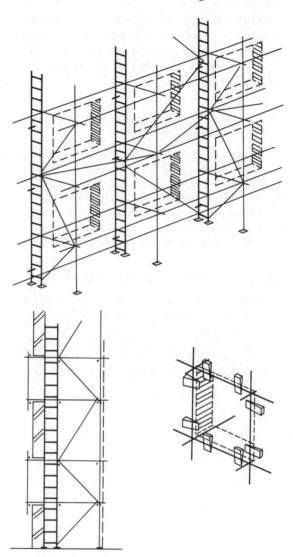

Figure 22.2 The use of ladder beams as stiffeners in demolition scaffolds.

22.10 Scaffolds required to shore the building wall for reuse

The demolition process sometimes requires a wall to be saved during demolition. Raking shores for walls that are to be maintained in position are engineering scaffolds and are not dealt with in this volume. Figure 22.1 gives some idea of the form of these scaffolds. The critical feature of the scaffold shown in Figure 22.1 is that as the wall is demolished from the top downwards, the effectiveness of the shore is not lost. The coaxial joints in the rakers are in pre-planned locations, and all the temporary lintel supports are in place.

When the shore does not have to be dismantled, it is not as necessary to pre-plan its joint locations. It may be necessary to design the shore as a push-and-pull buttress to hold the wall in place while the internal part of the building is removed.

22.11 Hazards with the removal of ties in demolition scaffolds

- A great hazard with demolition scaffolds is that the ties of the external scaffold are taken away as the wall is dismantled, leaving the external scaffold free-standing.
- The wall may be dismantled to a level lower than the scaffold, which will be left projecting well above any of the old building wall and so exposed to the exaggerated wind forces acting on its netting or dust sheets.

22.12 Hazard in the design of the shore

- A raking shore can work only as a shore, i.e. provide a horizontal resistance to an applied horizontal force, if the top of the raker is weighed down with a critically greater force. This vertical force can be calculated from the triangulated forces. It can be supported by the weight of the wall entering the shore through the lintel beams or through the needles. If these elements are demolished, the shore loses its effectiveness.

22.13 Hazard in evaluation of the load to be resisted by a shore

- At the time the demolition is planned, the walls being shored are probably standing vertically without any tendency to fall outwards or inwards. The whole of the demolition process can sometimes be carried out without the verticality of the wall being disturbed and without it applying any horizontal load on the shore.

What, then, is the horizontal load that the external scaffold and incorporated shore should be designed to resist if called upon to act as a shore? This question is one that can be answered only by the building owner's engineer, who will have investigated the verticality of the wall, its attachment to the floors, its condition and the order of demolition of the building. The scaffolding contractor should be told what horizontal force and at what level he has to provide against.

Typical forces are 10% to 15%, or even 20%, of the vertical weight of the wall above the level of each chord of the raking shores or above the flying shore.

23 Weather protection and sheeted stone-cleaning scaffolds, temporary roofs and canopies

23.1 The problem involved

The working platform on a sheeted scaffold is subject to the normal rules given elsewhere in this volume, and in the Construction (Working Places) Regulations 1966 and the Code recommendations. The rules with regard to toe boards and guard rails must be complied with, even if there is a sheet tied down the full outside face of the scaffold.

The problem with a sheeted scaffold lies in the wind forces to which it is subjected. It is worthwhile comparing the face area of an unclad scaffold with one clad with tarpaulins. Consider a single 2 m bay and a single 2 m lift.

The effective face area of the tubes at right angles to the wind is 9.33 m (length) $\times$ 0.049 m (diameter) $\times$ 1.2 (tube shape factor), giving 0.538 m^2. The clad scaffold has a face area of 4 m^2, i.e. 7.5 times greater. If the location under consideration is a working platform with a guard rail, a platform edge and a toe board, with their appropriate shape factors, a further 0.942 m^2 is added to the tube area, giving a total of 1.48 m^2 for the unclad scaffold. The 4 m^2 for the clad scaffold is still 2.7 times greater.

The consequence of this is that the bracing system of a sheeted scaffold has to deal with up to seven times more distorting force than an untied one, and the tying system has to deal with up to seven times more shear or tension between the scaffold and the building than the same scaffold without the sheeting.

23.2 Safe working loads on scaffold members when loaded by wind forces

The current practice is to allow 1.25 times the normal safe working load when a member is loaded mainly by wind forces. This is in accordance with the practice allowed in the current BS 449 for structural steelwork.

Because members are usually loaded by the self weight of the structure, the imposed load and the wind force, it is difficult to interpret how much extra load can be allowed on a member that has the triple loading. The most appropriate interpretation for scaffolding is to allow the 1.25 factor only on that part of the load due to the wind.

Thus the strut tables given in Chapter 31 will have to be used for the standards of a tied access scaffold because the wind load is transferred mainly to the building but, for the wind bracing and the bracing elements of a hoist tower, the values in the table can be multiplied by 1.25 for bracing elements only.

The same applies to right-angle couplers. The normal applied safe slip load is 635 kg (1400 lb) against self weight and imposed load, but the coupler used to connect the horizontal diagonal wind brace referred to in the previous paragraph may be designed to be loaded to 794 kg (1750 lb).

23.3 The nature of the wind effect on a sheeted scaffold

Reference should be made to Chapter 37 for details of wind forces. Three conditions need consideration in practice:

- A canopy, i.e. a roof without a skirt.
- A roof with a skirt without a decked lift at the bottom.
- A fully sheeted wall with or without a roof.

The rules for dealing with these types of structure are dealt with in CP3 Chapter V, Part 2. Variations in the building or partly erected building, the sheeted or partly sheeted scaffold and the nearness of other buildings makes it impossible to formulate a set of simple rules for universal application.

For canopies, experience has shown that both uplift and downward forces must be expected, together with horizontal shear.

If there is a skirt without a bottom, there will be uplift by pressure inside and suction from the top. This will depend on the length of the building, which governs how much wind can escape around the side and how much is swept up the skirt.

If the wall surface and roof are fully sheeted, the wind calculations are as for an ordinary building, and CP3 Chapter V, Part 2, can be applied as for buildings.

23.4 Tying to the building

It can generally be assumed that the building is strong enough to accept the wind forces applied to the sheeted scaffold servicing it. The building resisted the wind forces before the scaffold was put there, so the only problem for the building is that the wind forces will not be distributed over its face area but will be concentrated at the tie points.

The general rule of making the transoms butt up against the wall wherever possible helps, but in a stone-cleaning job this is an awkward requirement. If the wind load is taken through the ties, the part of the building to which the ties are attached must be able to resist a force of at least 1 tonne at each point.

The 1981 Code recognises that more ties are needed when the scaffold is sheeted. The recommendation that there should be one tie every 22 m^2 gives a load of 660 kg for a wind pressure of 30 kg/m^2 (6 lb/ft^2). This is within the safe working load of a coupler subjected to wind forces. The building should have some emergency strength, say a factor of 1.5, which increases 660 kg to 1 tonne.

The Code also recognises that 30 kg/m^2 (6 lb/ft^2) is not the maximum wind pressure that may be applied to the sheeted scaffold. Where the site and the degree of exposure indicate that greater pressures are possible, more frequent ties will be needed. In new structures being built, it is frequently possible to use full box ties and so obtain two couplers at each tie point. In existing structures being cleaned, it is usually only practical to use lip ties, and these have only one fitting inside. Supplementary couplers should be considered, and drilled-in anchors may be necessary.

23.5 The bracing systems

The ledger bracing rule, i.e. bracing at alternate pairs of standards, must be strictly adhered to.

Longitudinal bracing should be doubled in sheeted scaffolds, i.e. one set every 15 m (50 ft) instead of every 30 m (100 ft), which is acceptable for unclad scaffolds. The longitudinal bracing should be complete from top to bottom. This is important because the sheeting is on the outside face of the scaffold, where it has the width of the scaffold as its lever arm for moments resisted by the ties.

Plan bracing should be added in extreme cases and the transoms projected into the window reveals as soon as the progress of the work permits.

23.6 Hazards with protection sheeting

- Applying vertical weather protection sheeting to a scaffold not intended to carry such sheeting frequently results in the scaffold being forced off the building by horizontal wind force or lifted off the ground by wind uplift. The number of ties must be checked and supplemented where necessary.
- For the general case, the rules for ties are adequate, but for extreme exposures or when the building strength is suspect, the actual strength of the ties should be ascertained by tests and the appropriate number installed.

Ties not at the top

- The portion of the scaffold sheeted over is frequently at or near the top of the job and catches wind in both directions. It can exert large forces levering the scaffold off the wall against a level of ties not fixed near the top of the building. This situation must be made the subject of calculations.

23.7 Temporary roofs and canopies

These are engineering structures and should be calculated and designed by specialists. They must be designed in accordance with the British Standard Code of Practice CP3, Chapter V, Part 2, detailing wind forces. Because of their frequent use in the access scaffolding industry, a brief description is included in this work. Reference should be made to Chapter 38 on wind loads.

23.8 Temporary roofs bearing on an existing roof

In the case where the span is too large to be cleared without any central support, some of the downward load or some of the wind uplift will have to be carried by the existing structure over which the temporary roof is to be placed. This entails placing bearing pads and fixings on and to the existing roof structure. There is then a divided responsibility between the building owner and the scaffolding contractor for the stability of the whole assembly.

The scaffolding contractor must first study the roof structure and then design his temporary roof to bear on and be tied to the most favourable locations of the original roof, e.g. at the ridges, at locations of internal support for the main roof or at node points in the roof trusses. He will have to do this with little knowledge of whether the main roof structure is adequate to carry the loads to be applied to it, so his design will be provisional.

The scaffolder should then submit his proposals to the building owner or his professional team, or to the main contractor, accompanied by a schedule of the forces and their direction that he intends to impose on the main roof, the location of these forces, the means of transferring them to the roof and the local alteration in the roof necessary, e.g. lifting of tiles or placing spreader plates over rafters or asphalt.

The scaffolding contractor cannot assess the strength of the main roof for himself because he will not have the details of the roof design or knowledge of its condition. Responsibility for the bearing or tie points should be taken by the building owner or his contractor.

The building owner or his professional team or contractor must then assess the roof with regard to its ability to sustain the proposed loads from the temporary roof, and agree to the scheme proposed by the scaffolding contractor and to the means of bearing on or attachment to the roof and any modification needed. The owner cannot make this assessment without receiving the values and directions of the forces from the scaffolding contractor, so the details of the temporary roof are clearly a matter for engineering design and calculations. Once the owner has agreed to the scaffold construction proposal, with such modifications as are necessary, he is responsible for the strength of the main roof, and the scaffolding contractor is responsible for the stability of the temporary roof and its attachment to the main roof.

This is a simple procedure that, if followed, clearly defines liabilities. Unfortunately, in practice the owner usually does not know the strength of his roof. There may be no data about it, or it may be old and of uncertain strength. Neither party can proceed with the work until the existing roof has been reassessed in its present condition. The owner has no choice other than to obtain new advice concerning his roof.

If he asks the scaffolding contractor to carry out the assessment of the main roof, and the task is accepted by the scaffolding contractor, this agreement should be by a formal contract separate from the scaffolding contract. The relevant insurances will need to be considered, with design liabilities, indemnities and warranties properly set down and agreed by both parties.

23.9 Temporary roof fixed to existing walls

Where the old roof can be spanned without intermediate support, but the side scaffolds to which the temporary roof is fixed cannot be founded on the ground, they must be bracketed from the outside walls of the main structure. The division of responsibilities is similar to that described in the previous paragraphs, except that it is the strength of the walls and the places and means of attachment to them that must be the subject of strength analysis.

23.10 Hazards with temporary roofs

- The most noticeable shortcoming in the construction of temporary roofs is that one sees the same design of roof in Scotland as in the south of England. Referring to the wind speed map, Figure 38.1, London has a basic wind speed of 38 m/s, and Glasgow and Inverness 51 m/s. Because the wind pressure is proportional to the square of the wind speed, wind pressure in these northern cities is 1.8 times greater than in London. The hazard is that the temporary roofs are built without design.
- Temporary roofs are part of the temporary works for a building project and so are not in the bill of quantities as a priced item. They are the responsibility of the main contractor, with the costs incorporated in the price. This sometimes leads to hazardous economies.

23.11 The ideal solution to a temporary roof and wall scaffold

If a building is to be more or less gutted and re-roofed, a complete temporary building will be required. The design of a temporary structure affording a complete seal around the building is a job for the scaffolding contractor's engineering design office.

The ideal solution is to provide an external access scaffold fully tied to the walls with calculated ties and projecting above the eaves to receive a roof over the top. The combined structure of wall scaffolds and roof should be designed either as a fixed-angle or a pin-angled portal, and its construction must be correspondingly supervised.

In the ideal solution, there is an internal birdcage to service the internal wall surfaces. This birdcage should be carried up through the building roof and interlocked with the scaffold roof so that the whole temporary structure is one large unit vertically and horizontally linked through the window openings. During reconstruction of the building, the linking elements can be removed one at a time as the building work progresses and refixed elsewhere if necessary.

24 Free-standing scaffolds and power access scaffolds

24.1 Free-standing scaffolds

Chapter 13 deals with one type of scaffold that may be free-standing, i.e. not attached in any way to a permanent structure or by guys and anchors to the ground. Stability against wind overturning is achieved only by the self weight of the structure. The working imposed load cannot be taken into account because it may sometimes not be there. If the structure is rectangular or L-shaped, its stability in both directions must be analysed to ensure that the worst case is studied.

Building in extra tubes and couplers increases the weight, but they also catch more wind. Extra tubes can be placed and coupled in the base lift. Kentledge or any other material can be placed on a mattress of tubes in the base lift.

For structures of tubes and couplers taller than 6 m, joints will be required in the uprights, and the holding-down weight of the part of the structure above the joints is only the self weight of that portion of the structure. The tension strength of the joints in the standards may be nearly zero if joint pins are used, and inadequate if sleeve couplers are not correctly assembled and staggered.

For structures made of prefabricated frames with slip-on joint pins, the top of the structure is jointed across its full width at every lift height. This means that at the top levels, the self weight may be very small. The top few lifts may need holding down on to the lower lifts by vertical tubes and couplers. Wind analysis will be necessary at every lift from the top to the bottom because there will be a critical level that cannot be guessed when the weight stability compared with the wind overturning is least. At all levels, a stability factor of 1.5 must be maintained.

24.2 Hazard

- Ignoring the uplift effect of the wind on the boarded working platform, which has to be deducted from the self weight of the platform. The uplift is sometimes greater than the self weight of the top few lifts, which are lifted bodily off their joint pins.

24.3 Types of power access scaffold

Under this heading are included:

- Ground-based scaffolds on which there is a working platform that is raised and lowered mechanically, other than on suspension wires either manually or by power operation.
- Long platforms that climb masts attached to the building have been used for the last 20 years and are well proven.
- Short platforms that climb masts erected on trailers are also well proven.
- Hydraulic platforms raised on hinged arms mounted on motor cranes are now commonplace.
- Scissor lifts operated by screw threads or hydraulics are available mounted on either trailers or motor chassis.

24.4 Mechanical platforms on two masts

This system incorporates two masts set up the face of a building up to 20 m apart. The masts are tied to the building surface. Bolted racks are fitted up the masts, and a pinion system climbs these.

The system has a cost advantage where the work to be done on the building surface is of short duration, but each level has to be visited several times. Such a situation is the inspection, removal, replacement or repointing of brick slips on the edges of floor slabs in multi-storey blocks. Another case is the replacement of intermittent cladding materials such as mosaic.

Heights of 75 m (250 ft) and platform lengths of 20 m (65 ft) are available.

24.5 Single-mast systems

On jobs that are not as extensive as those referred to in the previous section, a system is available where the mast is mounted on a trailer or motorised chassis. The platform climbs the mast. The mast is extended as necessary and taken down from the platform at the end of the job.

Heights of 25 m (80 ft) are available, at which height the mast will require tying to the building.

Mast units are usually 1.5—2 m and some machines have platforms up to 8 m long.

24.6 Hydraulic arms

Hydraulic platforms with hinged arms are available in all sizes and heights up to 30 m. The supplier's rules must be obtained and followed.

24.7 Scissor lifts

Scissor lifts are available up to a height of 10 m. The supplier's rules must be obtained and followed.

24.8 Loads

In all the above cases, the safe working loads must be ascertained from the supplier of the equipment. The rules for use internally and externally must be obtained and followed precisely, especially with regard to any requirement for stability ties to the main structure and outriggers to the base.

24.9 Hazard due to mobility

• Many free-standing scaffolds and power access scaffolds are mounted on castors, wheels or a power-driven truck or trailer. This necessitates the correct reconstruction of the scaffold or its correct attachment to a building everywhere that it is to be used.

It is a common experience that the free-standing and mobile scaffold is properly erected, tied, stabilised and used on the first few occasions, but these essential precautions for its stability are partly or completely omitted during the later applications of the scaffold. The chargehand scaffolder must attend to this point with the greatest diligence if he is not to be caught out.

PART III
Administration and operation

25 Scaffolds requiring calculations, drawing office practice and planning for overall stability

25.1 Basic principle

If a scaffold can be built by a gang consisting of a charge-hand, an advanced scaffolder and a labourer, and it does not need any special knowledge of loads except those classified in Table 2.8, it does not need to be designed. It can be built by rule of thumb.

Calculations become desirable when there is any unorthodox feature in the scaffold, e.g. extraordinary height, no facilities for the attachment of ties, imposed loads outside the standard tabulated values, the scaffold incorporates special through passages or is subject to special wind forces, it is to be in place for many years, or it is in a marine or river environment.

25.2 The Code requirement

BS 5973 1981 recommends that unsheeted scaffolds higher than 50 m be subject to calculations. Special scaffolds, sheeted scaffolds and any that have peculiarities of size, shape or loading should be calculated.

25.3 Scaffolds generally calculated

- Access scaffolds above 50 m high.
- Scaffolds where the imposed load is greater than 3 kN/m^2 (300 kg/m^2).
- Scaffolds where the dimensions do not comply with Table 2.8, or where the bracing or tie patterns are unorthodox.
- Scaffolds with attachments such as loading bays and fans.
- Pavement frames and protection scaffolds.
- Hoist towers projecting more than 10 m above the scaffold.
- All sheeted scaffolds, canopies and temporary roofs.
- All slung and suspended scaffolds.

- Large birdcages and those supporting heavy loads.
- Scaffolds utilising the strength of the timber boards.
- Demolition scaffolds and dangerous structure scaffolds.
- Shoring scaffolds.
- Marine and river scaffolds.
- Bridges, gangways and ramps.
- Scaffolds incorporating prefabricated beams.

25.4 Drawing office practice

As with all structural engineering design work, the calculations should be checked by a section leader who has not personally had a hand in them.

The drawings following the calculations should also be checked by a section leader who did not do the drawing.

The drawings from a company's design section to its own workforce should always be in the same style so that the erectors habitually do what is required by observation of the drawing. In normal nuts and bolts structural engineering, the elements are numbered on the drawing and numbered as they leave the fabrication works. It is very unlikely that a mistake will be made in erection. Each element is of a specific length, and its bolt holes match up with bolt holes in the next member. There is little opportunity for the erectors to get it wrong or to leave pieces out.

In scaffolding, there are no set places for the joints and there are no set lengths of tube. The erector is empowered to move a column from its designed place. He may find that it stands on a fire hydrant when he comes to set out the job in the street. There are no precise lift heights. They may need slight modification to avoid services such as telephone wires or to be more suitable for attaching to ties. There is no set location for the bracing. The longitudinal bracing on the drawing may interfere with a schoolchildren's crossing. The ledger bracing may block a fire escape.

These sorts of problems have to be dealt with by the erector. If the chargehand is always given the same style of drawing, there is a better chance he will be able to accommodate these variables correctly into a sound completed structure. Moreover, the draughtsman and the checker will do their jobs more accurately if there is a regular drawing office practice.

25.5 The drawings

A scaffold is a structure with a large number of like-shaped frames, but they may have different bracing in them; they may have a varying tie pattern; they may be of different heights and at different spacings and standing on the ground at different levels. At the end of the day, the erector must be given every instruction necessary for him to produce what the designer wants.

The drawings should be elevations and plans. They should show all the uprights, ledgers and transoms by lines and dots on the drawings. They should show the longitudinal bracing, the ledger bracing and the plan bracing. They should show or refer to the tie pattern.

Some drawing offices use fairly thick lines to represent the building façade and slightly thinner line for the uprights and horizontal members of the scaffold. Failures are mostly due to deficient bracing, so the opposite emphasis in the drawing is better, i.e. thin lines for the building façade, medium-width lines for the uprights and horizontal members, and thick lines for the bracing and ties.

Bracing is an intermittent requirement that is difficult to show on a two-dimensional drawing. To overcome this difficulty, full elevations should be shown.

25.6 Planning for overall stability

Scaffold structures frequently have to be in place for a long time. When tall buildings are being built, the lower part of the scaffold structure may have to be in place for a year. During this time, it will be exposed to all the seasons with their varying winds, to all the various degrees of ground stability, settlement, freezing and thawing, to variations of length due to diurnal and annual temperature change, and to the possibility of being accidentally damaged.

It is a fundamental requirement, irrespective of the dictates of calculation, that the structure has sufficient strength and stability to sustain not only its self weight and the imposed loads but also a measure of abuse, local and general overloading, and some degree of alteration during its life. It should also be strong enough and sufficiently braced and tied to sustain the loads occurring during erection and dismantling, even though these may be critical for only a short period of time.

While ensuring the above characteristics, the structure must be capable of being built economically and without excessive materials, especially those that make no contribution to the stability of the structure.

When the structure is in a hazardous location, e.g. over the sea, at high altitude, or over the side of a tall building or natural ravine, the risk of injury to those building the structure and to those working on or near the structure during its life must be the prime consideration, taking precedence over productivity in erection and economy of materials.

If the structure is susceptible to damage by other plant or traffic, it must be capable of staying in place and of sustaining its imposed load, even if parts of it have been struck out by accident or removed during dismantling.

The consequence of failure, both in injury to people and in terms of financial loss, must be taken into account.

It should be realised that the difference between a good sound structure and a weak unstable one may well be only a few extra tubes and couplers, which cost very little to add to a structure already nearly completed.

Planning and design in advance of erection will always pay dividends, ensure extra stability and reduce the hazards. The planning should be applied not only to the handling and erection of the materials in the scaffold, but also to the safe procedures for building it and at a later date dismantling it. It must also include the duties and performance of persons, and have regard to their ability and actual experience of building a structure of the type contemplated and their susceptibility to error.

During the life of the scaffold, it must be maintained, even though it is a temporary structure. Its temporary nature induces work people to alter it, abuse it and leave it unmaintained. Its emergency strength above that necessary for its construction and use must not be allowed to decline to a hazardous level. Its lateral stability from bracing and its transverse stability from ties must not be allowed to decrease during the period of work on the building.

Special structures need special consideration. For example, scaffolds for cleaning boilers must be kept clear of debris on the platform; stone-cleaning scaffolds must be kept clear of dust, which becomes slippery when wet; high-rise structures in high locations must be well tied and braced against the wind forces.

Regard must always be paid to the structural maxim that it is better to deal with each force or load properly and separately than to have a hotch-potch of a structure without clearly defined load paths to the ground.

Scaffold structures must always be recognised as highly flexible. Even a properly built access scaffold is longitudinally braced on only one face, the outer one, and not on the inner one. Transverse bracing is generally only applied to alternate structure frames. There are joints at random places in nearly all the members.

The designer must take all these features into account and, when he cannot draw what he wants in the elevation and plans, he should draw attention to them with notes.

25.7 Planning for security

There are two aspects of security to be considered. One concerns the building serviced by the scaffold and the other concerns the scaffolding materials themselves.

When a scaffold is built against a finished building, there is relatively easy access for burglars through the upper storey windows or through the roof lights into the roof spaces above the loft hatches. If the building is occupied, it is undesirable for the windows to be left open for the installation of through ties. Drilled-in anchors should be used wherever possible. Reveal ties should be used, and if no other ties are desirable, twice the normal number of reveal ties will overcome any uncertainties about their safety after a long period on site.

Access to the scaffold should be made difficult by removing the access ladder, or by raising it each night and weekend until it stands or lies on the first boarded lift.

Excess material should not be left on the ground on site for others to steal. It should not be left loose resting in the scaffold framework. If spare materials, e.g. extra transoms and couplers for extending the platforms, have to be left on site, the items should be coupled into the structure where it is expected they will be wanted.

In the early stages of planning, security consideration must be given to night lights, electrical earthing, lightning earthing, etc.

26 Construction, dismantling, alteration and adaptation of scaffolds

26.1 Management

It is not intended to describe in this book the actual methods of handling materials on site.

Persons erecting scaffolds must know their duties before the work starts. Permission may have to be obtained from the local authority. In some locations, police permission may be required. There may be time restrictions on the parking of lorries for unloading and loading. In many cases, there are special technical requirements to be complied with, as in the case of scaffolding associated with railways, electrical power transmission, or oil or gas production. Many similar industries have internal rules to be obeyed. In some types of premises, there are special rules, e.g. those concerned with solvents or other chemicals, in mines, in docks and in offshore work. Many of these industries have their own codes of scaffolding practice, which supplement the British Standard Code.

It is important to realise that the legislation, the industrial rules, the special codes and the various construction regulations are not set down to hinder the work but to guide it along lines that experience has proved are sound and safe. Compliance assists the scaffolder, while non-compliance may be disastrous for him, his work people, his business and the public.

26.2 Construction

The essence of good site practice is in attention to three factors:

- Risk to the public.
- Risk to the scaffolders and subsequent trades.
- Risk to the scaffold itself or to the structure it serves.

Attention to these three details requires not only a careful observation of the site activities as they are carried out, but also an anticipation of the circumstances that will occur in both the immediate and the more distant future. This anticipation is particularly important in dismantling scaffolds. A good scaffolding manager, like any other manager, is a man who thinks ahead as well as relying on his staff to do it for him.

There is no necessity at any stage in any scaffold to create a hazardous structural situation. If a degree of instability is anticipated, the necessary raking tubes, braces or ties can be inserted to overcome the problem. If these means cannot be left in the final job for one reason or another, other means must be inserted before any temporary measures are removed.

A case is the construction of a scaffold where a temporary means of achieving stability may have to be used and subsequently removed, e.g. the stabilising of prefabricated beams by knee braces or chord stiffeners to hold them in place until the structure above them achieves this object. When this superstructure has to be dismantled, the same situation will at some time be reached from the other direction. The temporary instability will recur. In this there is a hazard for the dismantling gang, but the expert manager and supervisor will anticipate this. It is best if the same gang erecting a difficult structure can dismantle it because they know how they overcame the difficulties in erection.

26.3 Dismantling

There are frequent problems of instability during dismantling. This may be due partly to the fact that the dismantling process is twice as quick as erection and so receives less attention. It may also be due partly to the fact that the scaffolder does not expect the structure to become less stable as he reduces its height and so does not look out for a hazard being created.

The problem with dismantling often lies with taking out a key structural member such as a brace, which not only braced the immediate lift or bay being dismantled but also

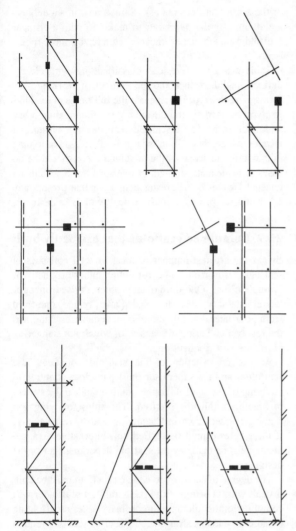

Figure 26.1 Hazards during dismantling.

braced other parts of the structure. Ties have to be taken out as they are met during dismantling and this may result in a condition where there are insufficient left, particularly in the lower 6–7 m of the job.

Structural reliance is not usually placed on joint pins or sleeve couplers during erection because they are placed at the ends of members already fixed, and new members are added that are subsequently fixed. However, in dismantling, if the members are not taken out in the reverse order, the joint pin or sleeve coupler may be called upon to carry a substantial load while another section of scaffolding is being taken away. If they fail there will be an accident.

Figure 26.1 contains two examples of the failure of sleeve couplers during dismantling. It also includes an example

of the consequences of removing ties and braces without properly stabilising the scaffold.

The scaffolder has to be very watchful to spot these circumstances in advance. A typical case occurs in a canti-levered loading bay when a projected transom of the main scaffold is coupled to the inside upright of the main scaffold just above a joint pin. Attachment of the cantilever at this point during erection is of no consequence because the whole scaffold will be in place before the loading bay is added, but in dismantling, the top part of the scaffold may have been taken away. The outside ledger of the loading bay cannot be taken away until the rakers that are attached to it have been removed. There is apparently no hazard in taking away the raker because there is no load on the bay. When the rakers and the full weight of the main scaffold have gone, and the scaffolder goes out on boards on the outrigger to take away the outside ledger, his weight breaks the joint in the inside upright of the main scaffold, which in tension cannot hold him safely. This is an easy case to spot if one has been alerted to the problem, but easily missed otherwise. There are many more cases where the trap is difficult to predict.

Storage of materials on a scaffold during dismantling has led to many problems. It is easier to load a lorry with materials lying on the first lift of a scaffold than to take it to the ground first and pick it up again. When the scaffold has been reduced in height to this last lift, its stability either at right angles to the building or parallel to it may be reduced to the danger point.

Removal of a scaffold in vertical slices can create a hazard. Such an operation is associated with glazing a building from one end towards the other or a progressive stone-cleaning job. Dismantling in vertical slices may result in the loss of a whole vertical line of ties. It may result in the removal of the bay in which the zigzag façade bracing has been fixed to stabilise a length of the scaffold.

Towards the last corner of a scaffold being dismantled, there may remain only a narrow vertical slice formerly stabilised by the return scaffold or possibly with no ties in place originally because it was stabilised by the corner.

Dismantled material should be handed to the ground or lowered on a gin wheel, not thrown down. It should not be leaned up against the partly dismantled scaffold or against the building, from which it may slide sideways. The last few tubes should be uncoupled carefully using a pre-planned system that will not put any person at risk.

26.4 Modification

When a scaffold has served its purpose for some time and its stability is accepted by all on site, it must not be assumed that it can be altered and retain its stability. Frequently it does not, and hazards must be anticipated.

If the modification was pre-planned when the scaffold was built, there should be no problem because provision will have been included for a safe modification. It is when the modification is unplanned or due to a suddenly arising need that problems arise. Examples are:

- Removal of the corner standard of a scaffold because there is insufficient room for a forklift truck to pass, followed by cantilevering the platform at the top.
- Incorporation of beams in a structure not designed for them when an unexpectedly bulky lorry load arrives.
- Adding a short extension to a scaffold that was built with insufficient length to begin with.
- Taking out pieces to allow the handling of pipes or duct work.
- Changes to the façade by the architect that require a change in the format of the scaffold.

All modifications may create hazards, sometimes simply by the removal of structural members, sometimes by causing joints to be all in one plane, and sometimes by creating discontinuity in the bracing pattern.

Additions to scaffolding such as extra cantilevered loading bays, storage bays, fans, vertical safety nets on the outside face and protection sheeting may all upset the stability.

Removals can similarly disturb the stability, as can removal of parts of the main structure in a demolition contract.

26.5 Adaptation

When a scaffold has to serve two or three functions, it may require adaptation. For instance, lifts may be required every 2.7 m to erect a concrete frame. For brick cladding, the lifts may then have to be reset at 1.35 m. Finally, for rendering or roughcast, the lifts may have to be reset at 2 m.

To make these changes, the ledger bracing will have to be taken out and reset, the longitudinal bracing will have to be taken off and reset, and the ties if originally set on the lifts will have to be modified.

These activities carried out continuously throughout the job cause a steady deterioration in the quality of the scaffold. The closest site management is needed. The original design should make the changes easy and provide for them to be made without leaving large areas of the structure minus some of its structural elements.

For instance, if half of each bracing system is fixed on the uprights with swivel couplers, this will not need to be taken off while the adaptations are made.

26.6 Completed scaffolds that are well built but of inadequate strength

This case is often caused by a lack of communication, e.g. the purchaser does not specify what he wants well enough, or the scaffolder does not ask. Sometimes, it is a deliberate attempt by the purchaser to make do with a cheaper scaffold for a job that he knows should have a more robust one.

Table 2.8 gives five load classifications with clearly defined loading values and the number of working platforms. The five types encompass the full range of loadings likely to be needed, and the table gives the maximum bay width for each. These scaffolds can be built on normal façades to a height of 50 m, so nearly all access scaffolding falls within the scope of the table and there should be no cases of inadequate strength or overloading. Scaffolds on unusual façades or of greater height, or that present any other unusual problem, must be dealt with by a designer.

26.7 Completed scaffolds that are badly built

By far the greatest number of local or total collapses of scaffolds result from bad construction either originally or through misuse. One-third of all problems arise from poor lateral stability due to inadequate tying, one-third from poor lateral stability due to inadequate bracing, and the rest from a variety of causes, of which not staggering the joints is a frequent one.

As the construction of the scaffold continues, the scaffolder and his supervisor must pay close attention to the tying of the scaffold to the building and to the system of bracing within the scaffold. The rules for tying and bracing are set down in Chapters 5 and 6, respectively. If these rules cannot be carried out, alternative arrangements must be made by the scaffold depot manager or the designer.

The degree of accuracy to which a scaffold must be built is dealt with in Section 30.4. Generally, a scaffolder will comply with the tolerances in ordinary access scaffolding because they are not stringent. Engineering scaffolds require much tighter tolerances.

Finally, with regard to staggering joints, there is a tendency to start off a job with tubes of the same length, usually the longer ones. This results in standard joints being in the same lift and ledger joints being in the same bay. There is no doubt that a scaffold loses much of its emergency strength when the joints are not staggered. While the coincidence of joints may not be a frequent prime cause of local or total collapse, one frequently observes that the unstaggered joints have failed to hold the structure together when some other effect has started a collapse.

26.8 The sleeve coupler hazard

- Figure 26.1 shows two examples of elements of a scaffold being taken away and the failure resulting from sleeve couplers losing their grip. There are many such cases each year. It is the clear duty of a scaffolder to make a structure safe before he starts to dismantle it and to see that it is modified to keep it safe as he dismantles it. The two examples given demonstrate that the scaffolder must pay particular attention to the tight-ness of the sleeve couplers. Because the sleeves are used mainly in bending in ledgers and mainly in compression in uprights, they do not receive full and proper tightening during erection.

27 Safety

27.1 Lifting scaffolding materials

On large jobs, materials will usually be lifted by the site crane and slings will be used with the tubes or boards horizontal. On smaller jobs, materials will be raised by gin wheels in loads of 30–50 kg.

Figure 27.1 shows bundles of tubes, some not tightly packed. Bundles varying from a single 6.4 m tube to five 1.5 m transoms are shown. In (a) and (b), the bundle cannot collapse, but in some of the other cases the bundle may collapse and the lifting hitches will momentarily release their grip. Group (c) may collapse into (d). Group (e) may collapse into (f) or (g). Groups of five tubes can collapse from (h), (i), (j) or (k) to (l), (m) or (n). In each case, there is a loosening of the hitch of 40–50 mm of rope.

Different types of materials, e.g. tubes and boards, should not be hitched together. Bags of fittings must be lifted separately from other elements of the scaffold.

27.2 Hazards with lifting tubes

Three hazards in the raising of tubes in a more or less vertical position are revealed by Figure 27.1:

- The hitch loosens momentarily when the tubes rearrange themselves.
- Despite the geometric fact that some arrangements have the same length of hitch, one of the tubes in the arrangement has only tangential contact with the rope. Because of the low friction of steel against steel, the tubes can slip easily out of the hitch and fall to the ground. In cases (c), (f), (i), (j) and (m), one tube is not retained by friction with the rope.
- The tubes are usually not exactly straight, and the top hitch may be around one arrangement of the tubes and the lower hitch around another.

27.3 The use of scaffolding in inclement conditions

It is not possible to lay down set rules or to fix limits when work or access should cease. The number of times that one has to perform the duty in question also affects the limit of severity that is acceptable.

Wind (see also Chapter 38)

The lowest 50 year, 3 s gust speed in the United Kingdom is in the southeast. On the wind map, it is marked as 38 m/s (85 mph). It is impossible to stand up in such a wind. At half this speed, it is possible to stand but impossible to walk and impracticable to work.

The safe limit for walking along a scaffold is about 16 m/s (35 mph), but working in this wind velocity is not practical.

To do one job on a single occasion, e.g. make an emergency change to a lighting system or turn off a valve, the practical limit is about 11 m/s (25 mph), but if the job has to be done every day, this is unacceptable. For a daily task, the reasonable limit is 8 m/s (18 mph). This is the velocity of the wind on the part of the scaffold being used, which may be much greater than the velocity in the general area of the site.

In the case of slung scaffolds under partly built roofs, the daily limit should be 7 m/s (15 mph) unless roofing sheets are being carried across the decking, in which case a more reasonable figure is 5 m/s (11 mph).

When working in cradles and bosun's chairs, there should be a limit for daily tasks of 5 m/s (11 mph), depending on the restraint system and on the closeness of the platform to the actual point of restraint.

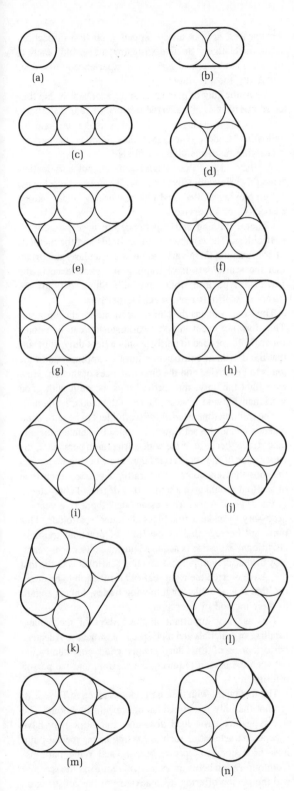

27.4 Snow, ice and cold, and high temperatures

Work should not be carried out on scaffolds covered with snow. This should be cleaned off first and all traces of ice removed. The Construction (Working Places) Regulations 1966 reflect this in Regulation 30, which requires appropriate steps, as soon as is reasonably practicable, to remedy any slippery conditions.

Low temperature as such is not a reason for stopping work, but the attention paid by a workman to his surroundings declines as he spends more time trying to keep warm; the hazard potential rises. A temperature of about −7 °C (below freezing) can be tolerated for a one-off job but is unacceptable for repeated jobs. For daily tasks, a temperature limit of −5 °C should not have an unacceptable hazard potential.

At the other end of the scale, a single short job can be carried out in a temperature of 45 °C. Continuous work cannot be carried out at 40 °C, but work on the scaffold can be done for up to two hours. At 30 °C, continuous work can be carried out, unless the scaffold is in a confined space without much air circulation, as may occur in boiler cleaning work.

27.5 Darkness

Scaffolds that are to be used as a means of access or as working platforms should not be used without adequate lighting of the means of access and at the location of the job.

Scaffolds that are used constantly at night and have lighting systems on them must be earthed. In this category are scaffolds inside places with temporary artificial lighting such as electrical generation boilers, chimney stacks, lift shafts, mine shafts and tunnels.

27.6 Hazards with scaffolds in dark places

* A person may receive an electric shock if a light bulb placed to illuminate it breaks and the filament contacts the scaffold, if this is founded on sole plates. This is a particular hazard on an indoor scaffold or a film set.
* A scaffold that is usually illuminated may become unlit and present a hazard, because the reason for the loss of light cannot be seen.
* Standards in places that obstruct the public may become dangerous when the illumination fails.

Figure 27.1 Various arrangements of tubes in a bundle being hoisted. Arrangement (c) may collapse into arrangement (d). Group (e) may collapse into (f) or (g). Groups (h), (i), (j) and (k) may collapse into (l), (m) or (n). All these readjustments loosen the hitch by 40–50 mm.

27.7 Earthing scaffolds against lightning discharges

Some topographical areas are known to be subject to lightning, and scaffolds on the roofs of high buildings are susceptible to being struck. Scaffolds in these locations must be earthed. Butting on to the building surface is not adequate to ensure that the lightning will not pass through a person's body if he is in contact with the metal framework.

Scaffolds associated with power line construction or adjacent to power lines or electrical transmission feeders must be earthed.

27.8 Morning dampness and rime

On days that otherwise would be expected to be good working days, the first hour in the morning may present hazards. Dampness may make steelwork slippery and put scaffolders at risk. Rime may destroy the customary non-slip surface of raw sawn scaffold boards and put both scaffolders and workmen at risk.

27.9 Friday afternoons and holidays

Irrespective of climatic conditions, experience shows that accidents happen frequently in the last hours of work before the weekend and before holidays, probably due to workmen hurrying to finish the work in hand, or perhaps because their minds are no longer on the job.

27.10 The duties of the erector

The erector should know something about how scaffolds are used, particularly the one he is building at any time. The user should know something about how scaffolds are built, of what they are made and what weights they are capable of supporting.

Both the erector and the user must keep in mind that there are statutes governing the construction and use of working places, and that the contents of these statutes are matters of law. The requirements of the statutes are not matters that can be overridden by engineering judgements or personal experience, or for reasons of expedience or economy. The erector and the user must both accept the Regulations, which on close study they will find neither unreasonable nor financially burdensome.

The recommendation in the Code that the erector, if in doubt on any point concerning scaffolding, seeks expert advice, has been specially included because scaffolding is not a precise subject. It permits a variety of methods of construction and its supporting ability, and its stability or weaknesses, are not always apparent on first inspection. Being a relatively flexible structure, a scaffold reacts to misuse and abuse, and it requires experience to assess its suitability for any duty.

The prime duty of persons erecting scaffolds is that they do so in a manner in which the erectors are safe at all times, and that the working platforms they make are safe for following trades. The same duty applies to persons modifying or dismantling scaffolds.

All the Codes of Practice, manuals and advisory leaflets have this prime objective in view. It can best be achieved by employing skilled scaffolders working under skilled inspectors or supervisors.

General training and experience are necessary for all scaffolders. The complexity of scaffolds and the methods of constructing them vary, and it is particularly important that the scaffolders have training and experience in the particular type of scaffold being built. Slung scaffolds and truss-out scaffolds need special expertise.

The Code contains the phrase 'adequate for the purpose'. This does not mean that the workmanship can be poorer for scaffolds needed to perform only a light duty. It means that the details of construction must be appropriate to the loads to be carried and the circumstances of the job. How ever light the duty, the scaffold must be built with good workmanship and in accordance with standard practice.

The training time needed for scaffolders to acquire competence cannot be set down, as indeed it cannot for any other trade, because of the wide variation in personnel, the experience of the supervisors they are working under and the jobs they encounter. Generally, a year on each type of scaffold should give a satisfactory degree of competence in that type of work. This means that about five years is necessary to make a first-class all-round scaffolder. This must not be regarded as the limit of practical training. Additional expertise is acquired during subsequent years.

A further duty of persons erecting scaffolds is to see that they have erected the right scaffold for the job to be done, i.e. that it is suitable for following trades, having regard to their method of operation.

This is very important in these days of mechanical handling of materials and high-speed construction. Adequate performance of this duty entails good communication between the person requiring the scaffold and the person offering it.

The five types, with their descriptions given in Table 2.8, are not the only five divisions of scaffold functions. The five types are basic load divisions. Each one may have special characteristics such as loading bays, special point load provisions or special bracing and tying patterns. Communication between the person requiring the scaffold and the person offering it, in advance of the construction period, is the only way these special matters can be dealt with.

Persons erecting scaffolds must consider the whole range of the construction team on the site, as well as the trades using the scaffold and the public at large. Many of the problems in this aspect of scaffolding are resolved by the communication referred to above, but the anticipation of problems needs imagination and forethought. Questions such as the following should be considered:

- How are workmen to enter the building?
- How are lorries to enter the building? What are their turning circles, and how closely will they come to corners of the scaffold?
- Where are the tower cranes to be, and what special provisions do they need?
- What means of loading will be used, and how many lifts can be loaded if the general planning of the job goes wrong?
- Do the public need protection?
- Is there a possibility of damage by vandals or children?

27.11 The duties of the user

The first obligation is that every person employing labour is responsible for their safety and the user himself must behave in a reasonable manner.

This responsibility of the employer entails the obligation to provide the right sort of platform for his work people in accordance with the statutory requirements. He must see that it is properly decked and guarded for any special function that he may want his work people to carry out.

It is not enough to assume that a scaffold on a job is suitable for all trades because it has been suitable for one. There is a special provision in the Construction (Working Places) Regulations for scaffolds to be used by several trades, and the employer is advised to acquaint himself with it.

The user must also see that he does not misuse the scaffold once he has accepted it as suitable. He must not alter it without authority, and in particular, he must not alter its bracing and tying pattern without properly consulting the scaffolder.

The user is responsible for inspecting the scaffold and signing the site register every week. The fact that the scaffold was constructed by others does not relieve him of this responsibility. He must recognise that the scaffolder cannot carry out these weekly inspections because he is probably not on the site after the initial handover of the scaffold.

The user must alert the responsible authorities on site to any defects that he sees in the scaffold that might cause a hazard to other users or to the public.

27.12 Deterioration of scaffolds

Scaffolds that are safe, neat and tidy when first constructed frequently become ramshackle and hazardous towards the end of the building operation because of repeated modifications, removal of ties and braces, impacts by crane loads and traffic, and sometimes by undermining of the foundations.

All these changes can be rectified as they occur if the site manager and general foreman are diligent in their attention to all the details of the scaffold. These persons will be experienced enough to know the consequences of a scaffold collapse.

Other causes of deterioration of scaffolds are: continual movement of ladders, interference with the boards near ladder holes, moving boards on to lifts with unsuitably placed transoms, and moving guard rails and toe boards.

27.13 Work people putting themselves at risk

Actions that put persons at risk are:

- Jumping down on to boarded lifts.
- Stepping from the floors of the building on to the scaffold when it is not properly boarded.
- Climbing down the scaffold framework instead of walking along to the ladder.
- Leaving debris on the working platforms.
- Rushing through the work in hand in the last few minutes of a shift.

27.14 Work people putting others at risk

This can result from:

- Throwing materials down or up.
- Leaving guard rails off after ladders have been moved.
- Interfering with hoist gates.
- Replacing ties and braces badly.

27.15 Safety handbooks

Companies operating as scaffolding contractors, and other companies employing work people to assemble scaffolding or staff to supervise the work, should prepare manuals giving general information on scaffolding and specific information about any special equipment peculiar to the company. These safety manuals should contain sections dealing with the following points.

There should be a well-defined tree of responsibilities for every safety aspect of the company and in every geographical location so that any person of any rank and in any location can quickly find the proper place in the chain of command to report defects in materials or assemblies, dangerous practices and potential hazards that may not yet have caused an accident.

Every reader of the manual should be made aware of the relevant contents of the Construction (Working Places) Regulations and the Health and Safety at Work Act, and should also be told that the requirements of these documents are the law and must override any other consideration. This is not an onerous requirement. The law as it relates to scaffolding is very simple to comply with (though regrettably not always as simple to understand at first reading) and it requires no special techniques. Its application costs no more than that of any decent scaffold and compliance with it is as much a protection for the scaffold company as it is for the general public.

Attention should be drawn to the requirements of the various Codes of Practice, both national and industrial.

Management must know its responsibilities to see that all work people and staff have the necessary information, training and supervision to carry out scaffolding work. Management and work people should know that they as individuals and as a company have responsibilities for the safety of others.

Omissions may cause as great a hazard as wrong acts, so the manual should emphasise the diligence and attention to detail that all ranks must apply to scaffolding, which is the major tool of the construction industry. One can build without cranes and hoists, but one cannot build without working platforms. All places on or from which persons walk or work must be safe, and the object of the manual should be to achieve this.

There is also the requirement of staff and work people that, once means of ensuring safety have been built into the scaffold, these means are maintained in the structure thereafter. All safety devices must be maintained; they must not be misused, overloaded, removed from their correct places or interchanged.

With regard to the actual operation by work people, it is not desirable in a manual to give precise instructions as to how a scaffolder should carry out the work, e.g. where he should place his feet, what weights he can hold and how he should move from one area of a partly completed scaffold to another. Such detailed information may be correct for one man of a specific size and shape, but wrong for another.

Obvious bad practices should be prohibited, such as throwing equipment up or down from heights, or jumping from one level to another or from a roof edge on to a scaffold. Erectors must not walk along scaffold tubes without a handhold, nor should they attempt to get on to ladders or parts of the building that are too far away. They should not lift loads that are too heavy or badly tied. They should obey the instructions of their foreman or supervisor.

27.16 Scaffolders' responsibilities

Personnel behaviour must be disciplined. Many accidents have been caused by apparently harmless larks, and many have been due to work people not following the recommendations peculiar to the particular construction site. Drinking and drugs must be avoided, and the effects of any special medication must be considered. Inducing others to take risks or not preventing them doing so also leads to accidents.

Clothing should be suitable, particularly footwear, which should be non-skid even when wet and soft enough to enable the feet to grip the tube rather than slip off it. Helmets must be worn.

Failure to report inadequacies in the services that should be provided by other contractors, and failure to report bad equipment and bad practices by others, also leads to accidents.

Management should ensure that tasks are not set for persons who are experienced enough to carry them out, and work people should not hesitate to say if they have been set a task for which they have no experience.

Inspectors must be thorough in their work, tedious though this can be on big scaffolds. Failure to see the omission of a critical coupler in a key place can result in a collapse.

Work people should provide themselves with enough boards to work on. They should not rely on ledges on buildings, terracotta decorative work, cornices, guttering, or brittle or fragile materials. They should not build into the scaffold any material that is defective or that may become defective during the life of the scaffold.

Chargehands, foremen and supervisors should not take orders from others on site in matters that reduce the safety of their operations. They should not alter structures at the request of site agents, general foremen or even the police if such requests will weaken the scaffold structure. If such controversial orders are received and they cannot be dealt with by discussion, they should be referred to headquarters for settlement.

Structures must never be left unstable over the lunch hour or overnight. Structures accessible to young children in playgrounds and parks must be fenced or boarded off so that there is no hazard, and ropes and gin wheels must be concealed. Ladders should be boarded over or, better, not left resting on the ground at all.

Paint and rust from scaffolding should not be left to damage the clothing of the public, and grease and slippery materials must be cleaned off the platforms or anywhere they may cause a hazard. In chemical works, boilers and other noxious environments, care must be taken by work people to wear the correct masks and goggles.

Debris from building operations should be cleaned from platforms, and loose couplers, boards and tubes from scaffolding operations should be cleared away. Nails in timber or any other trip in the platform or its toe boards and ladders must be removed.

All staff and operators should complete the paperwork required by the company because this can give safety protection in any circumstances. If headquarters has failed to send enough of the correct equipment to the site to do the job in safety, it should be notified through the paperwork, by telephone or by direct reporting.

Before the job starts, hazards may be created by badly stacked materials or by the bad loading or careless unloading of lorries. Instability can be caused in the scaffold by stacking material in it or leaning tubes against it. Instability can occur in the very early stages of a scaffold construction before there is enough material in it to make it firm. This situation must be watched very carefully by the chargehand.

Transoms, ties, roof crawling boards and saddles may damage the main structure, and this must be avoided. If the main structure has to be altered, e.g. tiles removed from roofs, glass taken out for through ties or facing stone drilled for anchors, the client's permission must be obtained. The work must not proceed to a dangerous condition while such permission is being obtained.

Pavement lights, traffic fenders, pedestrian walkways, use of electricity, use of welding gear, use of hoists and cranes and use of lighting should be dealt with in the planning stage before deliveries to the job have started. Other matters requiring pre-planning are hoardings; traffic lights; avoiding manholes, fire hydrants, water cocks, gas cocks and telephone manholes; and obstructing access to buildings, particularly domestic buildings, which are entered more frequently in the dark than industrial buildings. Hospitals, children, schools, zebra crossings and road junctions need special attention.

Signs should alert the public to 'men working overhead', alert the contractors to 'unfinished scaffolding' and alert factory workers to 'narrow corridors ahead'.

Where there are environmental dangers to the scaffolds, special precautions must be taken, e.g. working over water must be with harnesses, belts, safety nets and safety boats. Working in railway sites requires watchmen. Working in chemical works, mines and excavations may need watchmen. On larger sites, there may be special rules concerning first aid, meal arrangements and surrender of personal goods such as petrol lighters, matches, steel items, cameras and radios. Such site rules must be followed precisely. Permits and identity tags are necessary in many jobs, and these must always be obtained, worn and surrendered according to the safety rules of the location.

27.17 Scaffolds used by several trades

Building and construction works are now frequently carried out by specialist subcontractors, each contributing the special skills of his own trade. He expects to be provided with access. Even if one subcontractor provides his own scaffolding, another subcontractor will expect to use it. Herein may lie a hazard.

A brickwork contractor will require only one heavily loaded platform and possibly one lightly loaded one. A glazier or mullion fixer may need all the levels boarded. The lifts for building a concrete frame are different from those required to clad it later.

The Construction (Working Places) Regulations recognise this problem and include Regulation 23, which makes each trade responsible for seeing that the working platforms provided for others are suitable for them.

27.18 Hazards with scaffolds used by various trades

- Removing the guard rails to facilitate the passage of heating and ventilating plant into the building and not replacing them.
- Removal of toe boards by a trade that has to move a ladder along the job several times each day.
- Fast-moving finishing trades working faster than the scaffold can be modified for them.
- Leaving the top platform covered in bits of tiles and roofing debris ahead of the painting trades.
- Providing a cleaning and painting scaffold for work on the gutter and walls, and allowing it to be used as edge protection for roofers, for which it may be unsuitable.
- Fixing a gin wheel to a scaffold not specially prepared for it.
- Using ladders fixed to a scaffold for a previous trade, who have moved the platform along the job.
- Using crawling boards set out for roof sheeters for access to roof vents and flues when they are not suitably placed.

Many other cases will come to the mind of any builder that will demonstrate the importance of Regulation 23.

27.19 Collapse statistics

Throughout this text, attention has been drawn to the hazards existing in each type of scaffold structure. Failure to recognise these hazards and deal with them may lead to accidents and collapses, even if the scaffold is fairly loaded.

The deficiencies that cause collapses are given in Table

Table 27.1. Causes of structural collapse of access scaffolds

Dominant cause of the collapse or partial collapse, as deduced by analysis or actual observation:			
Main grouping and percentage		Subdivision	Percentages for each subdivision
Lateral instability	64%	Inadequate bracing on the original structures	24%
		Inadequate ties in the original structure to utilise the stiffness of the building	18%
		Ties removed during the life of the scaffold	12%
		Bracing, rakers or buttresses removed during the life of the structure	10%
Inadequate support for the boards	15%	Transoms not repositioned properly after the platforms have been moved	9%
		Transoms wrongly placed in the structure	6%
Overloading by the user	12%	Disregard of the loading limits	5%
		Attachment of vertical wind catching sheets not allowed for in design	4%
		Accidental impacts	3%
Dismantling errors	4%	Removal of structural integrity as elements are removed	2%
		Premature removal of ties	1%
		Premature removal of braces	1%
Defective materials	3%	Boards	1%
		Couplers allowed to go rusty, preventing proper tightening	1%
		Wrong type of couplers used or couplers not tightened even when oiled	1%
Others	2%	Traffic accidents	1%
		Floods, foundation failures, etc.	1%

27.1. Those observed most frequently are at the top of the table. The percentages are the results of observations over a period of 30 years.

Table 27.1 relates to structural failures and not to accidents to personnel caused by shortcomings in the working platforms that make them operationally unsafe.

These shortcomings are usually breaches of statute, i.e. the platform does not comply with the Construction (Working Places) Regulations. These Regulations deal adequately with the procedures necessary to make a working place safe for construction work to be carried out. It is not possible to predict the effect of a breach of the statutory requirements. A guard rail may be left unerected on one job for months and no one will fall through the gap. On another site, a man will fall through the gap within an hour of the start of the job.

The structural inspector can use Table 27.1 to guide him on the features to concentrate on but, when considering the adequacies of the actual working platform, he must give every requirement of the Regulations equally strict attention because it is impossible to predict where or when or how an accident will occur.

After an inspection has shown that a scaffold is in good condition, accidents, collapses and partial collapses often happen some days later. This has been observed more frequently towards the end of the construction period of the main building. The obvious reason for this is that glazing and cleaning requires ties to be interchanged and, in consequence, frequently left off, but there is also the regrettable deterioration in the attention given to the correct construction and use of the scaffold as the job proceeds. Items such as uniform tie patterns, regular bracing patterns and staggering of internal joints become scrappy. Control of the loads applied and their distribution declines. A combination of circumstances may then result in a failure.

27.20 Compliance with the statutory regulations

The statutory regulations of a country are part of its criminal law and must be complied with, irrespective of the fact that in some cases a lower standard is not unsafe. Working platforms on scaffolds are by definition above the ground, and the construction of these platforms requires a discipline, a uniformity of end product and a maintenance of quality over the whole period of the job. Compliance with the Construction (Working Places) Regulations 1966 in the United Kingdom ensures this discipline. Other countries have comparable documents.

27.21 Compliance with codes of practice

Codes of practice are advisory and deal with the technical aspects of scaffolding to supplement the legal requirements. They cannot deal with every practical case, but they can give sufficient guidance to the designer, the erector and the user to keep him on proven safe paths.

Three types of codes of practice relate to access platforms and scaffolding:

(i) The British Standard Codes.
(ii) The codes and advisory notes prepared by trade associations, manufacturers of equipment, the Health and Safety Executive and scaffolding contractors.
(iii) Special codes relating to working in special places, e.g. railway sites, dockyards, mines, television, radio and film studios and outside sites.

These documents are not at variance. They are complementary to one another, and all of them should be studied when applicable because they are aimed not only at making safer scaffolds but also at getting the best out of the equipment in the very wide circumstances of its use.

27.22 Types of access scaffold requiring special safety precautions

Codes of practice do not deal with the more advanced types of structure tabulated below:

- A single line of uprights with transoms resting on the slabs of a concrete or steel building frame.
- Multiple-upright scaffolds used for very high buildings.
- Scaffolds with inclined standards.
- Geodetic arrangements.
- Curved structures using interwoven straight or curved tubes.
- Arch scaffolds for domes.
- Untied scaffolds such as those used in a chemical tower or atrium.
- Signboards and hoardings.

- Access scaffolds on curved surfaces such as ships and pressure vessels.
- Chimney and cooling tower scaffolds.

All the recommendations governing safety must be applied more diligently to cases such as these because of their unusual nature and infrequent occurrence.

27.23 Human error and the safety pyramid

A useful concept for managing the safety of a scaffolding job is to visualise it as a pyramid with four tiers. The bottom tier, being the biggest, represents the contribution to safety made by the site labour force. Good scaffolders cannot be made from inferior workers, so the prime duty of management in safety control is to obtain skilled workmen, train them further and when necessary train them specifically for the job allocated to them.

The second tier represents the safety department of the depot constructing the scaffolds. Here the chargehands and the local management need extra training and to be specially chosen for their diligence in supervision on the sites.

The third tier represents the main company headquarters team that comes into action when it is recognised that there are special hazards in the job under consideration. This team should be specially formed for detailed supervision of every aspect of the job in question and at every stage in its progress. The company chief engineer, the contract director and the safety director should all take a hand to see that no hazard slips past the net.

The fourth tier at the top of the pyramid may apply to perhaps only a few jobs each year. This tier should consist of a very deep probe team for special jobs that have been singled out for their complexity and hazard potential. The company estimator, the conditions of contract officer, the managing director and if necessary outside consultants should all lend a hand to see that nothing goes wrong with the job from start to finish.

Finally, in this chapter on safety, all personnel should be reminded that good scaffolds do not collapse by themselves or cause hazards to others. It is scaffolds that have been built by persons who have failed to carry out their responsibilities that cause problems. Thus lack of safety is mainly a human problem and as such, it is a solvable problem. There have been no scaffold failures for which we have not known the cause, and management and work people are mainly responsible.

28 Training

28.1 Skills

Scaffolding requires five skills:

(i) Identifying and handling the materials.
(ii) Planning the sequence of work.
(iii) Assembling the materials into a sound and stable structure that provides the facilities required of it.
(iv) Not working from precarious places, and where necessary providing a safe temporary place from which to build the scaffold.
(v) Being able to work at heights.

These skills are diverse and have to be acquired by one person. This person must also have the character to do a first-class job on every occasion because any bad scaffold may be a killer.

28.2 Managers

The belief that a manager does not need to have the skills of his work people — a proposition more frequently heard from management than scaffolders — is dangerous in scaffolding. It is difficult to recognise a potentially hazardous scaffold. If the manager cannot identify hazards, his clients, his insurers or the public may find them for him. Top management and area or depot management should be more highly trained than the site staff.

28.3 Supervisors

Supervisors should come up through the ranks and should have had training at every level. Scaffolding is an art as well as a science and it has to be executed as a craft, calling on the five skills listed in Section 28.1 for each and every job. Thus a scaffold supervisor cannot be trained overnight. He should have training beyond that of his scaffolders in the choice of scaffolding systems to be used, in the rates of progress that can be achieved for various types of work and in deciding what can be done with scaffolding and what is doomed to failure if attempted.

The supervisor must be very good in assessing the qualities of his scaffolders and in placing them in jobs within their various degrees of competence. He should also know when to call for help from designers or engineers. He should have special training in how to bring young people forward in the industry and how to develop their skills safely.

Many serious and tragic accidents occur in which young persons are killed or injured. This is sometimes due to the over-keenness of youth, to inexperience in the job and in what is potentially hazardous, and to the desire to prove oneself. The young person is not to be blamed for being over-keen, for inexperience, for failure to recognise hazards or for a desire to prove himself. Who, then, is to blame? That this paragraph is under the heading 'Supervisors' expresses the opinion that the supervisory and management ranks are as much to blame as the scaffolders on site for accidents to young persons.

This fact was recognised by the drafters of the Health and Safety at Work Act 1974, which has a requirement that all ranks be trained and supervised.

The supervisor must be trained in selecting suitable persons for the industry, then trained to train them.

The following section deals with selecting suitable persons for training and subsequently with a programme of training for young persons or trainees.

28.4 Persons suitable for training in the industry

Scaffolding involves the construction of platforms at a considerable height above the ground. These platforms are to be used by persons other than those building them and who, having other jobs to do, cannot be continually watching for hazards. The platforms are frequently situated over places to which the public has access.

These three features necessitate a high degree of attention to detail in the construction of scaffolds and continuous watchfulness that errors do not occur at any time. The work therefore requires a conscientious type of person who is not likely to 'go off the rails' at any time.

Some persons are unsuitable for training as scaffolders. How ever much effort is spent in their training, their sense of responsibility will not improve. Such persons should not be trained for the industry. If they have entered the industry before their seniors have perceived their unsuitability, steps should be taken to find other employment for them. An inability to work at heights will probably be revealed early enough in a training scheme for the trainee to be diverted to other duties.

Of those persons suitable for the industry, the duration and breadth of the training scheme depends, as one would expect, on the person involved. Only a general training recommendation can be set down.

For simplicity, the full training of a scaffolder can be divided into three steps:

(i) Trainee scaffolder.
(ii) Basic scaffolder.
(iii) Advanced scaffolder.

In general, each of these stages of training takes one year. It is advantageous if during each of these years, the trainee can have specialised and formal instruction at a training establishment for about two weeks sandwiched between normal site work, which should also be under supervision.

28.5 Trainee scaffolders

There should be a period of about two months during which the trainee works at ground level acquiring experience of the various components used in scaffolding and their handling, raising and lowering, transportation, storage and fixing.

A further two months should be spent in gaining experience setting the scaffold out on the ground and starting off the erection process.

Working at increasing heights can then be permitted at one lift higher every month.

All the trainee's work must be under the closest supervision by a senior person.

The first year should include instruction on erection tolerances, accidents, lifting and lowering, dress, selection and rejection of materials and proper construction techniques.

28.6 Basic scaffolders

A basic scaffolder should have all the above matters revised, with close supervision of any special types of scaffold that

are encountered during the year.

The year should include instructions in the special hazards of dismantling, altering and adapting scaffolds, and in building access scaffolds in difficult locations such as on slopes, over water and in the high street.

Special structures involving prefabricated beams, pavement gantries, traffic and pedestrian openings, towers and bridges, protection fans and nets and cantilever and truss-out scaffolds should be included.

28.7 Advanced scaffolders

This third year should include all complicated structures and those difficult to build, such as temporary roofs, temporary buildings, public stands, slung scaffolds, footbridges, stairways and ramps, and roof scaffolds and roof edge protection.

The advanced scaffolder should also be able to supervise less experienced scaffolders, trainees and basic scaffolders. He must be able to recognise hazardous situations and to arrange the work so that it can be carried out safely. He must recognise dangers in the erection and dismantling of more difficult scaffolds.

He should also have knowledge of the Construction (Working Places) Regulations, the Codes of Practice and his own company's rules and standing orders.

28.8 Chargehands

Above the status of advanced scaffolders, most scaffolding companies have chargehands, foremen or working supervisors. All these persons should have training. It should be more intense than that detailed above for the scaffolders themselves.

The chargehand must be able to inspect the scaffold built by his gang and see that it complies with the Construction (Working Places) Regulations before he hands it over to the user. He will be responsible for issuing a handover certificate, if one is required.

28.9 Scaffold inspectors and safety officers

These persons have special responsibilities to locate other people's mistakes. This is a difficult task because those who have made the mistakes are not trying to make them. They are trying not to make them, so the mistakes are not obvious. Also, mistakes cannot be found by a sense of movement of the scaffold. If this were so, the scaffolders would rectify the errors themselves. To become an expert inspector or safety officer therefore requires a high degree of skill, experience, technical knowledge and training.

The training of scaffolders may contain only about two weeks in a formal training centre per year. Inspectors and safety officers should have a further two weeks of special training under the instruction of an engineer.

The suitability of a person to be an inspector or safety officer must be carefully assessed. He must be the type of person who will not become slack with time. He must not begin to condone bad practices because they have not produced failures for a year or two.

He should be trained to make formal reports on the condition of scaffolds, as detailed in Chapter 30.

To make a judgement on the tolerances in construction acceptable for each job, he must be able to recognise misuse and overloading by the user.

28.10 Recording training

Frequently in the investigation of a scaffolding failure or accident, there arises a suspicion that the personnel at some or perhaps several levels in the management and operational sides of the industry have been inadequately trained. The management claims that it is highly trained, while the labour force claims that it is the best in the land.

There is evidence that these claims may be false – the scaffolding structure is in a heap on the ground – but only rarely is there evidence that the claims are true. All training activities should be recorded, including the name of the person trained and every detail of the training given.

These records are of more use to management than to the personnel trained. Management cannot control any activity unless it has full knowledge of the activities of the fixing teams. The records also act as a prompt to management to keep its training up to date and in accordance with some previously thought-out plan.

29 Scaffolding depot activities

29.1 General functions

The main function of the scaffolding depot is as a stock yard for materials coming in from jobs and waiting to go out on new jobs. Scaffolding is carried out with second-hand material. Materials coming back from building contract works have frequently been badly used, and they are not in a suitable condition to send out to a new contract site without servicing.

The second duty of the depot is to maintain the equipment in a suitable condition for reuse, to destroy those elements that cannot be maintained and to record by goods despatched and goods returned sheets, the data necessary for the materials administration of the company.

The third function is to act as an interchange point for work people to collect their instructions, wages and other documentation necessary for the administration of the company.

29.2 Tubes

Three types of tubes are in general use: galvanised, untreated and painted.

Galvanised tubes have proved to be very durable and need very little maintenance. However, the ends wear down, both on the extreme outside end due to handling on pavements and in the storage racks, and on the inside due to repeated assembly on spigot pins. The depot foreman can deal with this problem by cutting 50 mm off defective ends in the tube saw.

The ends of tubes occasionally split up the welds, and this defective length can easily be removed by the saw. If the tubes have filled with water and frozen, the welds may split, or sometimes a longitudinal crack not on the weld line may appear. The defective part should be cut out, leaving two shorter tubes.

Many attempts have been made to weld short lengths of tube end to end to form more useful lengths. No process has proved entirely satisfactory, probably due to reversals of stress in the various uses of the tube and to the severe shock waves generated in the tube when it is handled on hard surfaces. Exposure to the elements is likely to cause deterioration of the weld.

The external surface of the weld should be ground down to the original tube surface level so that the amount of weld material left is small. Various codes of practice recommend that tubes less than 4−5 mm in wall thickness be welded without an inside sleeve being used, but this has not proved durable in use.

If an occasion occurs where end-to-end welding is essential, an inside sleeve should be used. Each end of the tube being joined should be welded to the sleeve before the gap is finally bridged over.

There is a good reason why welded tubes should not be used at all. Sooner or later, a tube will be overstressed on site, or in a state of total or partial collapse. It will then break at the weld rather than elsewhere. The cause of the fracture may be nothing to do with the weld. It may be due to misuse of the scaffold, but the blame will be wrongly placed on the fractured weld.

Untreated tubes will probably be delivered new with a thin layer of black bituminous paint on the surface. This coating will not last, and the tubes rust.

Rusty tubes are not necessarily significantly weakened. Rust 1 mm thick is derived from only one-tenth of this thickness of metal. The rusty surface increases the friction grip of the couplers used on the tube. If tubes have been stored for a long time in the open, water may remain along the contact lines of the tube and cause severe rusting, which should be carefully inspected.

Bent tubes can be reeled or crowed straight. A deviation from straight of less than 15 mm should be achieved in each 6.4 m length. Creased tubes cannot be straightened without leaving a bad dent. The affected part should be cut out, leaving two shorter tubes.

Storage should be in racks under a roof. The tubes should

be in bins of various lengths, with one end facing the depot yard so that the tubes can be handed out directly on to the tail of a lorry. The racks must be very firmly built because they will hold a very large weight. They will be struck accidentally by lorries and at various times will be eccentrically loaded.

29.3 Couplers

Couplers should be cleaned and the threads oiled so that the nuts run freely.

It is usually uneconomical to attempt to reshape a coupler that has been deformed. It should be scrapped.

Storage should be in bins under a roof facing the depot yard or in sacks on platforms at lorry deck level.

29.4 Prefabricated elements

Beams and frames should be stored under a roof according to type, with attention being paid to the stability of tall stacks.

Each unit should be inspected for deformation. Welds must be inspected before and after any deformation has been rectified.

29.5 Boards

Timber is a natural material and as such does not have the uniformity of man-made tubes and couplers and other factory-made elements. This imposes the extra duty to see that the material is correct when received from the timber supplier on the scaffold depot. No amount of maintenance can improve the quality of scaffold board once it has been accepted into the depot stock. Chapter 33 is devoted to the properties and specifications of boards suitable for scaffold platforms. The depot must see that all boards have the qualities specified in Chapter 33 when they are delivered.

Boards should be stacked under a roof on timber battens to keep them off the ground. They should be stacked about ten deep, over which another set of battens can be placed and subsequently a further ten boards. This will allow slow drying out to take place.

The end band and nail plates should be refixed if necessary.

Older boards should be sawn in half for sole plates. Boards that have been used for sole plates must never again be used for platform boards.

29.6 Ropes

Ropes must receive special attention. They must be inspected for deterioration.

Natural fibre ropes should be free from deterioration resulting from contact with chemicals such as stone-cleaning liquids or with chemicals manufactured or used in the works of the company being scaffolded.

Man-made fibre ropes should be inspected for burns.

Both types should be labelled and identified with a serial number. Man-riding ropes should be entered in an 'in and out' register, which should contain a certification of the inspection of the rope. Lifting ropes must be certified in the manner required by the Construction (Lifting Operations) Regulations 1961.

Wire ropes should be similarly inspected, certified and recorded.

All ropes should be stored under cover. The rope store should have the means of fixing the various rope terminations. Bulldog grips should be kept for every size of rope used.

Safe working load tables should be readily available for scaffolders to study.

29.7 Lifting gear

All mechanical devices, even pulleys and gin wheels, must be maintained mechanically. The inspections and tests required by the Construction (Lifting) Regulations, 1961 must be carried out and the necessary certificates kept at hand and recorded in a register.

29.8 Identification

Most scaffolding contractors use some means of identification on their components. This can be an indented marking or a special coloured paint. The depot should ensure that the marking will last for the duration of the contract.

29.9 Inspections

The inspection necessary to maintain good equipment should take place twice between each use. All goods coming in from building contracts should be inspected as they are handed from the lorry. Defective materials should be passed through a maintenance bay. Good materials can go directly into the racks.

On despatch, the equipment should be inspected a second time as it is loaded. This double check should ensure that no materials that may cause a hazard on the job are allowed to leave the depot.

29.10 Data sheets

Scaffolding depots should maintain a stock of data sheets giving details of the rules for assembling all components. These sheets should be prepared by the company's headquarters and should be freely available for use by all scaffolders of whatever rank or experience.

These sheets should also contain examples of problems, failures, collapses and hazards.

30 Inspection of scaffolds

30.1 Important features

The inspection of access scaffolds must cover four unrelated features:

(i) The structural stability of the scaffold, its effect on the building it serves and the adequacy of that building to keep it rigid. The effect of any unplanned attachments such as fans, hoist towers, sheeting, extra working lifts, loading bays, alterations and adaptations.

(ii) The adequacy and safety of the working platforms and their compliance with the Construction (Working Places) Regulations 1966. Compliance with the Construction (Lifting Operations) Regulations if any elements in the scaffold are subject to these. Compliance with other relevant regulations, and with the requirements of the local authority and police.

(iii) The safety of access for persons to and from the working platforms. The safety and access for persons and vehicles to the site through the scaffold. The safety of the public passing near to or under the scaffold.

(iv) The correct use of the scaffold by all workmen, contractors and subcontractors, and the correct loading.

30.2 Philosophy of inspection

Before dealing with the practical method of inspection, consideration has to be given to the nature of the scaffolding industry, its purpose, and the limitations set by its equipment and the behaviour of its users.

It should be borne in mind that, in general, a scaffold is not a designed structure. It is built by experience and craftsmanship. One gang of scaffolders will construct working platforms one way, while another gang will construct them a different way. Both structures will be safe, although they may not look alike.

The gang undertaking the job is sent out from the scaffold depot with only three types of material, i.e. tubes or tube

components, couplers and boards, and they have to produce safe working platforms of the right load-bearing capacity and in the correct place. The nature, preciseness and appearance of the structure supporting the platform is immaterial, provided that it performs its function and causes no danger to any person or property.

Therefore, in inspecting structures one asks the following questions: Is the structure under the platform safe? Is the platform itself safe? Is the access to the platform safe? Is the user of the platform using it in a safe manner?

It is necessary for the inspector to assess the degree of safety of the scaffold in all four respects and to decide whether the emergency strength is adequate for the particular circumstances of the job.

With a safety factor of just over 1, the structure will not fall down, but it may not have sufficient emergency strength to resist unforeseen impacts or overloads. A minimum safety factor of 1.65 is recommended for the loads in tubular material (based on the yield stress of the metal and 2.00 for the loads on the couplers).

The degree of safety of the working platform and the access to it cannot be expressed in figures because it depends on human behaviour, which is very variable, but there are accepted standards that have been found sensible in practice. These are set down in the Regulations, and the scaffold must not fall short of these requirements.

The user must not overload the structure or impact it beyond its proper use, because this will reduce its emergency strength needed to deal with any unforeseen forces. Thus there should be no relaxation in the limits on loading by the user.

30.3 Method of inspection

There are several ways of carrying out an inspection of a scaffold. Some inspectors will look at all the bases, then all the first lifts, then all the second lifts and so on. Others prefer to start at the bottom of a standard and look from the bottom to the top, repeating this for every standard. Others will seek out the bracing and the ties first and look at the remainder of the structure afterwards.

Whichever method is adopted, experience dictates that two functions must be carried out: every member and every joint in the whole structure must be seen separately, and all these items must be set out on a standard report form and ticked off as the inspection proceeds, because the process is tedious and it is easy to miss a defect if the inspection is carried out without the disciplining effect of a report form.

In advance of the inspection, it is preferable to allocate grid letters and numbers to each upright, and numbers to each lift so that the report can fix the location of any defect precisely and enable it to be found again, e.g. on standard G6, lift 6 − wrong fitting used, or G6/6 to G7/7 − brace missing.

In this way, clear instructions can be left with the scaffolder on site, with the scaffolding company and with the main contractor. The site can be revisited and the correction of defects seen on the second visit.

Figure 30.1 is an example of an access scaffold inspection sheet prepared for the method described of inspecting each standard from the bottom to the top.

The inspector needs a list of acceptable tolerances to enable him to report reliably. Such a list is given below.

The frequency of inspections should be governed by the consequences of a failure and by the processes being carried out by the main builder. There have been many cases where removal of ties during glazing or cladding has reduced a scaffold during a period of 14 days to a condition that resulted in a major collapse. It is better to have a programme of inspections related to the works in hand, even if this requires irregular visits rather than visits at regular intervals.

Inspectors and safety officers should pay no respect to construction staff or work people who will not heed their advice on the state of the scaffolding or its use, or make the necessary changes. They should report these persons to their superiors and get things straightened out as soon as there is any obstruction to rectifying errors.

30.4 Tolerances

Standards

Experience has shown that trouble does not arise from a deviation out of plumb of 50 mm in access scaffolding. By coincidence, this rule is simple to observe because the tube itself is nearly 50 mm in diameter, and this dimension can be seen at any range of vision. Whatever the height of a scaffold being inspected, and wherever the inspector is standing, he can always see what 50 mm is by using the

diameter of the tube. Falsework and engineering scaffolds should have 25 mm as the norm.

Ledgers

Inclination of the lifts from the horizontal is not critical, except that the platform may be disturbed and stepped if the ledgers are not set horizontally. The base lift, i.e. foot ties, can follow the ground and so be inclined to keep the bottom unbraced leg length short.

Lift height should not exceed that intended or designed by more than 150 mm.

Transoms

Transoms should be horizontal so that the platform is horizontal across its width.

Bays

The length of bays should be within 200 mm of their designed length. This tolerance need not be applied to every bay. It is to enable various lengths to be incorporated to finish off the length of the whole scaffold at the desired place.

The width of the scaffold must always contain the correct number of boards required for the project. The width of the framework will vary according to the type of couplers being used and according to whether the toe boards are to rest on the transoms outside or on top of the boards. The tolerance in the width of the framework should be ±20 mm. The negative tolerance may be impracticable because it is impossible to lay the boards on the transoms. The positive tolerance is tight so that large gaps do not form between the boards.

Working platforms

Gaps at the ends of boards should not exceed 25 mm.

Guard rail heights should be ±25 mm. This figure allows one rail to overlap the next where necessary, for a rail to be hinged on a swivel coupler, or to allow rails to be set one over the other when the platform is set around a corner.

The distance of the whole scaffold from the wall it is serving should be agreed between the contractor and the user, and it should be within ±25 mm of the agreed dimension.

Node points

The structural nodes, i.e. the places where the tubes meet and cross, or start and finish, should be as closely assembled as possible. Any brace or raker should be connected to the

ACCESS SCAFFOLD
INSPECTION REPORT

SITE: . . . .   REPORT No.:

. . . .   DEPOT:

LOCATION: . . . .

DRAWING No.: . . . .

COPIES TO		
	MANAGER	
	AGENT	
	SUPERVISOR	

INSPECTED BY: . . . .   DATE: . . . .

ITEM		GRID LOCATION OF DEFECT
SOLEPLATES		
BASEPLATES		
ADJ. BASES		
LIFT HEIGHTS		
Standards	SPACING	
	PLUMB	
	JOINTS	
ADJUSTABLES		
FORKHEADS		
Ledgers	SPACING	
	JOINTS	
Transoms	SPACING	
	JOINTS	
FOOT LACING		
HEAD LACING		
Long Bracing	BOTTOM	
	TOP	
	INT'SEC'N	
	LAPS	
Transverse Bracing	BOTTOM	
	TOP	
	INT'SEC'N	
	LAPS	
RAKERS		
PLAN BRACING		
PUNCHEONS		
BUTTRESSES		
LINK BARS		
X BRACES		
TIES		
BUTTING		
ANCHORS		
GUYS		
FITTING TIGHTNESS		
APPLIED LOAD		

Figure 30.1 An access scaffold inspection report form.

framework so that there is not more than 100 mm of tube showing between the couplers. The BS Code gives a tolerance of not greater than 150 mm between the centre lines of the couplers. This is a good way to specify the tightness of the nodes. If there are three bracings coming to the intersections between a ledger, a standard and a transom, it may be impracticable to fix the six tubes involved in the node so closely together.

Summary of the tolerances in Figure 30.2

(a)   25 mm bend.
(b)   50 mm out of place.
(c)   50 mm out of plumb.
(d)   Not greater than 500 mm change in lift height.
(e)   Not greater than 100 mm slope of the lift.
(f)   Variation not greater than 200 mm change in leg length for scaffolds up to 30 m high.
(g)   In scaffolds taller than 30 m, height must not vary more than +0 or −20 mm from calculated height.
(h)   Not greater than 450 mm.
(i)   300 mm to allow for toe boards.
(j)   150 mm.
(k)   150 mm.
(l)   150 mm.
(m)   250 mm.

30.5 Hazards, accidents and failures

A scaffold inspector must not only concern himself with the accuracy and adequacy of the structure he is looking at. He must also anticipate what hazards, accidents and failures may occur on it from every conceivable cause. He should visualise every mode of failure, then check that the structure cannot fail that way.

Two levels of hazard should be in the inspector's mind. One is the general hazard that affects all scaffolds. The other is the specific hazards for the structure being inspected.

General hazards include inadequate numbers of ties, braces or anchors, poor joint distribution, bad foundations, overloading and impacts. Specific hazards include dangers in construction methods and dismantling, unbalanced loads, cantilever bending, lateral instability of beams, electrical discharges, massive corrosion, excessive wind forces, latent defects in couplers, errors in calculation and eccentric loads.

Accidents specific to the job may be the result of working over water, working on steep roofs, inadequate edge protection for workmen working near the edges of slabs and roofs, poor access for the scaffolder and fragile materials. Every possible chance for an accident on any job being inspected must be investigated and the possibility of the accident happening removed.

Most failures in scaffolding are caused by lateral

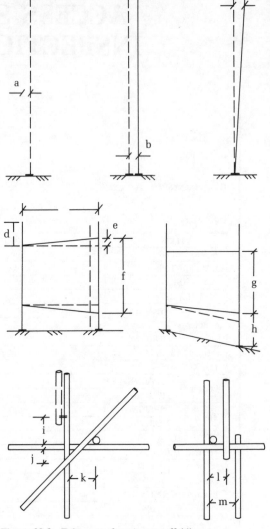

Figure 30.2 Tolerances for access scaffolding.

instability. A few are due to excessive axial loads, and some are due to eccentric loads, especially if the eccentricity is repeated in many adjacent members in the scaffold and all the eccentricities have their effect in the same direction.

The analysis of a failure after it has happened is easy compared with predicting the problem before it happens. However, this is the inspector's job.

30.6 Features that an inspector may overlook

Under the Construction Regulations, a scaffold must be inspected every week and a document signed certifying the inspection and the defects. This is done on well-controlled construction sites and, because it is done so frequently, the personnel doing it will be experienced. This raises the question: why, with this level of skill applied so frequently, can there be a failure or a collapse? Experience points to two areas where the inspector has not foreseen a hazard:

- He has not foreseen external influences.
- He has not foreseen reckless behaviour.

External influences are not taken into account in the design of the scaffold or not included in calculations. Suppose that a site vehicle or a public vehicle hits the corner standard of the whole structure. If the face bracing runs upwards along the two façades from the corner in question, it will be knocked out with the lower lifts of the standard. If it runs upwards from some distance along each façade towards the corner in question, it will hold up that corner, even if the corner standard is knocked away.

Where inward or outward services have to be installed for a building, trenches will be cut below the scaffold. An A frame of bracing over this point may overcome the effect of removal of the foundations by this trench work.

Acceleration of the wind as the height of the building increases is a predictable hazard. The storage of excess material is predictable if the delivery schedules for building components are studied.

Reckless behaviour is often repetitive and becomes predictable; for instance, the non-replacement of guard rails after a series of rough-terrain forklift loadings. The stacking of roof tiles along a gutter painting scaffold is also predictable.

The scaffold inspector should not deem it beyond his responsibility to issue a timely warning of a potential hazard to the manager of even the most thoroughly organised and controlled site.

PART IV
Materials and technical data

31 Scaffold tubes

31.1 General

The British Standards dealing with scaffolding materials refer to materials when new. The materials used in scaffolding construction have usually been used previously and therefore may not comply with the appropriate standard at the time of use.

For scaffolds that use standard couplers, the outside diameter of the tube is 48.3 mm ($1\frac{29}{32}$ in). Data are given for three different steel tubes and one aluminium tube with this external diameter.

31.2 Section properties of four types of scaffold tube when new

In the United Kingdom, scaffold tube is traditionally made of steel with a yield stress of 210 N/mm² and a 4 mm wall thickness.

Two other steel tubes are used in other European countries. One has a yield stress of 235 N/mm² and a 3.2 mm wall thickness, and the other has a yield stress of 235 N/mm² and a 4 mm wall thickness.

Aluminium tube has a 255 N/mm², 0.2% proof stress and a 4.47 mm wall thickness.

It is recommended that an allowance be made on the strength when new to take into account any deterioration or change in characteristics that have taken place during continued reuse of the material. Accordingly, eight tables of strengths have been included in the text, and the designer is cautioned to make sure that he is using the appropriate table.

Table 31.1 gives the structural section properties of the four tubes. Table 31.2 gives other specified details.

It will be observed that the United Kingdom BS 1139 tube has the lowest allowable stresses, but it is not the weakest tube. The 3.2 mm wall thickness, 235 N/mm² yield steel is slightly weaker in standards at the higher lift heights. Aluminium tube is weaker than any of the steel tubes at the higher lift heights.

Table 31.1. Section properties of various scaffold tubes when new (48.3 mm outside diameter)

Property	BS 1139 Steel	EN 39 Steel	EN 244 Steel	BS 1139 Aluminium
Wall thickness				
SWG	8	10	8	7
mm	4.0	3.2	4.0	4.47
in	0.16	0.126	0.16	0.176
Weight				
kg/m	4.405	3.58	4.37	1.674
lb/ft	2.96	2.41	2.94	1.12
Cross-sectional area				
cm²	5.67	4.60	5.57	6.15
in²	0.878	0.715	0.863	0.953
Second moment of area				
cm⁴	14.03	11.82	13.80	14.90
in⁴	0.337	0.284	0.331	0.358
Modulus of section				
cm³	5.78	4.88	5.70	6.18
in³	0.354	0.208	0.349	0.377
Radius of gyration				
cm	1.57	1.600	1.57	1.56
in	0.620	0.630	0.620	0.614
Yield stress				
N/mm²	210	235	235	255 (0.2% PS)
T/in²	13.6	15.2	15.2	16.5
Max. allowable stress in compression				
N/mm²	127	155	155	139
T/in²	8.22	10.0	10.0	9.0
Max. allowable stress in bending				
N/mm²	137	165	165	154
T/in²	8.87	10.68	10.68	9.97

Table 31.2. Other specified properties of various scaffold tubes when new

Property	BS 1139 Steel	EN 39 Steel	EN 244 Steel	BS 1139 Aluminium
		Tube type		
Tolerance on outside diameter	±0.5 mm	±0.5 mm	±0.5 mm	±0.5 mm
Tolerance on mass per unit length	−7.5% on any one piece	±10% for any one piece	±10% for each piece	−7.5% on any one piece
	±4% on quantities over 150 m	±7.5% on batches of 10 tonnes	±7.5% on batches of 10 tonnes	±5% on batches over 150 m
Tolerance on length specified	±6 mm	—	—	±6 mm
Tolerance on wall thickness	+0.8 mm −0.4 mm	−12.5%	−12.5%	±0.56 mm
Straightness	$\frac{1}{600}$th of the length	Reasonably straight	Reasonably straight	$\frac{1}{600}$th of the length

The modulus of elasticity, E, can be taken as:

Steel   13 600 T/in^2, 3.05 × 10^7 lb/in^2,
            2.1 × 10^5 N/mm^2.
Aluminium   4500 T/in^2, 10^7 lb/in^2, 6.9 × 10^4 N/mm^2.

31.3 Tubes: after use

The words 'when new' require some explanation. They have been inserted in the Code because scaffolding structures are not made with new materials. They are built, used and dismantled, and the material stored and reused repeatedly.

A scaffold tube may be up to 50 years old, i.e. it may have had between 1 and 500 uses.

Its condition at the time of use, therefore, is not covered by the properties set down in the standard specifications. This can be taken into account in two ways. A safety coefficient is proposed in Section 31.21 to limit the height to which a scaffold can be built when calculated on the allowable stresses listed in Tables 31.9 to 31.12, or a reduced axial load can be used for the tubes.

Used tubes should be straightened if they have been bent by more than 15 mm in the full length.

The wall thickness should not have been reduced by rusting to a degree whereby the bending strength of the tube is less than 85% of that originally specified. If tubes are to be loaded to their allowable maximum in compression, the ends should not have been significantly reduced in thickness on the outside by wear and tear in handling or by contact with sleeve couplers, or on the inside by contact

with internal spigots. A loss of 15% of the cross-sectional area of the tube metal is a fair limit to apply.

The surface of tubes can be untreated, painted or galvanised. The latter is preferred, but there is no objection to the use of untreated tubes, provided that the consequent rust stains will not cause damage.

Pavement frames, footbridges and walkways should not be made with untreated tubes because the public will inevitably come into contact with them, and in damp conditions, clothes and hands will be stained with rust.

31.4 Corrosion and environmental considerations in relation to steel tubes

In addition to the three surface treatments referred to above, i.e. untreated, painted and galvanised, there are three types of environment to be considered: marine, inland and areas with high levels of chemical pollutants.

In the normal course of events, untreated steel scaffold tubes rust initially and the rust is scoured off by repeated use, stacking and restacking, and constant handling. Deterioration due to rusting slows down with time and tubes rarely reach a condition where they have to be discarded because of rusting. The loss of weight and of sectional area is not significant.

Exceptions to this general rule are:

(i) When tubes have been left stacked and undisturbed for a long time and water has been retained along the contact lines between the tubes.

(ii) When the structure has been exposed to the atmosphere in chemically polluted surroundings for a long time, say 10 years, without treatment. In this case, corrosion may occur where rainwater cannot run off, e.g. at intersections and at couplers.

In marine environments, there is usually no trouble with untreated steel tubes in constant use and reuse on land. If they are over the sea, they should be checked at six-monthly intervals in the expectation that they will need replacing after two or three years. Tubes under water, as in a temporary jetty, a repair job for a sea wall or a drilling tower for a bridge abutment soil survey, should be inspected every month. In clean sea water, there will be little trouble during most jobs lasting up to six months.

In tidal estuaries where there is a large silt and sand content in the water, the products of corrosion are abraded by the solid content of the tidal stream, and the life of the tubes may be reduced to a few months.

In areas where there are chemical works, or coal and gas plants, and where there are known corrosive elements in the air, tubes should be inspected every year — or more frequently in known extreme cases.

Galvanised tubes rarely give any trouble in any of these circumstances, but they must be monitored closely in the tidal case referred to above.

The author has no records of electrochemical deterioration of tubing or couplers due to differences between the metals.

Rusting does not usually appear uniformly over the surface of a tube; it creates small pits. In serious cases, these pits are about 5 mm in diameter and 1 mm deep, and they occur in close contact. This results in a loss of the outer skin metal to an average depth of 0.5 mm.

The loss in cross-sectional area is about 24 mm^2 on an original value of 567 mm^2 for BS 1139 tube, i.e. a loss of 4.2%. With a wall thickness of 3.2 mm, the loss is 5%.

The loss in modulus of section is about 14%, and the bending strength is also reduced by 14%.

It will be evident from this that the recommended reduction of 15% from the strength of new steel will cater for the corrosion described above.

31.5 Hazards with steel tubes

- The worst case in the author's records was for 4 mm wall thickness unprotected steel tube in a drilling tower at the edge of the Firth of Forth, in the narrowest section where the tide was fast and the silt content high. The tube deteriorated to the danger point in three months. On the other hand, 4.9 mm wall thickness unprotected steel tube driven into the beach off Cromer 44 years previously was found to be in good condition. The behaviour is thus unpredictable and must be constantly monitored.
- Steel tube at very low temperatures in refrigeration plants and in ground-freezing processes has fractured due to impact, e.g. falling equipment, but it has carried its normal load satisfactorily.
- In petroleum and inflammable chemical environments, steel tube may spark when dragged across flint, gravel or concrete floors.

    In underground coal-mining works, advice must be taken about steel and flint sparking.

31.6 Corrosion of aluminium tubes

Aluminium tubes rarely give any trouble in normal circumstances, but they should not be used in chemical environments or under the sea without specialist metallurgical advice.

They should also not be used in mines without specialist advice.

The durability of aluminium scaffold tube is satisfactory in normal use. It corrodes more rapidly when new but at a slower rate with increasing time, provided that there are no corrosion accelerating conditions.

Table 31.3. Susceptibility to corrosion of various metals

Metal	Electrical potential (negative)
Galvanised iron	1.14
Zinc	1.05
Aluminium	0.9−0.8
Mild steel and cast iron	0.78
Lead	0.55
Tin	0.5
Stainless steels	0.2−0.4
Brass	0.33
Copper	0.22

In marine locations, the products of corrosion are brushed away by the water flow and particularly by suspended solids in the water. Corrosion is thus accelerated. In heavily silt-laden and fast-flowing estuaries, tube life may be reduced to a few months.

Exposed to fire, aluminium tube is far inferior to steel and may disintegrate in minutes, depending on the severity of the fire. Steel tubes will not disintegrate at such a rate unless furnace temperatures are reached. In contact with other metals in dry conditions, there is no electrolytic or bi-metallic action, but there is in damp conditions. This is accelerated either by continuous washing away of the products of corrosion or by the trapping of water between or in contact with the two metals so that it does not escape but becomes more concentrated in its chemical composition.

To check the various building metals in order of their susceptibility to corrosion when paired, Table 31.3 is useful.

Metals near the top of Table 31.3 corrode when paired with one lower down. Corrosion is generally more robust towards the top of the table and when the difference in negative potential is greater.

Thus if galvanised mild steel is used in contact with aluminium, the zinc will corrode slowly. The zinc coating on the mild steel will thus deteriorate and, if it becomes porous or broken, the aluminium will then corrode under the influence of the mild steel.

Brass and copper also cause deterioration of aluminium.

From this, it will be seen that there is no safe pairing of any other metal with aluminium for a long-term project in a damp environment. In this case, protection should be applied.

It is usual to apply the protection to the aluminium element of the pairing. It can be by insulation using waterproof fibre washers and sleeves, or by painting with various chromatic or zinc chromate varnishes or paints. Bituminous paint applied in several coats is frequently satisfactory.

31.7 Hazards with aluminium tubes

- When using aluminium tube, problems arise through failure to remember that its deflection under load is three times greater than that of steel tubes under the same load on the same span.
- The flexibility of an aluminium scaffold is greater than that of a steel structure.
- Where mass is important, aluminium tubes are only one-third as heavy as steel tubes, but resistance to wind overturning is reduced and extra counterweights must be used.
- In aluminium smelters, munitions factories and storage sheds, and other places where explosive dust or vapour may be in the air, the sparking potential must be considered. For adverse circumstances, wooden scaffolds should be considered, using poles and fibre rope lashings in the pre-1920 style, bearing in mind the hazard of sparks from the use of hammers and nails.
- Where there is a fire risk, aluminium tube should not be used if the platforms are to be used as escape gantries.

31.8 Scaffold tubes as uprights and struts

There are two types of strut to consider:

- Axially loaded.
- Eccentrically loaded.

It is sometimes contended that because the uprights in a scaffold receive their load from ledgers, which are coupled to the side of the upright tube, the uprights are eccentrically loaded. Therefore, the normal strut tables for tubes cannot be applied.

This does not appear to make much difference in practice. There are very few cases of properly braced and tied access scaffolds failing by buckling of the uprights due to overload alone, without there being some other cause of the failure.

This is because a transom spanning across the two ledgers of a scaffold acts as a beam, with some rigidity at the ends achieved by the rotational strength of the grip of the fittings on the ledger. This moment relieves the bending in the upright at the position of the ledger to upright coupler.

Figure 31.1 demonstrates this. Figure 31.1(a) shows a true cantilever with a root moment of 0.5 wL, which when distributed between the upper and lower parts of the upright creates a moment of 0.25 wL in each part. Figure 31.1(b) shows the end of the cantilever rotated back to the horizontal by end propping. For this condition, the fixed end moment is 0.125 wL. Distributing this for the normal proportions of a scaffold gives 0.0444 wL in the upper and lower parts of the upright. Figure 31.1(c) shows the normal scaffold transom with a fixed end moment of 0.0833 wL, which

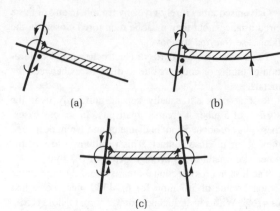

Figure 31.1 The bending moments in the uprights from cantilevers and transoms.

distributed gives 0.0296 wL in the upper and lower parts of the standard.

This last value of 0.0296 wL is only about one-tenth of the moment in the upright caused by a transom as a cantilever.

If the transom is not coupled to the two ledgers but just rests on them without the end stability afforded by the couplers, the moment applied to the uprights is about 0.021 wL, derived from the 50 mm offset of the ledger from the upright. This distributes to 0.01 wL in the upper and lower parts of the uprights. This value is also very small.

After dealing with axially loaded tubes in the following section, eccentrically loaded struts will be considered in Section 31.10.

The following sections give the section properties and load-carrying capacity of uprights and axially loaded struts for four types of new scaffold tube and for the same four tubes de-rated to a typical used condition.

31.9 Load tables for uprights and axially loaded struts

Eight load tables (Tables 31.5−31.12) are relevant to axially loaded uprights or struts in scaffolding.

Engineers wishing to use other maximum allowable compressive stresses should ensure that all the pieces of tube in the structure have the appropriate strengths.

For stresses other than those given, the permissible axial load can be interpolated or extrapolated from the tables given.

The values for used tubes are about 15% less than those for new tubes. It is recommended that calculations for all scaffold standards and struts be based on the used material tables. The 15% will make an allowance for joints in the tube, eccentricities of load, corrosion, badly seating ends to uprights and slight irregularities in the load applied.

Table 31.4. Load tables for uprights and axially loaded struts

Table	Material		Wall thickness	Maximum allowable compressive stress (N/mm^2)
31.5	BS 1139	New steel	4 mm	127
31.6	EN 39	New steel	3.2 mm	155
31.7	EN 244	New steel	4 mm	155
31.8	BS 1139	New aluminium	4.47 mm	139
31.9	BS 1139	Used steel	4 mm	108
31.10	EN 39	Used steel	3.2 mm	132
31.11	EN 244	Used steel	4 mm	132
31.12	BS 1139	Used aluminium	4.47 mm	118

Table 31.5. Permissible axial stresses and loads for new steel, 4 mm wall, scaffold tubes manufactured in accordance with BS 1139 Part 1, 1982

Effective length, l (mm)	Slenderness ratio, l/r	Permissible axial compressive stress Pc (N/mm^2)	Permissible axial load		
			kN	kg	tonnes
0	0	127	70.7	7207	7.21
250	15.9	123	68.5	6983	6.98
500	31.8	119	66.2	6748	6.75
750	47.8	113	63.0	6422	6.42
1000	63.7	104	57.7	5882	5.88
1250	79.6	90.3	50.3	5127	5.13
1500	95.5	75.4	42.0	4281	4.28
1750	111.5	61.4	34.2	3486	3.49
2000	127.4	50.0	27.9	2844	2.84
2250	143.3	40.9	22.8	2324	2.32
2500	159.2	34.0	18.9	1927	1.93
2750	175.2	28.7	16.0	1631	1.63
3000	191.1	24.2	13.5	1376	1.38
3250	207.0	20.9	11.6	1182	1.18
3500	222.9	18.1	10.1	1030	1.03
3750	238.8	15.9	8.8	897	0.90
4000	254.8	14.1	7.9	805	0.81
4250	270.7	12.5	6.9	703	0.70
4500	286.6	11.2	6.2	632	0.63
4750	302.5	10.1	5.6	571	0.57
5000	318.5	9.1	5.1	520	0.52
5250	334.4	8.2	4.6	469	0.47
5500	350.3	7.5	4.2	428	0.43
5750	366.2	6.9	3.9	398	0.40
6000	382.2	6.4	3.5	357	0.36

Table 31.6. Permissible axial stresses and loads for new steel, 3.2 mm wall, scaffold tubes manufactured in accordance with EN 39

Effective length, l (mm)	Slenderness ratio, l/r	Permissible axial stress pc (N/mm^2)	Permissible axial load		
			kN	kg	tonnes
0	0	155	70.2	7156	7.16
250	15.6	151	68.4	6972	6.97
500	31.3	146	66.1	6738	6.74
750	46.9	138	62.5	6371	6.78
1000	62.5	126	57.1	5821	5.82
1250	78.1	112	50.7	5168	5.17
1500	93.8	88.4	40.0	4077	4.08
1750	109.4	66.8	30.3	3089	3.09
2000	125.0	55.3	25.1	2559	2.56
2250	140.6	47.5	21.5	2192	2.19
2500	156.3	37.6	17.0	1733	1.73
2750	171.9	31.2	14.1	1437	1.44
3000	187.5	26.5	12.0	1223	1.22
3250	203.1	23.2	10.5	1070	1.07
3500	218.8	19.9	9.0	917	0.92
3750	234.4	17.0	7.7	785	0.79
4000	250.0	15.2	6.9	703	0.70
4250	265.6	13.8	6.3	642	0.64
4500	281.3	12.2	5.5	561	0.56
4750	296.9	11.2	5.1	520	0.52
5000	312.5	9.9	4.5	459	0.46
5250	328.1	9.1	4.1	418	0.42
5500	343.8	8.2	3.7	377	0.38
5750	359.4	7.6	3.4	347	0.35
6000	375.0	5.2	2.4	245	0.25

The table for used aluminium tubes is not to take into account corrosion, which in scaffolding is generally insignificant, but to take into account factors such as reduced mass inertia, increased eccentricities due to the higher flexibility of aluminium tube and increased damage due to misuse.

31.10 Scaffold tubes eccentrically loaded

It was explained in Section 31.8 that the transoms coupled across two ledgers do not cause the uprights to be loaded with an eccentricity of about 50 mm, i.e. the distance apart of the centre lines of the ledgers and the uprights. This is because the couplers joining the ledgers to the uprights and the transoms to the ledgers together form a rigid joint, as shown in Figure 31.2(a). The only moment transferred to the upright is from the change of slope of the end of the relatively short and stiff transom, as shown in Figure 31.2(b).

However, if there is no coupler on the end of the transom to restrain the rotation of the ledger, there will be a 50 mm eccentricity in the load applied to the upright. This occurs in some prefabricated scaffold systems that have loose transoms or transoms with one end loose. It also occurs when the transoms are replaced by timber battens in a pedestrian walkway or rostrum.

Table 31.7. Permissible axial stresses and loads for new steel, 4 mm wall, scaffold tubes manufactured in accordance with EN 244

Effective length, l (mm)	Slenderness ratio, l/r	Permissible axial compressive stress Pc (N/mm²)	Permissible axial load		
			kN	kg	tonnes
0	0	155	86.3	8803	8.80
250	15.9	150	83.6	8527	8.53
500	31.8	145	80.8	8242	8.24
700	47.8	137	76.3	7783	7.78
1000	63.7	125	69.6	7099	7.10
1250	79.6	111	61.8	6304	6.30
1500	95.5	87.4	48.7	4967	4.97
1750	111.5	65.7	36.6	3733	3.73
2000	127.4	54.8	30.5	3111	3.11
2250	143.3	46.5	25.9	2642	2.64
2500	159.2	37.0	20.6	2101	2.10
2750	175.2	30.3	16.9	1724	1.72
3000	191.1	25.0	13.9	1418	1.42
3250	207.0	22.1	12.3	1255	1.26
3500	222.9	18.7	10.4	1061	1.06
3750	238.8	16.5	9.2	938	0.94
4000	254.8	14.3	8.0	816	0.82
4250	270.7	13.2	7.4	755	0.76
4500	286.6	12.0	6.7	683	0.68
4750	302.5	11.5	6.4	653	0.65
5000	318.5	11.0	6.1	622	0.62
5250	334.4	8.9	5.0	510	0.51
5500	350.3	7.8	4.3	439	0.44
5750	366.2	7.1	4.0	408	0.41
6000	382.2	6.6	3.7	377	0.38

Table 31.8. Permissible axial stresses and loads for new aluminium, 4.47 mm wall, scaffold tubes manufactured in accordance with BS 1139, Part 1, 1982

Effective length, l (mm)	Slenderness ratio, l/r	Permissible axial stress Pc (N/mm²)	Permissible axial load		
			kN	kg	tonnes
0	0	139	84.4	8603	8.60
250	15.9	132	80.1	8165	8.16
500	31.8	118	71.6	7299	7.30
750	47.8	103	62.5	6371	6.37
1000	63.7	82	49.8	5076	5.08
1250	79.6	54	32.8	3344	3.34
1500	95.5	38	23.1	2355	2.36
1750	111.5	27.5	16.7	1702	1.70
2000	127.4	21.0	12.7	1295	1.30
2250	143.3	17.0	10.30	1050	1.05
2500	159.2	13.4	8.16	832	0.83
2750	175.2	11.12	6.75	688	0.69
3000	191.1	9.24	5.61	572	0.57
3250	207.0	7.96	4.83	492	0.49
3500	222.9	6.77	4.11	419	0.42
3750	238.8	5.98	3.63	370	0.37
4000	254.8	5.24	3.18	324	0.32
4250	270.7	4.66	2.83	288	0.29
4500	286.6	4.15	2.52	257	0.26
4750	302.5	3.72	2.26	230	0.23
5000	318.5	3.36	2.04	208	0.21
5250	334.4	3.05	1.85	189	0.19
5500	350.3	2.80	1.70	173	0.17
5750	366.2	2.55	1.55	158	0.16
6000	382.2	2.31	1.40	143	0.14

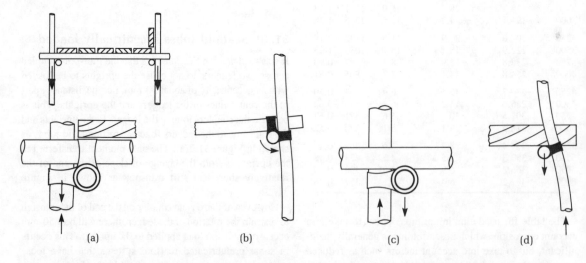

Figure 31.2 Axial and eccentric loads in uprights. The standards in (a) and (b) can be considered axially loaded when the transoms are coupled to the ledger. The standards in (c) and (d) are eccentrically loaded because there is no coupling between the horizontal tube or board and the ledger.

Table 31.9. Proposed axial stresses and loads for used steel, 4 mm wall, scaffold tubes manufactured in accordance with BS 1139 Part 1, 1982

Effective length, l (mm)	Slenderness ratio, l/r	Proposed axial compressive stress Pc (N/mm^2)	Proposed axial load		
			kN	kg	tonnes
0	0	108	60.1	6126	6.12
250	15.9	105	58.2	5933	5.93
500	31.8	101	56.3	5739	5.74
750	47.8	96.2	53.6	5464	5.46
1000	63.7	88.1	49.1	5005	5.01
1250	79.6	76.8	42.8	4363	4.36
1500	95.5	64.1	35.7	3639	3.64
1750	111.5	52.2	29.1	2966	2.97
2000	127.4	42.5	23.7	2416	2.42
2250	143.3	34.8	19.4	1978	1.98
2500	159.2	28.9	16.1	1641	1.64
2750	175.2	24.4	13.6	1386	1.39
3000	191.1	20.6	11.5	1172	1.17
3250	207.0	17.8	9.9	1009	1.01
3500	222.9	15.4	8.6	877	0.88
3750	238.8	13.5	7.5	764	0.76
4000	254.8	12.0	6.7	683	0.68
4250	270.7	10.6	5.9	601	0.60
4500	286.6	9.5	5.3	540	0.54
4750	302.5	8.6	4.8	489	0.49
5000	318.5	7.7	4.3	438	0.44
5250	334.4	7.0	3.9	398	0.40
5500	350.3	6.4	3.6	367	0.37
5750	366.2	5.9	3.3	336	0.34
6000	382.2	5.4	3.0	306	0.31

Table 31.10. Proposed axial stresses and loads for used steel, 3.2 mm wall, scaffold tubes manufactured in accordance with EN 39

Effective length, l (mm)	Slenderness ratio, l/r	Proposed axial stress Pc (N/mm^2)	Proposed axial load		
			kN	kg	tonnes
0	0	132	54.7	6083	6.08
250	15.6	128	58.1	5926	5.93
500	31.3	124	56.2	5727	5.73
750	46.9	117	53.1	5415	5.42
1000	62.5	107	48.5	4948	4.95
1250	78.1	66.4	43.1	4393	4.39
1500	93.8	75.1	34.0	3465	3.47
1750	109.4	56.8	25.8	2626	2.63
2000	125.0	47.0	21.3	2175	2.18
2250	140.6	40.4	18.4	1863	1.86
2500	156.3	31.9	14.5	1473	1.47
2750	171.9	26.5	12.0	1221	1.22
3000	187.5	22.5	10.2	1040	1.04
3250	203.1	20.0	8.93	910	0.91
3500	218.8	16.9	7.65	779	0.78
3750	234.4	14.5	6.55	667	0.67
4000	250.0	12.9	5.87	598	0.60
4250	265.6	11.7	5.36	546	0.55
4500	281.3	10.4	4.68	477	0.48
4750	296.9	9.5	4.34	442	0.44
5000	312.5	8.4	3.83	390	0.39
5250	328.1	7.7	3.49	355	0.36
5500	343.8	7.0	3.16	320	0.32
5750	359.4	6.5	2.89	295	0.30
6000	375.0	4.4	2.04	208	0.21

When there is this 50 mm eccentricity of applied load to the standards, the allowable loads in the standards are very much reduced.

The bending stress due to the eccentricity of 50 mm must be added to the axial compression stress. The combined stress evaluations should be such that $\dfrac{f_c}{P_c} + \dfrac{f_{bc}}{P_{bc}}$ does not exceed 1, or $\left(\dfrac{1}{AP_c} + \dfrac{50}{ZP_{bc}}\right)$ does not exceed $\dfrac{1}{W_e}$.

Where f_c = the applied axial average stress = $\dfrac{W_e}{A}$.

A = the cross-section area.

W_e = the allowable eccentric load.

f_{bc} = the applied bending stress = $\dfrac{50W_e}{Z}$.

Z = the section modulus of the tube.

P_c = the allowable stress in compression.

P_{bc} = the allowable stress in bending.

Tables 31.13–31.16 give the strut values for the load applied at an eccentricity of 50 mm. Other cases can be worked out from the formulas given above.

31.11 Hazard with strut loading

- Tables 31.13 to 31.16 give the load capacity of scaffold tube struts in various materials and loaded in various ways. It should be noted that the range of values is from a maximum of about 7 tonnes to a minimum of 0.1 tonnes. These loads have to be transferred into the tubes by couplers with a safe working load of only 0.625 tonnes.

  Failure to recognise the great reduction in strength due to increased length and the limits imposed by the couplers has led to accidents. Inspectors who have not carried out the drawing office calculations for the structure they are inspecting may not be advised of the loads being dealt with and, in consequence, will not check the struts as rigorously as they should.

Table 31.11. Proposed axial stresses and loads for used steel, 4 mm wall, scaffold tubes manufactured in accordance with EN 244

Effective length, *l* (mm)	Slenderness ratio, *l/r*	Proposed axial stress *Pc* (N/mm²)	Proposed axial load		
			kN	kg	tonnes
0	0	132	73.4	7482	7.48
250	15.9	128	71.1	7248	7.25
500	31.8	123	68.7	7005	7.01
750	47.8	116	64.9	6615	6.62
1000	63.7	106	59.2	6034	6.03
1250	79.6	94.4	52.5	5358	5.36
1500	95.5	88.3	41.4	4222	4.22
1750	111.5	55.8	31.1	3173	3.17
2000	127.4	46.6	25.9	2644	2.64
2250	143.3	39.5	22.0	2246	2.25
2500	159.2	31.5	17.5	1786	1.79
2750	175.2	27.8	14.4	1465	1.47
3000	191.1	21.3	11.8	1205	1.21
3250	207.0	18.8	10.5	1066	1.07
3500	222.9	15.9	8.8	902	0.90
3750	238.8	14.0	7.8	798	0.80
4000	254.8	12.2	6.8	694	0.69
4250	270.7	11.2	6.3	642	0.64
4500	286.6	10.2	5.7	581	0.58
4750	302.5	9.8	5.4	555	0.56
5000	318.5	9.4	5.2	529	0.53
5250	334.4	7.6	4.3	434	0.43
5500	350.3	6.6	3.7	373	0.37
5750	366.2	6.0	3.4	347	0.35
6000	382.2	5.6	3.1	321	0.32

Table 31.12. Proposed axial stresses and loads for used aluminium, 4.47 mm wall, scaffold tubes manufactured in accordance with BS 1139 Part 1, 1982

Effective length, *l* (mm)	Slenderness ratio, *l/r*	Proposed axial stress *Pc* (N/mm²)	Proposed axial load		
			kN	kg	tonnes
0	0	118	71.7	7313	7.31
250	15.9	112	68.1	6940	6.94
500	31.8	100	60.9	6204	6.20
750	47.8	88	53.1	5415	5.42
1000	63.7	70	42.3	4315	4.32
1250	79.6	45.9	27.9	2842	2.84
1500	95.5	32.3	19.6	2002	2.00
1750	111.5	23.4	14.2	1447	1.45
2000	127.4	17.9	10.8	1101	1.10
2250	143.3	14.45	8.76	893	0.89
2500	159.2	11.39	6.94	707	0.71
2750	175.2	9.45	5.74	585	0.59
3000	191.1	7.85	4.77	486	0.49
3250	207.0	6.77	4.11	418	0.42
3500	222.9	5.75	3.49	356	0.36
3750	238.8	5.08	3.09	315	0.32
4000	254.8	4.45	2.70	275	0.28
4250	270.7	3.96	2.41	245	0.25
4500	286.6	3.53	2.14	218	0.22
4750	302.5	3.16	1.92	196	0.20
5000	318.5	2.86	1.73	177	0.18
5250	334.4	2.59	1.57	161	0.16
5500	350.3	2.38	1.45	147	0.15
5750	366.2	2.17	1.32	134	0.13
6000	382.2	1.96	1.19	122	0.12

31.12 Scaffold tubes as braces

The stabilising force necessary to maintain a rectangular structure in its original square profile is quite small if the structure is built truly. If a scaffold is built off plumb by 20 mm in a lift height of 2 m, with all its uprights inclined the same way, the horizontal force it exerts is only about 1% of the vertical load.

BS 449 suggests a figure of 2.5% for permanent structural steelwork made and erected with fabrication shop accuracy. This is not sufficient for scaffolding for either access or support work.

After hearing evidence from all parts of the world that lateral instability is the cause of most temporary work failures, the Bragg Committee recommended an increase from 2.5 to 3% for structures that are correctly designed, and built with a high degree of supervision and checking.

For ordinary scaffolding work, a horizontal restraint against lateral instability of 5% is appropriate. In practical terms, this means one 45° diagonal brace for every 16 standards in a free-standing structure. The horizontal force derived from a 45° brace is about 70% of the inclined force.

In the case of a scaffold attached to a building, it is assumed that the ties provide the lateral stability for the inside line of standards, leaving the longitudinal brace to support the outside line. The traditional rule is one longitudinal brace for every 30 m of scaffold on the outside, which corresponds to one brace for 14 or 16 bays.

The brace is a longer tube than either the upright or the ledger in the panel that it braces. This must be taken into account in calculations and particularly in checking structures on site, where the brace crosses two bays and two lifts without a coupler at the mid-point.

When the brace is attached to the horizontal tubes by right-angle couplers, the joint shown at the top right-hand corner of Figure 31.3(a) is comparable to that shown in Figure 31.2(a), i.e. the brace is axially loaded because the rotational strength of the coupler compensates for the offset of 50 mm. Tables 31.5–31.12 can be used to give the strength of compression braces coupled this way. In tension, the number of couplers will govern the strength.

It is sometimes contended that a brace coupled to the same side of the two horizontals, as in Figure 31.3(a), is weaker

Table 31.13. Permissible loads applied parallel to the axis of a strut at a distance of 50 mm from the axis on the same side at each end. New and used steel, 4 mm wall thickness, manufactured in accordance with BS 1139 Part 1, 1982

Strut length (mm)	New tube Permissible load offset 50 mm			Used tube Permissible load offset 50 mm		
	kN	kg	tonnes	kN	kg	tonnes
0	12.66	1291	1.29	10.76	1097	1.10
250	12.60	1285	1.29	10.71	1092	1.09
500	12.50	1275	1.28	10.63	1084	1.08
750	12.34	1259	1.26	10.49	1070	1.07
1000	12.20	1244	1.24	10.37	1057	1.06
1250	11.76	1200	1.20	10.00	1020	1.02
1500	10.98	1120	1.12	9.33	952	0.95
1750	10.64	1085	1.09	9.04	922	0.92
2000	9.90	1010	1.01	8.42	859	0.86
2250	9.17	935	0.94	7.79	795	0.80
2500	8.40	857	0.86	7.14	728	0.73
2750	7.57	772	0.77	6.43	656	0.66
3000	6.90	704	0.70	5.87	598	0.60
3250	6.25	638	0.64	5.31	542	0.54
3500	5.68	579	0.58	4.83	492	0.49
3750	5.10	520	0.52	4.34	442	0.44
4000	4.72	481	0.48	4.01	409	0.41
4250	4.26	435	0.44	3.62	370	0.37
4500	3.91	399	0.40	3.32	339	0.34
4750	3.58	365	0.37	3.04	310	0.31
5000	3.28	335	0.34	2.79	285	0.29
5250	2.99	305	0.31	2.54	259	0.26
5500	2.76	282	0.28	2.35	240	0.24
5750	2.55	260	0.26	2.17	221	0.22
6000	2.36	241	0.24	2.01	205	0.21

Table 31.14. Permissible loads applied parallel to the axis of a strut at a distance of 50 mm from the axis on the same side at each end. New and used steel, 3.2 mm wall thickness, manufactured in accordance with EN 39

Strut length (mm)	New tube Permissible load offset 50 mm			Used tube Permissible load offset 50 mm		
	kN	kg	tonnes	kN	kg	tonnes
0	11.76	1200	1.20	10.00	1020	1.02
250	11.70	1193	1.19	9.95	1014	1.01
500	11.63	1186	1.18	9.89	1008	1.01
750	11.49	1172	1.17	9.77	996	1.00
1000	11.36	1159	1.16	9.66	985	0.99
1250	10.99	1121	1.12	9.34	953	0.95
1500	10.75	1097	1.10	9.14	932	0.93
1750	9.90	1010	1.01	8.42	859	0.86
2000	9.17	935	0.94	7.79	795	0.80
2250	8.58	875	0.88	7.29	744	0.75
2500	7.81	797	0.80	6.64	677	0.68
2750	7.04	718	0.72	5.98	519	0.52
3000	6.37	650	0.65	5.41	553	0.55
3250	5.81	593	0.59	4.94	504	0.50
3500	5.26	537	0.54	4.47	456	0.46
3750	4.78	488	0.49	4.06	415	0.42
4000	4.35	444	0.44	3.70	377	0.38
4250	3.94	402	0.40	3.35	342	0.34
4500	3.60	367	0.37	3.06	312	0.31
4750	3.30	337	0.38	2.81	286	0.29
5000	3.02	308	0.31	2.57	262	0.26
5250	2.76	282	0.28	2.35	240	0.24
5500	2.54	259	0.26	2.16	220	0.22
5750	2.35	240	0.24	2.00	204	0.20
6000	2.17	221	0.21	1.84	188	0.19

than that coupled to opposite sides, as in Figure 31.3(b). A close study of the length in bending will show it to be the same in both cases, so the strength will be the same.

When the brace is attached to the uprights with swivel couplers, it must be considered as pin-jointed at each end, and because of the tolerance in the swivel pins, it must be considered as eccentrically loaded. Tables 31.13–31.16 can be used to give its compression strength.

31.13 Scaffold tubes used as tension members

Where a purely tension member is required, the strength of the tube is much greater than that of the couplers at each end.

The various types of tube have the safe tension values set out in Table 31.17 when new.

In practice, the limitation of the load depends on the end connections. Engineering bolts can be used at each end. Welding is sometimes used, in which case its effect on the tensile strength of the material must be taken into account.

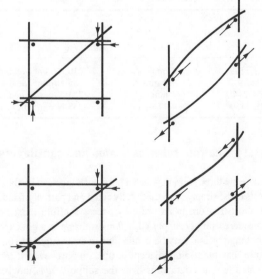

Figure 31.3 The length in bending of braces.

Table 31.15. Permissible loads applied parallel to the axis of a strut at a distance of 50 mm from the axis on the same side at each end. New and used steel, 4 mm wall thickness, manufactured in accordance with EN 244

Strut length (mm)	New tube Permissible load offset 50 mm			Used tube Permissible load offset 50 mm		
	kN	kg	tonnes	kN	kg	tonnes
0	15.45	1576	1.58	13.09	1335	1.34
250	15.35	1566	1.57	13.05	1331	1.33
500	15.26	1557	1.56	12.97	1323	1.32
750	15.09	1539	1.54	12.83	1308	1.31
1000	14.81	1511	1.51	12.59	1284	1.28
1250	14.42	1471	1.47	12.26	1250	1.25
1500	13.57	1384	1.38	11.53	1177	1.18
1750	12.43	1268	1.27	10.57	1078	1.08
2000	11.64	1187	1.19	9.89	1009	1.01
2250	10.90	1112	1.11	9.27	945	0.95
2500	9.84	1004	1.00	8.37	854	0.85
2750	8.93	911	0.91	7.59	774	0.77
3000	8.00	816	0.82	6.80	694	0.69
3250	7.44	759	0.76	6.32	645	0.65
3500	6.84	698	0.70	5.81	593	0.59
3750	6.18	630	0.63	5.25	536	0.54
4000	5.60	571	0.57	4.76	486	0.49
4250	5.29	540	0.54	4.50	459	0.46
4500	4.93	503	0.50	4.19	427	0.43
4750	4.78	488	0.49	4.06	414	0.41
5000	4.62	471	0.47	3.93	400	0.40
5250	3.92	400	0.40	3.33	340	0.34
5500	3.53	360	0.36	3.00	306	0.31
5750	3.27	334	0.33	2.78	284	0.28
6000	3.08	314	0.31	2.62	267	0.27

Table 31.16. Permissible loads applied parallel to the axis of a strut at a distance of 50 mm from the axis on the same side at each end. New and used aluminium, 4.47 mm wall thickness, manufactured in accordance with BS 1139 Part 1, 1982

Strut length (mm)	New tube Permissible load offset 50 mm			Used tube Permissible load offset 50 mm		
	kN	kg	tonnes	kN	kg	tonnes
0	14.93	1523	1.52	12.69	1295	1.30
250	14.71	1500	1.50	12.50	1275	1.28
500	14.49	1478	1.48	12.32	1256	1.26
750	14.08	1436	1.44	11.97	1221	1.22
1000	13.33	1360	1.36	11.33	1156	1.16
1250	11.76	1200	1.20	10.00	1020	1.02
1500	10.70	1091	1.09	9.10	927	0.93
1750	8.77	895	0.90	7.45	761	0.76
2000	7.58	773	0.77	6.44	657	0.66
2250	6.62	675	0.68	5.63	574	0.57
2500	5.68	579	0.58	4.83	492	0.49
2750	4.85	495	0.50	4.12	421	0.42
3000	4.24	432	0.43	3.60	367	0.37
3250	3.76	384	0.38	3.20	326	0.33
3500	3.28	335	0.34	2.79	285	0.29
3750	2.94	300	0.30	2.50	255	0.26
4000	2.62	267	0.27	2.23	227	0.23
4250	2.35	240	0.24	2.00	204	0.20
4500	2.12	216	0.22	1.80	184	0.18
4750	1.92	196	0.20	1.63	167	0.17
5000	1.74	177	0.18	1.48	150	0.15
5250	1.63	166	0.17	1.39	141	0.14
5500	1.46	149	0.15	1.24	127	0.13
5750	1.33	136	0.14	1.13	116	0.12
6000	1.21	123	0.12	1.03	105	0.11

Table 31.17. Tension strength in scaffolding tubes

	Tube type		Safe tension value (tonnes)	
			New tube	Used tube
BS 1139	4 mm	210 N/mm^2 yield steel	7.3	6.2
EN 39	3.2 mm	235 N/mm^2 yield steel	7.1	6.0
EN 244	4 mm	235 N/mm^2 yield steel	7.3	6.2
BS 1139	4.47 mm	255 N/mm^2 0.2% proof stress aluminium	8.7	7.4

31.14 Scaffold tubes as beams and cantilevers

There are three ways in which a horizontal tube can be fixed at its ends; simply supported, fully fixed and partially fixed. A short transom in an ordinary access scaffold must be considered simply supported. A long transom in a birdcage or slung scaffold may be continuous over its supports, but there may be a joint adjacent to one support, even if the jointed tube is continuous over the support. It is unwise to assume that any horizontal tube can ever have a fully fixed end. A partial end fixing must be assumed.

A ledger is sometimes simply supported in a short scaffold or a tower, or it may be a continuous tube, but it should only be considered as partially fixed at the end.

These two end conditions affect the bending stresses in the tube. The manner in which the load is applied also affects the bending stresses. The least severe effect is from a uniformly distributed load. The most severe is from a centre point load. Between these extremes are conditions such as one-third or quarter point loads.

Table 31.18. Permissible total loads that can be applied on various spans of scaffold tubes used as beams. Values are given in kg. Divide by 102 to give kN. Divide by 1000 to give tonnes.

Material type and allowable stress	End fixing	System of loading	Span											
			(mm)						(m)					
			500	750	1000	1250	1500	1750	2000	2500	3000	4000	5000	6000
BS 1139 4 mm 137 N/mm² new steel	SS	CPL	646	431	323	259	215	185	162	129	108	81	65	54
		$\frac{1}{3}P$	969	646	485	388	323	277	242	194	162	232	97	81
		UDL	1292	862	646	518	430	370	324	258	216	162	130	108
	PF	CPL	1454	647	485	389	323	278	243	194	162	122	98	81
		$\frac{1}{3}P$	1211	808	606	485	404	246	303	243	203	290	121	101
		UDL	1615	1078	808	678	538	463	405	323	270	253	163	135
EN 34 3.2 mm 165 N/mm² new steel	SS	CPL	696	469	348	278	232	199	174	139	116	87	70	58
		$\frac{1}{3}P$	1044	696	522	418	348	298	261	209	174	131	104	87
		UDL	1392	928	696	556	464	398	348	278	232	174	140	116
	PF	CPL	1044	704	522	417	348	299	261	209	174	131	105	87
		$\frac{1}{3}P$	1305	870	653	523	435	373	326	261	218	164	130	109
		UDL	1740	1160	870	695	580	498	435	348	290	218	175	290
EN 244 4 mm 165 N/mm² new steel	SS	CPL	768	512	389	307	256	219	192	154	128	96	77	64
		$\frac{1}{3}P$	1152	768	576	461	384	329	288	230	192	144	115	96
		UDL	1533	1024	768	614	512	438	384	308	256	192	154	128
	PF	CPL	1152	768	584	461	384	329	288	231	192	144	116	96
		$\frac{1}{3}P$	1440	960	720	576	480	411	360	288	240	180	·144	120
		UDL	1916	1280	960	768	640	548	480	385	320	240	193	160
BS 1139 4.47 mm 154 N/mm² new aluminium	SS	CPL	777	518	388	311	259	222	194	155	129	97	78	65
		$\frac{1}{3}P$	1165	777	583·	466	388	332	291	233	194	146	117	97
		UDL	1554	1036	776	622	518	444	388	310	258	194	156	130
	PF	CPL	1166	777	582	467	389	333	291	233	194	146	117	98
		$\frac{1}{3}P$	1456	971	729	583	485	415	364	291	243	183	146	121
		UDL	1943	1295	970	778	648	555	485	388	323	243	195	163
4 mm 116 N/mm² used steel	SS	CPL	549	366	275	220	183	157	138	110	92	69	55	46
		$\frac{1}{3}P$	824	549	412	330	275	235	206	165	138	103	82	69
		UDL	1098	732	550	440	366	314	276	220	184	138	110	92
	PF	CPL	824	549	413	330	275	236	207	165	138	104	83	69
		$\frac{1}{3}P$	1030	686	515	413	344	294	258	206	173	129	103	86
		UDL	1373	915	688	550	458	393	345	275	230	173	138	115
3.2 mm 140 N/mm² used steel	SS	CPL	591	394	296	236	197	169	148	118	99	74	60	49
		$\frac{1}{3}P$	887	592	444	355	296	253	222	187	150	111	88	74
		UDL	1182	788	592	472	394	338	296	236	197	148	120	98
	PF	CPL	887	591	444	354	296	254	222	177	149	111	90	74
		$\frac{1}{3}P$	1109	740	555	444	370	316	278	223	188	139	110	93
		UDL	1478	985	740	590	493	423	370	295	245	185	150	123
4 mm 140 N/mm² used steel	SS	CPL	653	435	331	261	218	186	163	131	109	82	65	54
		$\frac{1}{3}P$	979	653	490	392	326	280	245	196	139	132	98	82
		UDL	1306	870	662	522	436	372	326	262	218	164	130	108
	PF	CPL	980	653	497	392	327	279	245	197	164	123	98	81
		$\frac{1}{3}P$	1224	816	613	490	408	350	306	245	174	165	123	102
		UDL	1633	1088	828	653	545	465	408	328	273	205	163	135
4.47 mm 131 N/mm² used aluminium	SS	CPL	660	440	330	264	220	189	165	132	110	82	66	55
		$\frac{1}{3}P$	990	660	496	396	330	282	247	198	165	124	99	82
		UDL	1320	880	660	528	440	378	330	264	220	164	132	110
	PF	CPL	990	660	495	396	330	284	248	198	165	123	99	83
		$\frac{1}{3}P$	1238	825	620	495	413	353	309	248	206	155	124	103
		UDL	1650	1100	825	660	550	473	413	330	275	205	165	138

Table 31.19. Permissible bending stresses and bending moments in scaffold tubes used as beams

Tube type			Permissible bending stresses (N/mm^2)	Section modulus (mm^3)	Permissible bending moments (kNm)
Material	Specification	Wall thickness (mm)			
New steel	BS 1139	4.0	137	5780	0.792
New steel	EN 39	3.2	165	4880	0.852
New steel	EN 244	4.0	165	5700	0.941
New aluminium	BS 1139	4.47	154	6180	0.952
Used steel	BS 1139	4.0	116	5780	0.670
Used steel	EN 39	3.2	140	4880	0.683
Used steel	EN 224	4.0	140	5700	0.798
Used aluminium	BS 1139	4.47	131	6180	0.896

Table 31.20. Formulas for calculating the applied bending moments on beams

W = total load L = length of tube between uprights

Type of loading	SS	PF
CPL	$wL/4$	$wL/6$
$\frac{1}{3}P$	$wL/6$	$wL/7.5$
UDL	$wL/8$	$wL/10$

Table 31.21. Permissible ledger loads for BS 1139 tubes

Span (m)		1.2	1.5	1.8	2.0	2.1	2.4	2.5	2.7
New BS 1139 tube	CPL (kg)	269	216	179	162	154	135	129	120
	UDL (kg)	539	432	358	324	308	270	258	240
Used BS 1139 tube	CPL (kg)	240	194	161	145	138	121	116	108
	UDL (kg)	484	388	322	290	276	242	232	216

Table 31.22. Weight of bricks (kg) that can be stacked along the two outside boards of a scaffold

Scaffold tube		Span (m)							
		1.2	1.5	1.8	2.0	2.1	2.4	2.5	2.7
New BS 1139 tube UDL	(kg)	593	475	394	356	339	297	284	264
Used BS 1139 tube UDL	(kg)	532	427	354	319	304	266	255	238

31.15 Load tables for horizontal tubes used as beams

Table 31.18 gives the permissible total loads on horizontal tubes in the various materials available. Each of the eight rows has two subdivisions that take into account the manner in which the tube is fixed at its end and three further subdivisions for the manner of loading.

In this and the following tables, the materials are designated by the standard number and wall thickness. The end fixtures are designated by the letters SS for simply supported and PF for partially fixed. The systems of loading are designated CPL for centre point load, ⅓p for one-third point load and UDL for uniformly distributed load.

Permissible bending stresses and section moduli are given in Table 31.19. Applied bending moments for various loadings on simply supported and partially fixed-ended beams are given in Table 31.20.

Table 31.18 is too cumbersome for quick use in the field.

In the Code, Table 31.21 of permissible ledger loads is given for BS 1139 tubes, new and used, assuming simply supported beams carrying centre point and uniformly distributed loading. No distinction is made between steel and aluminium.

There is a further complication that an access platform in an ordinary scaffold is frequently loaded more on the outside edge when bricks are hand stacked. This does not apply to bricks delivered on pallets by rough-terrain forklift trucks.

The bricks may occupy two out of five boards in the width, and some of this weight is transferred to the inside ledger. Similarly, some of the personnel imposed load near the inside ledger is transferred to the outside ledger. Making these allowances, the weight of bricks that is allowed can be increased slightly, as detailed in Table 31.22.

Irrespective of the span, the ledger load must not exceed 635 kg, which is the safe load of the right-angle coupler supporting the ledger.

Table 31.23. Permissible point loads that can be applied to scaffold tube cantilevers. The applied moment is *wL*.

Type of tube		Cantilever lever arm (mm)							
		300	600	900	1000	1200	1500	1800	2000
BS 1139	N	2640	1320	880	792	660	528	440	396
4 mm new steel	kg	269	135	90	81	67	54	45	40
BS 1139	N	2135	1117	749	670	559	447	372	335
4 mm used steel	kg	218	114	76	68	57	46	38	34
BS 1139	N	3172	1586	1057	952	793	639	529	476
4.47 mm new aluminium	kg	324	162	108	97	81	65	54	49
BS 1139	N	2699	1349	900	810	675	540	450	405
4.47 mm used aluminium	kg	275	138	92	83	69	55	46	41

It should be noted that the values in this table assume uniformly distributed loads. Some continuity over the transom near the standards is assumed, and some strength can be allocated to the boards.

31.16 Load tables for tubes used as cantilevers

Tubes used as cantilevers to support loads, to provide lateral stability, as that part of a scaffold standard between the ground and the lowest lift, or as a pavement frame leg, act in bending and are restricted to the same bending stresses. Table 31.23 gives the cantilever strengths of BS 1139 steel and aluminium tubes. The values in the table are for point loads in newtons and kilograms for various lengths of cantilever.

Where a horizontal cantilever is used to give lateral restraint to a scaffold, it should be capable of providing a resistance of 5% of the vertical load on the part of the scaffold that it is intended to stabilise. A tie tube may have to stabilise three pairs of uprights. In a tall, heavily loaded scaffold, these may carry 12 tonnes altogether, but 5% of this, 600 kg, can safely be transferred through one scaffold coupler. Table 31.23 shows that such tie tubes should not be longer than about 150 mm.

In practice, the tie tubes will be 300−400 mm long. At this length, their inadequacy to act unassisted indicates how important the longitudinal bracing is in an access scaffold. It must provide half to two-thirds of the lateral stability required for the whole weight of, and imposed load on, the scaffold. The remainder must be taken by the horizontal tie tube cantilevers.

31.17 Hazards in horizontal tubes

- Starting off the ledgers at the end of a scaffold with tubes all the same length results in all the joints being in one bay. Despite emphatic teaching on this point in training schools, one often sees every longitudinal tube − ledgers, guard rails and braces alike − jointed at 6.4 m from the starting end of the scaffold. Failures are frequent. To exaggerate the error, a hoist may be fixed to the end of the scaffold. The joints part under the tension force and are helped to do so by vibration from the hoist.

- A guard rail without a joint in a bay can be considered to make good a joint in the ledgers. If both ledgers have been fixed with joints in the same bay, the guard rail should be attached with right-angle couplers.

31.18 Lengths and straightness of scaffold tube available

The longest normal tube length is 6.4 m. Scaffolding contractors will stock all lengths shorter than this in stages of 300−500 mm.

To prefabricate special beams or trusses, longer lengths can be purchased by specification to the supplier.

The British Standard requires that deviation from a straight line is not more than $\frac{1}{600}$th of the tube length. This is approximately 10 mm in the longest tube.

31.19 Elastic extension and compression of tubes

The whole structure of a tubular scaffold is so flexible and able to accommodate minor distortions that elastic changes in the length of tubes under load can be ignored in access structures.

However, in high or long falsework, consideration may have to be given to it. A 10 m jointed steel standard carrying 2.5 tonnes will change in length by about 3 mm.

A 10 m jointed aluminium standard carrying 2.5 tonnes will change in length by about 9 mm.

31.20 Thermal extension and shortening of tubes

Thermal change in length cannot be ignored in cases where expansion is restrained and may cause buckling, or shrinkage may cause the scaffold to come adrift from some fixing.

A 10 m steel tube changing temperature by 20 °C expands or contracts by 2.3 mm.

A 10 m aluminium tube changing temperature by 20 °C expands or contracts by 4.6 mm.

In summer, such changes of temperature may occur daily.

31.21 The maximum height of a scaffold

For any size of bay and the loads thereon, it is possible to calculate the load that the bay applies to each of the standards supporting it. By adding up the number of lifts with their various loads, the total load on the standard can be calculated.

In a perfectly built scaffold, i.e. one that is fully braced in two directions and has been tied to the building so that every node point is fixed in space, the ability to resist the applied load is governed by the unsupported length of the standards.

This is the lift height in a perfectly built scaffold. At failure, the standards will 'snake' through the node points, each lift having an effective length equal to the lift height, i.e. acting as pin-jointed coaxial struts. The maximum load that the scaffold can carry is the strength of the tallest lift built into the system.

Unfortunately, in practice such an ideal scaffold cannot be built. It has spigot joints in all the members in irregular places. It has varying degrees of stiffness at the joints. It has widely spread node points, with the members often 300 mm apart and with the lines of force not intersecting at a point. The ties to the building will not be rigid but more or less hinged attachments, sometimes permitting many degrees of angular movement. With such a structure as this, the effective length of the uprights will vary widely between members that appear to be similarly fixed to each other and to the structure being served. Some scaffold frames will be diagonally braced, and some will not be braced at all.

While the calculations may be very reliable, the scaffold will not match this degree of reliability. There is no other way of calculating the load-carrying capacity of a scaffold, or the height to which any scaffold can be built, except by relying on the column formulas. To deal with variations in the scaffold, an empirical factor must be incorporated into the calculations to ensure that they represent the actual condition of the scaffold.

Because the deficiencies that make it variable are random in number, place and relation to one another, the only certainty about them is that the bigger the scaffold, the more deficiencies it will have.

If a scaffold has ten deficiencies in the lowest 10 m and in each 10 m of length and it is doubled in face area, it will have double the number of deficiencies in it on average. If the deficiencies are spread lengthways along a long, low scaffold, they will not compound one another significantly, and separately they may not do much damage to the overall stability of the scaffold. However, if the deficiencies are spread vertically up a tall, narrow scaffold, they will jointly weaken the same upright and combine, one on top of the other, to restrict its vertical load-supporting ability.

To correct for deficiencies, it is thus more important to restrict height than length, and a factor that limits the height of a scaffold will overcome the deficiencies. This factor needs to have more effect on a tall scaffold than a low one because it has to deal with more and compounding deficiencies.

Such a factor is $1 + \dfrac{h}{c}$, where h is the calculated height of the structure, assuming the tubes can carry their full column load, and c is a constant governing the overall value of the factor and hence the height of the structure to keep it within well-known and practically proved heights.

The designer has two options in applying the factor. He can calculate the maximum height to which a particular scaffold with a specified loading can be built, based on the allowable load for the standards at the lift height required, then reduce the height by applying the factor $1 + \dfrac{h}{c}$ as a divisor. Alternatively, he can apply the factor as a divisor into the maximum allowable load on the standard for the lift height required, then use this reduced value as the limit for the height of the scaffold.

Table 31.24 gives the allowable load in standards for three lift heights and for various heights of scaffold in terms of the number of lifts. The factor used in the table is $1 + \dfrac{h}{200}$, where h is the height of the scaffold in metres.

This value of the factor is empirical, but it has been found to result in scaffolds of a height that in practice has been found to be safe, even allowing for the unavoidable deficiencies described above.

It will be seen that, using the factor $1 + \dfrac{h}{200}$, a scaffold that by calculation can be built 200 m high will in practice be restricted to half that height. This does not mean that a scaffold higher than the restricted height cannot be built. To raise the scaffold higher, it must be strengthened by closing in the bay length or adding more standards to those in the original design.

Table 31.24. The allowable load (kN) in used scaffold tube uprights with various lift heights after application of the factor $1 + \dfrac{h}{200}$. BS 1139 tube

No. of lifts	Lift height		
	1.37 m kN	2 m kN	2.7 m kN
1	40.00	24.00	14.00
2	39.45	23.53	13.63
4	38.95	23.08	13.28
6	38.42	23.64	12.95
8	37.91	22.22	12.64
10	37.42	21.82	12.33
12	36.97	21.43	12.05
14	36.50	21.05	11.77
16	36.04	20.68	11.51
18	35.62	20.34	11.26
20	35.18	20.00	11.02
22	34.75	19.67	10.79
24	34.36	19.35	10.57
26	33.96	19.05	10.36
28	33.61	18.75	10.16
30	33.07	18.46	9.96
32	32.55	18.18	9.78
34	32.46	17.91	9.60
36	32.08	17.65	9.42
38	31.75	17.78	9.25
40	32.55	17.14	9.09

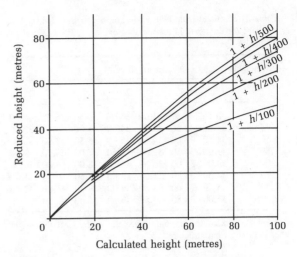

Figure 31.4 Height limitation coefficients for access scaffolds. The choice of the factor depends on the accuracy of the construction.

Figure 31.4 is a graph showing the height restriction resulting from five different factors. The most severe factor shown is $1 + \dfrac{h}{100}$, which reduces a calculated height of 200 m to one-third of that value. The least severe factor is $1 + \dfrac{h}{500}$, which reduces a calculated height of 200 m to 143 m.

BS 5973 1981 recommends the factor $1 + \dfrac{h}{200}$ as an arbitrary additional safety coefficient for high scaffolds. The designer should take into account the quality of supervision, the use of removable or non-removable ties, the number of functions the scaffold has to perform, and its length of life.

31.22 Data for calculating the height of a scaffold

Table 31.25 gives the load on one standard from the tubular framework of a normal scaffold, i.e. the self weight of half a bay one lift high. Multiplying the appropriate value by the number of lifts gives the self weight of the structure at its full height. Changes in width make very little difference.

Table 31.26 gives the weight to be added to one standard for each lift that is boarded, with toe boards and guard rails. Multiplying the appropriate value by the number of boarded lifts gives the self weight of the decking to be added to that for the framework.

Table 31.27 gives the imposed load to be added to one standard for each boarded lift in use. Multiplying this value by the number of platforms in use at any one time gives the imposed load to be added to that of the framework and the boarded lifts.

The sum of the three values is the load in the standard for the scaffold envisaged.

The sum of the three loads is compared with the appropriate load set down in Table 31.24 to see if the scaffold is safe in practice or overloaded.

For a scaffold of a fixed height that has to perform a fixed duty, the tables can be used to find the bay length, i.e. the spacing of the standards.

Example

A scaffold five boards wide is required to construct the brick facing of a 20-storey reinforced concrete column and slab building. The lift height for construction of the concrete frame is 2.7 m. Intermediate lifts at 2 m are to be inserted, ten lifts are to be boarded, with one working lift at 250 kg/m² and two lifts at 150 kg/m². A weather protection sheet weighing 1.25 kg/m² is to be added.

Try a standard spacing of 2.1 m.

Table 31.25. Load from half a bay one lift high = load per standard for the scaffold tube framework only, BS 1139 tube. Values in kg

Lift height (m)	Length of bay (m)						
	1.2	1.5	1.8	2.0	2.1	2.4	2.7
1.2	30.50	31.50	33.00	34.00	34.50	35.50	37.00
1.37	31.00	32.50	33.50	34.50	35.00	36.50	37.50
1.50	31.50	33.00	34.50	35.00	35.50	37.00	38.50
1.80	33.00	34.50	35.50	36.50	37.00	38.50	39.50
2.00	34.00	35.00	36.50	37.50	38.00	39.00	40.50
2.10	34.50	35.50	37.00	38.00	38.50	39.50	41.00
2.40	35.50	37.00	38.50	39.00	39.50	41.00	42.50
2.70	37.00	38.50	39.50	40.50	41.00	42.50	43.50

Note: The width of the scaffold makes little difference to the weight of the unboarded structure.

Table 31.26. Load from half a bay = load per outside standard from 38 m decking boards only (at 5 kg/m), including the toe board and guard rail. Values in kg.

No. of boards	Width (m)	Length of bay (m)						
		1.2	1.5	1.8	2.0	2.1	2.4	2.7
3	0.675	20.40	25.50	30.60	34.00	35.70	40.80	45.90
4	0.900	23.40	29.25	35.10	39.00	40.95	46.80	52.65
5	1.125	26.40	33.00	39.60	44.00	46.20	52.80	59.40
6	1.350	29.40	36.75	44.10	49.00	51.45	58.50	66.15

Table 31.27. Load from half a bay = load per standard from the imposed load only. Values in kg.

Scaffold width (boards)	(m)	Length of bay (m)						
		1.2	1.5	1.8	2.0	2.1	2.4	2.7
			Imposed load 75 kg/m²					
3	0.675	30.38	37.97	45.56	50.63	53.16	60.75	68.34
4	0.900	40.50	50.63	60.75	67.50	70.88	81.00	91.13
5	1.125	50.63	63.28	75.94	84.38	88.59	101.25	113.91
6	1.350	60.75	75.94	91.13	101.25	106.31	121.50	136.69
			Imposed load 150 kg/m²					
3	0.675	60.76	75.94	91.12	101.26	106.32	121.50	136.68
4	0.900	81.00	101.26	232.50	135.00	141.76	162.00	182.26
5	1.125	101.26	127.26	151.88	168.76	177.18	202.50	227.52
6	1.350	121.50	151.88	182.26	202.50	212.62	243.00	273.38
			Imposed load 200 kg/m²					
3	0.675	81.02	101.27	121.52	135.03	141.78	162.02	182.26
4	0.900	108.01	135.03	161.22	180.02	189.04	216.03	243.04
5	1.125	135.03	168.77	202.53	225.04	236.27	270.03	303.80
6	1.350	162.02	202.53	243.04	270.03	283.53	324.04	364.55
			Imposed load 250 kg/m²					
3	0.675	101.26	126.55	151.85	168.75	177.18	202.48	227.78
4	0.900	134.99	168.75	202.48	224.98	236.24	269.97	304.44
5	1.125	168.75	210.91	253.11	281.24	295.27	337.47	379.60
6	1.350	202.48	253.11	303.74	337.47	354.33	404.96	455.59
			Imposed load 300 kg/m²					
3	0.675	121.52	151.88	182.24	202.52	212.64	243.00	273.36
4	0.900	162.00	202.52	243.00	270.00	283.52	324.00	364.52
5	1.125	202.52	254.52	303.76	337.52	354.36	405.00	455.64
6	1.350	243.00	303.76	364.52	405.00	425.24	486.00	546.76

		Load per standard
Tubes:	20 lifts at 41 kg/lift =	820 kg
Weather sheeting:	20 lifts at 7 kg/lift =	140 kg
Decking:	10 lifts at 46.2 kg, with guard rail =	462 kg
Imposed load:	1 lift at 250 kg/m², 295.27 kg =	295 kg
Imposed load:	2 lifts at 150 kg/m², 177.18 kg =	354 kg
Add:	10 lifts of 1 ledger at 4.5 kg/m =	95 kg
		2166 kg
Allowable load for 20 lifts of 2 m =		2000 kg Not acceptable

Try a standard spacing of 1.8 m.

Tubes:	20 lifts at 39.5 kg/lift =	790 kg
Weather sheeting:	20 lifts at 6.08 kg/lift =	122 kg
Decking:	10 lifts at 44.1 kg =	441 kg
Imposed load:	1 lift at 250 kg/m², 253.11 kg =	253 kg
Imposed load:	2 lifts at 150 kg/m², 151.88 kg =	304 kg
Add:	10 lifts of 1 ledger at 4.5 kg/m =	81 kg
		1991 kg
Allowable load for 20 lifts of 2 m =		2000 kg Acceptable

31.23 Impact and dynamic loads

It is not within the scope of this book to deal with shock, dynamic, oscillatory or vibrating loads, but a designer may need to consider some of the consequences.

Let W be the actual or static weight of the body and P be its effective weight when subjected to a sudden or accelerating force.

The effective or increased weight P, i.e. the static equivalent of a body subject to an upward acceleration or a downward deceleration, is given by the expression:

$$P = W \times (1 \pm f/g),$$

where f is the actual acceleration and g is the normal acceleration due to gravity.

Loads that are applied suddenly but without any height of fall double the stresses in the members resisting the load. Loads that are applied suddenly from a falling weight that strikes a structure with the kinetic energy due to its fall may multiply the stresses up to one hundred times. These stresses may pass through the members so quickly that deformation may not reach critical proportions.

Let E be the modulus of elasticity (210,000 N/mm²), I the second moment of area of the member, h the height fallen by the weight, L the length of the member sustaining the impact and A its cross-sectional area. One can continue the analysis as follows.

Case 1: A weight W falling from a height h on to a column of height L

To find the ratio of P to W, evaluate the deflection that a static force P will produce. This is $\dfrac{PL}{AE}$, and the energy absorbed during the deflection is $\frac{1}{2}P \times \dfrac{PL}{AE} = \dfrac{P^2L}{2AE}$.

The energy of the falling weight is $W\left(h + \dfrac{PL}{AE}\right)$.

Equating these two functions gives:

$$\frac{P^2L}{2AE} = W\left(h + \frac{PL}{AE}\right), \text{ or } \frac{1}{2W}P^2 - P - \frac{AEh}{L} = 0.$$

This solves to

$$\frac{P}{W} = 1 \pm \sqrt{\left\{1 + \frac{2AEh}{WL}\right\}}.$$

It should be noted that if $h = 0$, $\dfrac{P}{W} = 2$.

Case 2: A weight W falling from a height h on to the centre of a simply supported beam of length L

To find the ratio of P to W, evaluate the deflection that a static force P will produce. This is $\dfrac{PL^3}{48EI}$, and the energy absorbed during the deflection is $\frac{1}{2}P \times \dfrac{PL^3}{48EI}$.

The energy of the falling weight is $W\left(h + \dfrac{PL^3}{48EI}\right)$.

Equating these two functions gives:

$$\frac{1}{2}P \times \frac{PL^3}{48EI} = W\left(h + \frac{PL^3}{48EI}\right), \text{ or } \frac{1}{2W}P^2 - P - \frac{48EIh}{L^3} = 0.$$

This solves to

$$\frac{P}{W} = 1 \pm \sqrt{\left\{1 + \frac{96EIh}{WL^3}\right\}}.$$

It should be noted that if $h = 0$, $\dfrac{P}{W} = 2$.

Case 3: A weight W falling from a height h on to the centre of a fixed-end beam of length L

To find the ratio of P to W, evaluate the deflection that a static force P will produce. This is $\dfrac{PL^3}{192EI}$, and the energy absorbed during the deflection is $\frac{1}{2}P \times \dfrac{PL^3}{192EI}$.

The energy of the falling weight is $W\left(h + \dfrac{PL^3}{192EI}\right)$.

Equating these two functions gives:

$$\frac{1}{2}P \times \frac{PL^3}{192EI} = W\left(h + \frac{PL^3}{192EI}\right), \text{ or } \frac{1}{2W}P^2 - P -$$

$$\frac{192EIh}{L^3} = 0.$$

This solves to

$$\frac{P}{W} = 1 \pm \sqrt{\left\{1 + \frac{384EIh}{WL^3}\right\}}.$$

It should be noted that if $h = 0$, $\dfrac{P}{W} = 2$.

32 Scaffold couplers and fittings

32.1 Background

Early scaffolds were made of round timber poles 7–8 m long. These were lashed together with 10 mm diameter natural fibre rope. The lashings could not contain any knots because these could not be undone when the rope was wet. Figure 32.1(a) shows how the lashing was started. A round turn was put on the upright at the appropriate height, and the loose end of the rope, about 750 mm long, was twisted with the main length of the lashing rope. An ordinary square lashing was then completed, as in Figure 32.1(b) and (c), which included the twisted length. Three complete cycles were made, and the end of the rope could then be pulled between the underside of the ledger and one set of turns. The weight on the ledger locked the end firmly. The same lashing was used for all angles of intersection of the poles, as in Figure 32.1(d).

Subsequently, when rectangular sawn timber was used, the joints were made with 100 mm nails driven in by a claw hammer, with the head left clear for subsequent withdrawal. This phase was short-lived in the United Kingdom but has persisted elsewhere in mainland Europe, where timber is cheaper.

Round pole scaffolds are still used in the United Kingdom where steel is forbidden. When it is permissible to use wire rope lashings, these have replaced the fibre rope of the early 1900s.

The rope used is a special soft wire 6 mm in diameter, which fits into the claw of a joiner's hammer. Lengths of the lashing rope can be purchased with an eye fixed in one end and a ferrule on the other.

Figure 32.1(b) and (c) shows the start of the lashing, with a clove hitch on the upright or a round turn using the eye. A three-cycle square lashing is then completed. The turns are moused to tighten them, and the spare end of the wire rope is finished off with a clove hitch on one or other of the poles. The claw of the hammer is used when necessary to tighten the rope.

It is recommended that the same form of lashing be used for any angle of intersection of the poles so that there are three parts of the rope in each of the four angles of the intersection. In this way, the lashing gives angular stability to the joint in both directions. This is shown in Figure 32.1(c) and (d), which also shows three mousing turns to finish off the lashing.

The next development was the chain plate, which consisted of a bolt passing through a large cruciform nut that had the corners turned back to act as hooks to engage the

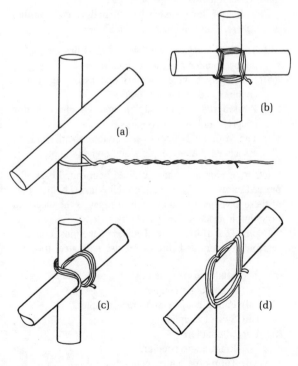

Figure 32.1 Fibre and wire rope lashings for timber scaffold poles.

links of a chain, which was usually crimped on to one of the hooks.

The centre bolt was screwed hard into the timber pole, and the chain was pulled tight on the turned-up corners of the plate. Both the indentation of the bolt into the pole and the friction grip of the chain contributed to the resistance to slipping.

32.2 Couplers and fittings for steel tubes

When steel tubes began to replace timber poles, the chain was replaced by the band and the chain plate by a new plate to match the band. The plate used today has the same characteristics that the old plate had in the 1920s. The centre bolt impinges directly on to the tube, and the turned-up edges lock into the band.

Subsequently, the double-flap fitting in the form of two pipe clamps was introduced. This was later made into a swivel form.

Finally, a lighter coupler was introduced to attach putlogs and transoms to ledgers, and to fix other tubes not requiring such a large resistance to slipping.

Other fittings and accessories associated with steel tube-and-coupler scaffolds have been introduced from time to time. All the main couplers and accessories have undergone continued change and improvement, and no doubt many more developments are yet to come.

Each line of development has created several families of couplers. The four main variants are:

(i) The SGB band and plate with its DH brace coupler, putlog coupler and swivel coupler.
(ii) Forged right-angle couplers and associated putlog couplers and swivel couplers.
(iii) Pressed right-angle couplers and associated putlog couplers and swivel couplers.
(iv) The Mills-GKN interlocked double coupler and its associated putlog couplers and swivel couplers.

Other developments have utilised bent round bars, flat pressed sheets, welded boxes, pipe clips and U bolts. Many of these have been tried out with wedges as the means of fastening instead of nuts and bolts.

Figure 32.2(a) to (q) are sketches of the most common scaffold couplers and fittings for use with steel tubes:

(a) The band and plate family.
    (i) Right-angle coupler.
    (ii) Putlog coupler and brace coupler.
    (iii) Swivel coupler.
(b) Forged couplers.
    (i) Right-angle coupler.
    (ii) Tulip-type putlog coupler.
    (iii) Swivel coupler.

(c) Forged couplers.
    (i) Right-angle coupler.
    (ii) Half wrapover-type putlog coupler.
    (iii) Swivel coupler.
(d) Pressed couplers.
    (i) Right-angle coupler.
    (ii) Half wrapover-type putlog coupler.
    (iii) Swivel coupler.
(e) Pressed couplers.
    (i) Right-angle coupler.
    (ii) Tulip-type putlog coupler.
    (iii) Swivel coupler.
(f) Mills-GKN couplers.
    (i) Right-angle coupler.
    (ii) Putlog coupler.
    (iii) Swivel coupler.
(g) Brace couplers.
    (i) Doubles with a putlog coupler.
    (ii) Doubles with a putlog coupler.
(h) In-line or coaxial couplers.
    (i) External sleeve coupler.
    (ii) Internal expanding joint pin.
    (iii) Combined internal and external coaxial pin (may have retaining pins fitted to the inside of the sleeve).
(i) Parallel couplers.
    (i) With a space between the tubes for other couplers to be placed.
    (ii) Parallel tubes in contact.
    (iii) With a space between the tubes for other couplers to be placed.
(j) Girder couplers.
    (i) To clip on to a girder flange.
    (ii) To fit through a hole in any steel section.
    (iii) To roll along a girder flange.
(k) Tie couplers.
    (i) To screw into an expanding anchor or a bolt hole.
    (ii) To screw into an expanding anchor or a bolt hole.
(l) Reveal pins.
    (i) Long type with a long tail inside the reveal tube (can double as an adjustable base or as a jack).
    (ii) Short type.
(m) Wire toe board and ladder clips.
    (i) End toe board clip.
    (ii) End-to-end toe board clip.
    (iii) Adjustable ladder clip.
(n) Flat strip toe board clips.
    (i) Double toe board clip (can double as a decking board clip).
    (ii) Hinged toe board or decking board clip.
    (iii) Toe board or decking board clip.
(o) Stair tread fittings and rail finials.
    (i) Fixed-angle stair tread fitting.

(ii) Adjustable-angle stair tread fitting.

(iii) Fixed right-angle finial.

(iv) Adjustable-angle finial.

(p) Putlog ends.

    (i) Flattened tube.

    (ii) Tube socket adaptor.

    (iii) Plate adaptor.

    (iv) Coupler adaptor.

(q) Miscellaneous.

    (i) Sheeting clip for roof sheets and wall sheets.

    (ii) Gin wheel.

    (iii) Base plate.

Each of the families of couplers contains the three basic types, i.e. the right-angle or double coupler, the swivel coupler and the putlog coupler. A brace coupler is included in some families. Each family has a spigot pin and a sleeve coupler, both of which enable tubes to be extended end-on.

Couplers from different families can be mixed on the same job because the basic types from any family have the same minimum load-carrying capacity defined by the British Standard.

Because of the different styles, the couplers have different practical characteristics. They have their own special tightening torques, different angular stiffnesses, and different ultimate failure and distortion loads.

Scaffolds built with the different families of couplers may have to be constructed slightly differently and be of slightly different dimensions, but this does not affect the stability of the structure. However, a scaffold inspector must study what has been done and he must also ensure that, if couplers from different families are used on the same structures, there are no resulting hazards.

The important properties of the right-angle coupler are:

- Its resistance to slipping down the tube.
- The factor of safety against slipping.
- The load at which distortion takes place.
- The load at which the ledgers will come adrift from the uprights.
- The angular stiffness of the coupler resisting lateral distortion of the scaffold in a plane parallel to the building façade.
- The rotational resistance of the coupler around the ledger resisting distortion of the scaffold in a plane at right angles to the building façade.
- The tightening torque required to achieve the strengths detailed above.
- The durability of the coupler against long-term exposure.
- The maintenance required on the coupler.
- The distortion of the tube, if any, caused by the coupler either during tightening or under the normal working load.

- The variation between one coupler and another.

Not all of these characteristics are specified in BS 1139.

32.3 Desirable performance for couplers

The characteristics that are necessary for successful couplers have been established over the years by usage. These are tabulated below.

Right-angle couplers

Slip: the coupler should not slip down the tube under a vertical load of less than 1270 kg so that it can be used safely at a vertical load of 635 kg.

Ultimate strength: the coupler should be able to hold the tubes together, even if the vertical load exceeds 1270 kg. It should be able to sustain a load of 3000 kg if prevented from slipping. At this load, it will have been subjected to accidental circumstances, and it is unimportant whether it has distorted or not, although BS tests require it not to distort.

Reversal of load: if the load alternates in direction, such as occurs in a coupler on a brace resisting wind loads from different directions, the same slip and ultimate strength values should apply.

Swivel couplers

Swivel couplers should have the same load characteristics as right-angle couplers. EN 74 allows a statistically determined lower limit of load-bearing capacity of 85% of that of a right-angle coupler. There should be consultation with and certification from the manufacturer on the failure loads. A safety factor of 1.6 can be applied to the statistically determined values at the specified deflections.

Parallel couplers

These can be used in series, and there is no need to have a nationally recognised strength. The manufacturer must declare the slip and ultimate strengths of each type of parallel coupler, and the user must fix an adequate number on to the lapped tubes to perform the required duty.

The slip strength should preferably be not less than 1270 kg so that the supporting value of one parallel coupler is equal to that of one right-angle coupler.

Brace couplers

Couplers for the attachment of braces are sometimes used at loads of 500 kg and so should have a slip value of 1000 kg.

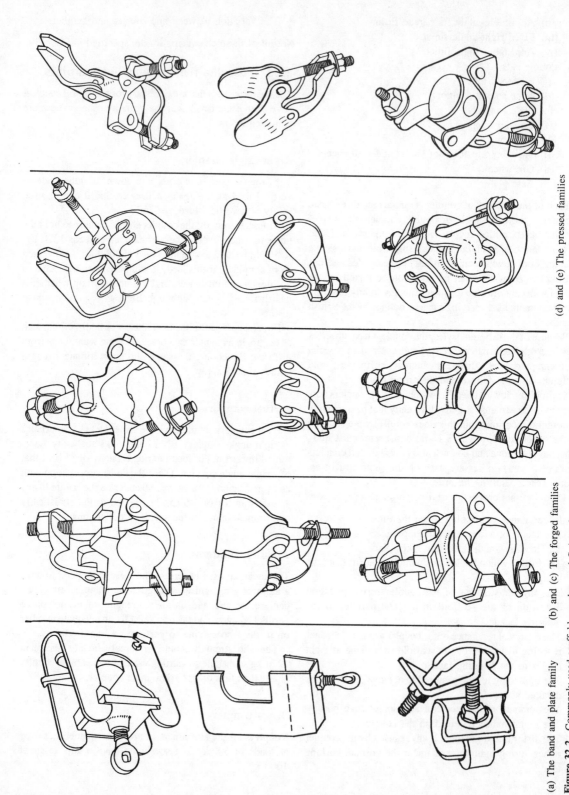

(a) The band and plate family   (b) and (c) The forged families

(d) and (e) The pressed families

Figure 32.2 Commonly used scaffold couplers and fittings for steel tubes.

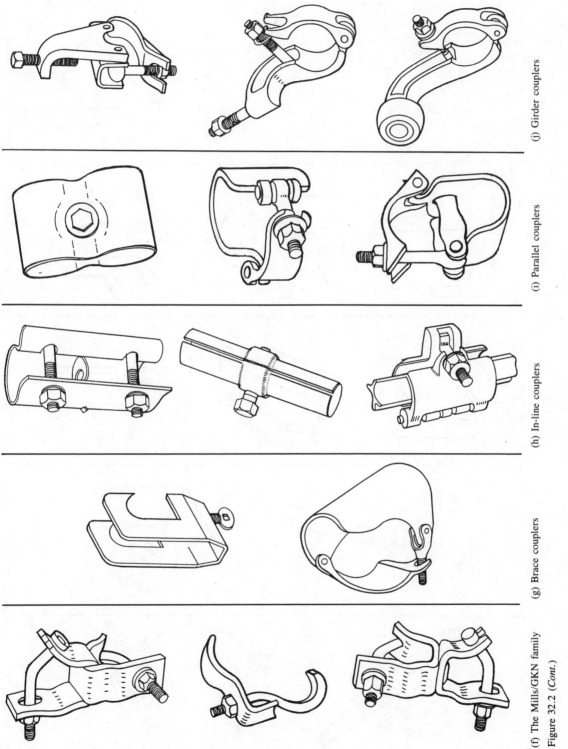

(j) Girder couplers

(i) Parallel couplers

(h) In-line couplers

(g) Brace couplers

(f) The Mills/GKN family

Figure 32.2 (*Cont.*)

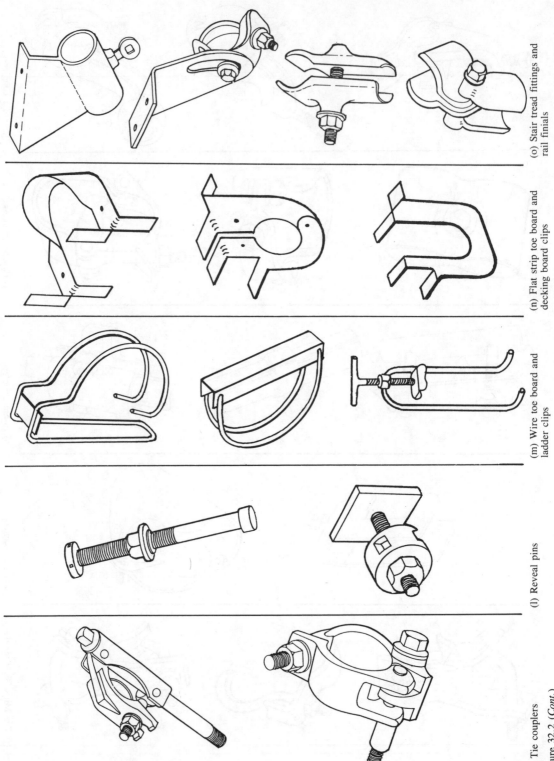

(o) Stair tread fittings and rail finials

(n) Flat strip toe board and decking board clips

(m) Wire toe board and ladder clips

(l) Reveal pins

(k) Tie couplers

Figure 32.2 (*Cont.*)

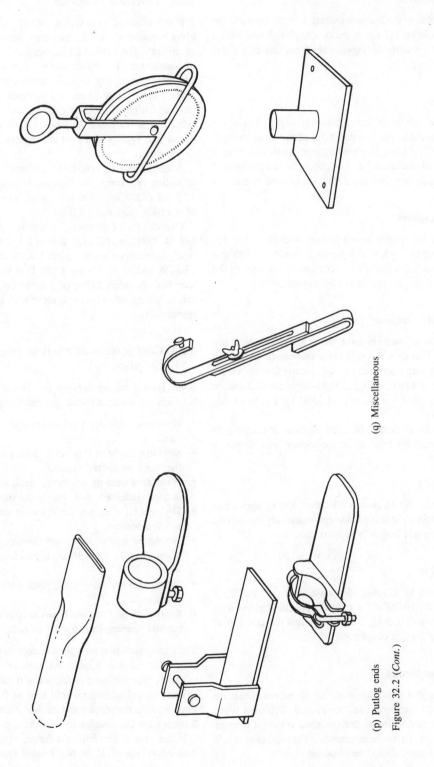

(q) Miscellaneous

(p) Putlog ends

Figure 32.2 (*Cont.*)

Putlog couplers

The slip value of the transom pulling at right angles to the ledger should be 127 kg. A putlog coupler should have a shape that is readily distinguishable from that of a right-angle coupler.

Joint pins

These need not have any tension strength. The shear strength should be not less than 80% of the shear strength of steel scaffold tube. The bending strength should be such that they will not break into two pieces before the ends of the tubes that they join are deformed beyond reuse.

Sleeve couplers

Sleeve couplers should have a tension strength of 635 kg. The bending strength in any position should be such that they will not break into two pieces before the ends of the tubes that they join are deformed beyond reuse.

Heavy-duty couplers

All the above couplers have been allocated minimum strengths. This does not rule out a manufacturer making any coupler with a declared strength greater than the values given. For instance, a forged right-angle coupler can be designed to resist slip at a load of 2500 kg, i.e. twice that given above.

Such couplers should be used at these loads only in designed scaffolds built under the closest supervision.

Putlog ends

A putlog end should sustain a load of 450 kg applied to the putlog tube near to the blade end when only the outside half of the blade length is supported.

Reveal pins

These should be capable of sustaining a pull-out force applied to the end of a 1.8 m tube opposite the end with the reveal pin of 635 kg, i.e. it will equal the safe slip of a single right-angle coupler.

Tightening torque

In all the above characteristics, it will be noted that no tightening torques have been mentioned. Different types of couplers and different designs need different torques to satisfy the above requirements. The manufacturer's recommendations should be obtained.

32.4 Tests for couplers

It is not intended to describe the test methods. BS 1139 gives test methods and the minimum performance in terms of strength. The EN and ISO codes give statistical methods of assessment. The buyer and the engineer in a scaffold contracting company and the manufacturer may need to make a close study of the test methods and results. The scaffold designer is not particularly interested in the technicalities. The scaffolder only needs to be able to recognise the various types and makes of couplers and know where to use them.

A safety factor of 2 should be applied to the early method of testing requiring, for right-angle couplers, no slip at 12.5 kN (1270 kg, 2800 lb), giving a safe working load of 6.25 kN (635 kg, 1400 lb).

Two classes of couplers, A and B, are referred to in EN 74, 1988, and BS 1139 Section 2.1, 1991, with a statistically determined lower limit of load bearing of 10 kN for Class A and of 15 kN for Class B at the test deflection specified. A safety factor of 1.6 can be applied to these values, giving safe working loads of 6.25 kN and 9.38 kN, respectively.

32.5 Coefficients of friction relevant to couplers

Some typical values for the coefficient of friction (μ) between the materials used in scaffolding are:

- Hard machined dry steel on hard machined dry steel.   0.15
- Steel that has been exposed to the atmosphere on the same weathered material.   0.25
- Weathered steel on weathered steel, when there is some indentation of one by the other.   0.4
- Dry steel on either dry aluminium or zinc without any indentation.   0.2
- Weathered steel on weathered aluminium or zinc, when there is some indentation of one by the other.   0.3
- Oiled steel on oiled steel, with both surfaces machined.   0.1
- Rusty steel on rusty steel, particularly rusty bolt threads running in rusty nut threads.   0.5

The significance of these values in couplers is obvious, and their wide range is a cause for concern.

Taking the coefficient of friction of the steel in a scaffold coupler on galvanised scaffold tube as 0.25, the contact force between the coupler and the tube must be 4 × 1.25 = 5 tonnes for the coupler to pass the slip test.

In the case of the flap of a forged right-angle coupler with a leverage of 2, the bolt tension must be 2.5 tonnes

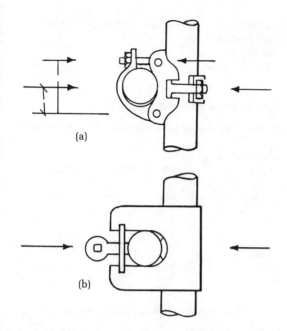

Figure 32.3 The tightening torque forces for two types of scaffold coupler: (a) A hinged gate coupler. (b) A band and plate.

(2500 kg). Equating this to the power of a screw thread enables the tightening torque to be calculated. Figure 32.3 shows the forces.

In the case of a band and plate, there is no leverage to assist the bolt, and a bolt thrust of 5 tonnes is required.

32.6 Tightening torque

Scaffold couplers are friction connections, i.e. they are essentially pipe clips relying for their grip on the tightness of the clamping bolts. If the bolts are not tightened by the correct amount, the couplers will slip at too low a load, and the platforms will not support the loads for which they are designed. If the couplers that connect the bracing tubes on to the structure are not tightened by the correct amount, the bracing may be inadequate to retain the structure in a rectangular pattern, and failure may result.

The maximum load that should be applied to a normal coupler in a scaffold is 0.625 tonnes (625 kg, 6.25 kN). This is the safe resistance against slipping that the British Standard requires of all right-angle couplers, whatever their design or origin. The coupler bolts must be tightened up to the torque specified by the manufacturer to attain this amount of grip with a safety factor of 2, i.e. the coupler must be capable of sustaining a load of 1.25 tonnes when done up to the manufacturer's specified torque.

Each manufacturer specifies his own requirement for the torque, depending on the design of the coupler. As stated, a small torque results in the coupler slipping below the required load, but a large torque may permanently distort the coupler and make it useless or dangerous for future use.

The coefficient of friction applies not only to the coupler on the tube but also to the bolt thread in the nut.

If the bolt thread is free-running in the nut and well oiled, the tightening torque will be converted into bolt tension much more effectively than if the bolt thread is tight or rusty.

For the sake of simplicity, three degrees of screw thread tightness will be considered: well-oiled and free, with a coefficient of friction of 0.1; normal looseness and dry, with a coefficient of friction of 0.25; and tight, rusty and dry, with a coefficient of friction of 0.5.

The bolt tension resulting from the tightening torque can be derived by representing the circular thread of the bolt by a straight ramp, as shown in Figure 32.4. This gives the following formula for the relation between the torque and the bolt tension or thrust.

Ignoring friction, the force up the ramp to create the tension T is given by:

$$F_1 = T\frac{p}{2\pi r},$$

where p is the pitch and $2\pi r$ is the circumference.

An amount to overcome the friction has to be added to the value of F_1 to create the tension. This is μT, i.e.

$$\text{total } F = T\frac{p}{2\pi r} + \mu T = T\left(\frac{p}{2\pi r} + \mu\right).$$

The torque Q to overcome the value of F is $F \times r$, so that

$$Q = T\left(\frac{p}{2\pi} + r\mu\right).$$

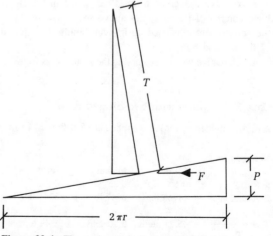

Figure 32.4 The wedge equivalent of a bolt thread.

If the three values of μ given above for the nut and bolt are inserted, and a bolt tension of 2.5 tonnes is required to correspond to a gripping force of 5 tonnes and a slip value of 1.25 tonnes in the coupler, the value of Q is as follows.

Case 1: Couplers with two hinged flaps

Assume 0.5 in Whitworth Imperial threads, for which $p = 0.083$ in, $r = 0.225$ in average, and $2\pi r = 1.414$ in. Bolt tension required $= 5600$ lb (2500 kg).

Nut and bolt condition	Oiled	Medium	Rusty
μ	0.1	0.25	0.5

$$Q = T\left(\frac{p}{2\pi} + r\mu\right) = 5600\,(0.0132 + 0.0225\,\mu).$$

$Q =$	199.92 in lb	388.92 in lb	703.92 in lb
	16.06 ft lb	32.41 ft lb	58.66 ft lb
	2.30 kgm	4.48 kgm	8.11 kgm

Case 2: Band and plate couplers

0.75 in Whitworth Imperial threads, for which $p = 0.1$ in, $r = 0.343$ in, and $2\pi r = 2.155$ in. Bolt thrust required $= 11200$ lb (5000 kg).

μ	0.1	0.25	0.5

$$Q = T\left(\frac{p}{2\pi} + r\mu\right) = 11200\,(0.0159 + 0.343\,\mu).$$

$Q =$	562.24 in lb	1138.48 in lb	2098.88 in lb
	46.85 ft lb	94.87 ft lb	174.91 ft lb
	6.48 kgm	13.12 kgm	24 kgm

The excessive tightening torques calculated in the right-hand column of the examples above show how much of the applied torque is used up in overcoming the friction of the nut and bolt.

An alternative way of expressing the results is as follows.

Case 1: Couplers with two hinged flaps

Assume a tightening torque is applied at 35 ft lb (420 in lb).

Nut and bolt condition	Oiled	Medium	Rusty
μ	0.1	0.25	0.5
Bolt tension	11765 lb	6047 lb	3341 lb

$$T = \dfrac{Q}{\dfrac{p}{2\pi} + r\mu}$$

	5.25 tons	2.70 tons	1.49 tons

Coupler grip	10.504 tons	5.4 tons	2.98 tons
Corresponding slipping value at $\mu = 0.25$	2.63 tons	1.35 tons	0.75 tons

Case 2: Band and plate couplers

Assume a tightening torque is applied at 70 ft lb (840 in lb).

Bolt thrust and coupler grip	16733 lb	8264 lb	4482 lb

$$T = \dfrac{Q}{\dfrac{p}{2\pi} + r\mu}$$

	7.47 tons	3.69 tons	2.00 tons

Corresponding slipping value at $\mu = 0.25$	1.87 tons	0.92 tons	0.50 tons

32.7 Hazards with poor bolts on couplers

- The calculations above clearly indicate the rapid decrease in the supporting value of a coupler when the bolts are allowed to become dry or rusty. Thread maintenance must be strictly controlled.
- The use of short spanners and cutting down orthodox spanners will obviously create a hazard.

32.8 Wedge couplers

Some couplers operate by driving a wedge rather than by screwing up a bolt. The recommended hammer weighs 1 kg and two heavy blows are required.

The same decrease in efficiency will result if the wedge surfaces are not maintained, but there is the saving grace that the surfaces are more self-cleaning than those of a threaded bolt in a nut.

32.9 Manufacturer's recommendations on tightening torques

The calculations given above are for the Whitworth system of threads because a lot of couplers now used have these threads.

For other threads and metric threads, the torque recommendations of the manufacturer should be obtained. The same applies for wedge couplers.

Where the nature of the coupler is such that it can lock on to the tube because of its shape and not rely entirely on the bolt tension, the tightening torque requirements may differ from those given above, and the manufacturer must be consulted.

When the threads are treated to resist corrosion by galvanising or sherardising, this will affect the bolt tension for any given torque, but the user must not overlook the effects of wear and tear when the surface has worn.

Other factors affect the manufacturer's recommended tightening torque appropriate to any particular coupler. A distortion of parts of the coupler under a large torque, and a progressive distortion resulting from many small distortions from repeated use, must be avoided.

32.10 The British Standard distortion test

In addition to resistance to slipping, couplers have to pass a distortion test. The intention of this is that if the slip load, which is twice the safe working load, is exceeded accidentally, the coupler will not fail completely.

The British Standard requires that a coupler, when prevented from slipping, be able to carry a load of 3 tonnes applied to the ledger without distorting.

There is a difficulty in interpreting this test because the term 'distortion' is not defined. Generally, it is taken to mean that the coupler is not so distorted as to become useless.

32.11 Right-angle couplers

The right-angle coupler is the cornerstone of all scaffold structures. It fixes tubes at right angles, but this does not mean that only structures with members in the three cardinal directions can be built using it. Diagonal bracing in vertical planes is fixed using right-angle couplers on the horizontals at right angles to those planes. Plan bracing is fixed using right-angle couplers on the uprights.

Cylindrical structures can be built by joining short ledgers to uprights or by joining alternate uprights with ledgers. Spherical structures can be built by interweaving straight tubes above and below one another or by curving the tubes.

The slip load required by the Code is 1.27 tonnes, giving a safe load of 635 kg. The distortion load must be 3 tonnes. Right-angle couplers are available that will give better performance than these requirements.

32.12 Swivel couplers

Swivel couplers are frequently made in two halves of the same form as the right-angle coupler of the same family and so have the same slip and distortion test values. The slip load required in the Code is 1.27 tonnes, giving a safe working load of 635 kg. The distortion test load must be 3 tonnes.

To achieve these values, the pivoting rivet joining the two halves of the swivel coupler together must be capable of sustaining a shear load of 3 tonnes.

The swivelling action is not impeded by tightening up, i.e. swivel couplers do not become fixed-angle couplers when they are fixed tightly to the two tubes they are joining. Because of this, they should never be used to perform the function of a right-angle coupler. If they are so used, a rectangular structure without bracing will become a mechanism, and dangerous instabilities will result.

Diagonal bracing is best fixed to the rectangular structure by right-angle couplers on to the horizontal tubes rather than by swivel couplers on to the standards.

The main use of swivel couplers is to attach bracing to standards when it is not practical to join it to horizontals with right-angle couplers.

One case, other than attaching braces for which swivel couplers have to be used as primary structural couplers, is where tubes are used as guys. Attachment of the guy tube to the anchor tube may require a swivel coupler, and if the guy tubes are set outside a rectangular tower in line with the diagonals of the tower, their attachment to the uprights will require swivel couplers.

A discussion of the use of swivel couplers to attach bracing tubes can be found in Section 5.5.

A half swivel coupler, i.e. one that has never been assembled with a rivet to another like half, is acceptable as a check or supplementary coupler.

If a countersunk tie bolt can be assembled through the rivet hole, the half coupler is acceptable as a tie fixing. If the half coupler is welded on to any other sort of plate, it can be used as a tie attachment, as a means of suspension or for fixing timber members to a scaffold.

Swivel couplers sometimes have to be used to fix the rakers in a truss-out scaffold or to lower a protection fan into position.

32.13 Brace couplers

Two types of brace are used in scaffolding. One maintains the rectangular framework in its upright or load-bearing position, and the other resists known horizontal loads or transfers them to the ground or to any firm point.

Section 31.12 discusses scaffold tubes used as braces. When the function of the brace is to maintain the shape of the structure, it may not be necessary to use a coupler as strong as a right-angle coupler. Some suppliers offer couplers that can sustain a load of 0.5 tonnes for this purpose.

32.14 Putlog couplers

The term 'putlog coupler' refers not only to the coupler used to attach the outside of a putlog tube to a ledger in a putlog scaffold, but also to the attachment of transoms to both ledgers in an independent tied scaffold. It also applies to the attachment of the board bearers to the ledgers in a birdcage scaffold, a loading bay or a slung scaffold.

Putlog couplers are not expected to transfer the large loads of the platforms on to the standards. They should be sufficiently strong to fix the board-bearing tubes firmly on to the ledgers on which they rest. No other capability should be expected of them unless they are specially rated for that purpose, such as performing the duty of a brace coupler or a guard rail coupler.

The slip load for putlog couplers is set at 0.127 tonnes in the Code.

32.15 Hazard with putlog couplers

- The design of most putlog couplers is such that there is no large projection of the coupler above the transoms to interfere with the boards resting on the transoms. This feature also makes them useful for fixing rails and purlins to which sheets are to be attached. Putlog couplers may not be strong enough to resist wind force suction and uplift in roofs. The manufacturer's recommendation and data must be obtained.

32.16 The strength of couplers by statistical analysis

The figures given previously are taken from the 1964 edition of BS 1139. They have been found satisfactory for general use in the construction industry. They have also been found satisfactory from a manufacturing point of view in that a cheap product can be made with the specified strengths. The term 'cheap' here means that the price enables it to sell in hundreds of thousands each year and to be made without the finely machined surfaces or other refinements that mechanical engineering components frequently require.

However, there are always schools of thought demanding increased strength, decreased deformation, and improved reproducibility of shape and load characteristics. Such refinements are inappropriate to the construction industry. There is a nearly zero record of failure of couplers when properly used in practice, and a nearly zero record of structural collapse or serious accident caused by a properly used coupler, going back 60 years. Refinements in the specification, taking an overall view of the construction industry, will probably do more harm than good.

The strength values given above and in the current British Codes are minimum values, and no statistical variation has been given. Nor have a specified number of tests had to be performed to justify the various types manufactured. In other words, these strengths are absolute minimum values.

Statisticians go to some lengths to point out that nothing can be manufactured in a continually reproducible form and that there will be a distribution of strength characteristics, even in one family of couplers. This should be taken into account in the specification.

The ISO and European Standards approach the strength of couplers from a statistical point of view and also relate the displacements of the two tubes joined by the coupler to the loads applied to the coupled tubes. These statistical requirements have been written around the behaviour of existing couplers and so have made no contribution to the art of scaffolding or of coupler manufacturing.

Statistical specifications require statistical control of production, and this has also been found unnecessary over the 60 years referred to. It adds to the cost of scaffolding while providing no benefit, and is thus of academic interest only.

The ISO and European Standards are summarised as follows:

Right-angle couplers
- For a slip distance of 0.5 mm, the load must be at least 10 kN (1000 kg, 1 tonne).
- For a ledger displacement of 7 mm, the load must be at least 7 kN (700 kg, 1543 lb).
- The breaking load must be not less than 20 kN (2000 kg, 2 tonnes).

Swivel couplers
The corresponding loads are:
- For a slip distance of 0.5 mm: 8.5 kN (850 kg, 1874 lb).
- For a ledger displacement of 7 mm: 6.0 kN (600 kg, 1323 lb).
- The breaking load: 17 kN (1700 kg, 3748 lb).

Sleeve couplers
For a 2 mm slip, the load must be not less than 3.0 kN (300 kg, 661 lb). The breaking load must be not less than 50 kN (5000 kg, 5 tonnes).

Parallel couplers
- Load capacity: 15 kN (1500 kg, 1.5 tonnes).

Test data
The above characteristics must be met by 90% of the samples in the batch, with a 95% confidence level.

The standard gives the batch size, which can vary from 25 to 100.

32.17 Loads on supplementary couplers

The safety factor normally used to determine the safe working load for a coupler is 2. This is applied to the specified minimum slip load of the coupler of 1270 kg.

If a load greater than 635 kg has to be transferred to a tube, a supplementary coupler must be attached to the

Table 32.1. Relaxation of the safety factor for couplers when more than one coupler supports the load

	Safety factor	Safe working load (kg)	Safe working load of the full joint (kg)
One coupler	2	635	635
Second coupler	1.8	706	1341
Third coupler	1.6	794	2135
Fourth coupler	1.4	907	3042
Fifth coupler	1.2	1058	4100

upright to prop up the coupler used as the main tube connector. The safety factor for the supplementary coupler can be reduced because it is unlikely that two defective couplers will be used in the same joint. A second supplementary coupler may have to be used to support the joint at a further reduced factor of safety.

Table 32.1 gives proposals for the relaxation of the safety factor. No reduction below the value of 1.2 should be used.

32.18 Girder couplers

Girder couplers are designed to attach a scaffold tube to the flange of a joist, a beam or column, or an angle or tee. They are load-bearing couplers as well as a means of attaching a tube.

A second type of girder coupler enables a slung scaffold to roll along a horizontal structural steel construction to give access to several parts of it without scaffolding the whole structure.

Examples of the use of girder couplers are given in Figure 17.4.

The load-carrying capacity is a matter for the manufacturer, and his rules for the use of the coupler to attain the declared safe working load must be strictly obeyed.

When a scaffold is slung below a series of beams, e.g. to provide a deck to paint the soffit of a bridge, this structure can be hung on girder couplers, but it must be treated as a slung scaffold and comply with the recommendations in Chapter 17. In particular, the vertical hanging tubes must have check couplers in place at both ends. This requirement raises the question that if check couplers are necessary on the vertical hanging tubes, the horizontal tubes should also be checked against breaking away from the bottom flanges of the bridge joists.

This can be achieved in the girder coupler itself by designing into its body a part that overlaps the joist flange. The application of direct tension to the flap bolt of the coupler must be avoided.

When these characteristics are not built into the girder coupler, a supplementary tube structure should be erected to safeguard the slung scaffold.

Girder couplers, in which direct tension is on the flap bolt or flange bolt, should only be used in such large numbers that, if one fails, the structure remains safe and the remaining couplers are not overloaded.

32.19 Calculating the load in girder couplers

A scaffold tube slung beneath two joists is frequently attached by four couplers, as indicated in Figure 17.4. This means that a sagging bending moment occurs in the centre of the tube and hogging bending moments occur at the ends. This enables a calculation to be made of the downward load on the two couplers nearest the centre span.

For a distributed load w over the whole span, the centre moment is $+\dfrac{wL}{24}$, and the end moments are $-\dfrac{wL}{12}$. The end moment is thus the limiting factor in the loading on the tube.

Taking as an example a span of 4 m and a safe moment of resistance of $f_b \times Z = 139 \times 5.7 \times 10^3 = 792300$ N/mm^2. The end moment is given by $\dfrac{w.4000}{12}$, so $w = \dfrac{12 \times 792300}{4000} = 2376.9$ N distributed over the span.

Now suppose that the distance between the two couplers on the end of the tube is 200 mm. The coupler force to give an end moment of 792300 N/mm^2 is 3961.5 N. This is very much greater than the end reaction of $\dfrac{2376.9}{2} = 1188.45$ N.

32.20 Hazards with girder couplers

- Failure to recognise that a girder coupler may be loaded with much more than the end reaction of a beam.
- Assuming that the beam is pin-jointed has caused failures.
- If a 6.4 m tube weighing 27.97 kg is coupled to a girder at only one end during the erection of a slung scaffold, it applies a moment at its fixed end of 884 kN/mm^2, which is greater than the safe bending moment of the tube. It may distort the coupler to failure or near failure before the scaffold is completed and loaded.

32.21 List of other scaffold fittings

- Base plates, adjustable base plates, swivel base plates.
- Expanding joint pins, plain joint pins, tension pins.
- Sleeve couplers, tension sleeves.
- Parallel couplers.
- Finials, swivel finials.
- Putlog tubes, putlog ends.
- Tie tubes, tie adaptors, reveal pins.

- Half couplers.
- Stair tread and adjustable stair tread couplers.
- Head plates and forkheads.
- Brick guards.
- Board retaining bars.
- Board clips, toe board clips.
- Tube-end protectors.
- Foot steps.
- Couplers for sheeted structures and temporary buildings.
- Prefabricated elements such as hop-up brackets, jib arms, gin wheel brackets, beams, components for buildings, components for towers and show stands.

32.22 Materials for couplers

Steel should comply with BS 4360, BS 970 or BS 1449.
  Bolts and nuts should comply with BS 916.

Aluminium should have a 0.1% proof stress of not less than 154 N/mm^2, which should not exceed 80% of the tensile strength of the material, and an elongation of not less than 8%, measured in accordance with BS 18.

32.23 European specifications and laboratory performance

The scaffolding contractor is not usually concerned on the building site with the laboratory performance of the couplers, but he should be aware that in some countries there are different grades of coupler, i.e. their safe working loads are different. He must be aware at all times which grade he is using and, if he is constructing a special scaffold that has been designed for a special duty, he must make sure that he constructs the scaffold with the couplers required by the designer.

The supplier of the couplers must be able to give the technical details to the user. These details can come only from the manufacturer.

EN 74 refers to two grades of coupler, Class A and Class B, of which Class B has a higher load capacity. There is a minimum requirement for Class B couplers for torsional rigidity, i.e. the angular stiffness achieved between the ledger and the upright. The test methods are described in EN 74 and BS 1139 Section 2.1, 1991. The manufacturer should have the test results available for the purchaser so that any batch of couplers purchased can be compared with the specified performance.

Class A right-angle couplers must have a load-bearing capacity of not less than 0.7 tonnes (700 kg, 7 kN), with a ledger movement of not greater than 7 mm when tested by the statistical method. Class B right-angle couplers must have a load-bearing capacity of 10 kN, with a ledger movement of not greater than 7 mm.

Class B right-angle couplers must have a load-bearing capacity of 1 tonne (1000 kg, 10 kN), with a ledger movement of not greater than 10 mm.

Swivel couplers must have a load-bearing capacity of 0.6 tonnes (600 kg, 6 kN) when tested by the statistical method.

Sleeve couplers must have a load-bearing capacity of 0.3 tonnes (300 kg, 3 kN).

Parallel couplers must have a load-bearing capacity of 1.5 tonnes (1000 kg, 15 kN).

The same safety factor of 2 should be applied to all these couplers in use in the field.

When the above test results are used, and the lower limit of load-bearing capacity has been determined by statistical tests, a safety factor of 1.6 can be used.

33 Scaffold boards and timber

33.1 Types of board

In tube-and-coupler scaffolds, the most common board is 38 mm thick, 225 mm wide and 3.96 m long. If the whole range of working platforms is considered from small towers to trestles, thicknesses range from 25–75 mm.

Various species of timber are supplied, including redwood, whitewood, fir, spruce, pine, larch and western hemlock.

The choice of thickness is governed by the clear span, or the span between supports is chosen for the thickness available. Some prefabricated scaffolds have fixed-length boards to match the bay lengths in the system.

Timber 75 mm thick is available in 5 m lengths. Above this length, e.g. a span between roof trusses, a lightweight fabricated stage is required.

33.2 Specifications of boards

The sizes and tolerances of the common scaffold board are:

- Width: 225 ± 5 mm.
- Length: 3.96 ± 50 m.
- Thickness: 38 ± 2 mm.

The thickness tolerance on 50 mm and 63 mm boards is ± 3 mm.

BS 2482 1981 gives the detailed specifications.

End splitting is prevented by end hoops or nail plates. The corners can be chamfered or square.

Boards may be visually or machine graded. They must be marked with the British Standard number, the supplier's identification, the letter M or V denoting machine or visual grading and the distance over which they can span.

In BS 2482, reference is made to BS 4978 for characteristics other than knots. BS 2482 specifically lists knots, fissures and resin pockets, wane, distortion, damage, decay, insect holes, abnormal features and slope of grain as defects that must be assessed.

This list of characteristics does not include density, which is particularly important. Wood that has grown rapidly has a lower density than wood that has grown slowly. This means that rapidly grown timber has less wood fibre per unit volume and hence is of a lower strength than more slowly grown and denser timber.

A scaffolding yard foreman or a scaffolder will be able to pick out a light board by feel and inspect that board more critically for other defects. A scaffold inspector or safety officer should likewise feel the weight of any boards he suspects for other reasons and make his judgement accordingly.

The approximate density of the various timbers of which scaffold boards are made are given in Table 33.2. Exact figures cannot be given.

The values given in Table 33.2 represent the quantity of actual wood in any given volume. An assessment of this

Table 33.1. The spans for British Standard boards

Thickness (mm)	Span (m)
38	1.50
50	2.60
63	3.25
73	5.00

Table 33.2. Approximate density of various timbers

Timber	Density (g/cm^3)	Weight of boards 225 mm wide and 38 mm thick per metre (kg)
Spruce	0.45–0.52	4.0–4.5
Pine and Fir	0.65–0.70	5.5–6.0
Redwood and Whitewood	0.52–0.60	4.5–5.2
Western Hemlock	0.55–0.65	4.8–5.6
Larch	0.70–0.75	6.0–6.5

quantity can also be made by counting the number of growth rings per centimetre observed in the end grain and measured at right angles to the rings. The closer the rings, the more actual wood there is in the specimen and hence the stronger the board.

The number of rings per centimetre ranges from 1.5 to 7. Boards with fewer than 2 rings/cm should not be used in scaffolding, but 2.5 rings/cm is a good value and has proved generally to give a satisfactory scaffold board. Any value above 3 rings/cm represents a density above the average required for scaffold boards.

Both the density and the number of rings per centimetre have been referred to in previous Codes on scaffold boards, and it is regrettable that the committee revising the Code for the 1981 edition did not think fit to retain these two very simple and practical means of assessment, as well as the other characteristics given above.

Visual grading

Face knots should not be larger than one-third of the width of the board.

Edge knots should not exceed three-quarters of the thickness of the board.

Arris knots, i.e. those on the corners of the board, which are visible both on the face and on the edge, should be measured on both surfaces. The face dimension of the knot should not be greater than one-third of the width of the board, and the edge dimension should not be greater than two-thirds of the thickness of the board.

Arris knots that are opposite each other when the face is observed are particularly dangerous because the remaining thickness of the board may average only about 62.5% of the full thickness. In consequence, if the knots come free from the rest of the wood, the bending strength of the board will be only 40% of the unblemished timber, because $0.625^2 = 0.39$.

Fissures

Fissures and resin pockets deeper than one-third of the thickness of the board and longer than 225 mm are hazardous.

Wane

The Code requirement is that the sum of the wane dimensions on both the face and the edge of a board must not exceed 25 mm, and the wane must not extend behind the end loop of the board and so cause a danger from the metal edge.

Distortion

Bow should not exceed half the thickness of the board in any 3 m length, twist should not exceed 10 mm over the full width of the board in any 3 m length, and cup should not exceed 5 mm. Of these, twist is the most critical. In a standard board length of 3.96 m, the maximum twist should be 13 mm. This is the height of the step in a fully boarded deck that a warped board will cause.

It is difficult to rectify a warped board. Storage on the flat may help for a short time, but the board will frequently warp again.

Some boards warp without any misuse, while others warp with wetting, and others with excessive drying when used in hot locations such as boilers or permanently dry locations such as the underside of factory roofs.

Damage

Mechanical damage is rarely visible, but over-straining may occur and not be visible. Visible mechanical damage will need assessing on its merits. Mechanical testing is probably the only way that the effect of observed damage can be assessed.

Decay

Decay is usually accompanied by a softening of the wood fibres. This can be assessed by pushing the point of a knife into the scaffold board. Any suspect reaction should be thoroughly investigated. A white colouring or deposit on the faces of a board is a tell-tale sign of decay.

Insect holes

These can be seen and need thorough investigation. Wood borers feed on rot fungus in the timber.

Abnormal features

Any abnormal features should be thoroughly investigated.

Slope of the grain

This is a very important characteristic. A slope of 1 in 10 on the face or the edge is the limit allowed in the Code. Sometimes, the slope is such that the grain when tested on the edge of the board goes from face to face in as little as 75 mm. This is a very dangerous condition, leading to the board snapping easily when loaded.

Inspectors must pay special attention to the slope of the grain. It should be measured by drawing a steel pointer along the grain and observing the trace so formed.

33.3 Machine grading of boards

Machine grading is a strength grading of new timber. It is based on the well-established fact that the ultimate strength is reasonably closely related to the dynamic modulus of elasticity. Boards are fed through a machine which deforms them and measures the relationship between the load and the deformation, i.e. the modulus of elasticity for the specific rate of loading. Previous tests have established the correlation between the measured modulus and the failure strength.

The machine can thus accept a board that may be rejected by visual grading. There is nothing in BS 2482 that gives either visual or machine grading preference over the other. The choice is a matter for agreement between the purchaser and the supplier.

The purchaser of machine-graded boards will not possess a grading machine, and thus will be unable to check the goods delivered to him. To overcome this difficulty, he should ascertain from the supplier what ultimate strength the boards are intended to have for the grading offered and take a selection of the boards delivered for strength testing.

33.4 Strength

The British Standard does not specify a minimum strength for boards in bending. Appendix D states that boards complying with the specification can be expected to have a failure bending strength of at least 15.2 N/mm² with a 27% moisture content (1.55 kg/mm², 2205 lb/in²). This figure is for the lowest one percentile.

Table 33.3 lists the working stresses and working moments of resistance given in Appendix E.

The average failure strength in bending of boards will be much greater than 15.2 N/mm². It should be 35—55 N/mm² for boards to behave on the job as expected.

33.5 Structural characteristics of boards

Table 33.4 gives some useful characteristics of scaffold boards evaluated for the nominal sizes.

Table 33.5 gives the moduli of section for the nominal thicknesses and widths, and also the values for boards cut to the minimum tolerances in thickness and width. The loss of strength at the minimum tolerances is given as a percentage of the value for the nominal sizes. The strength of the various boards relative to the standard scaffold board is also given.

Table 33.5 shows that when a board is cut to the minimum sizes allowed by the tolerances, there is a loss of strength of 10—14%.

Scaffold inspectors should bear this in mind when they are considering any other defects in the boards.

The intention of the tolerance in any specification is that there can be a local variation in the basic dimension. It is not intended that a whole consignment of boards can be supplied at the minimum tolerances.

If non-standard scaffold boards are produced by cutting 75 mm timber with a 4 mm saw, the two boards will have a thickness of 35.5 mm and there will be a 12.5% loss of strength in the boards produced. If the nominal 75 mm boards are at their lowest tolerance of 72 mm, the two scaffold boards will be only 34 mm thick and there will be a 20% loss of strength.

33.6 Boards used on edge

With the publication of Codes on tubular scaffolding and the training schemes in the craft, many of the older techniques are falling into misuse. In particular, the use of boards on edge is now rare.

An example is a birdcage through which a 3.5 m gap must be left for traffic access. Another example is a pier-and-beam platform for treating the inside of the roof of a structure when the piers need to be 3—3.5 m apart.

Boards on edge are excellent ledgers and, threaded into ladder beams every 1.2 m, they form an excellent support for the platform boards. With ladder beams as ledgers, the transom boards on edge can be fixed every 300 mm through the ladder beams.

The structural section properties of standard scaffold boards on edge are given in Table 33.6.

The bending moments are given in Table 33.8 for boards on the flat and boards on edge.

33.7 Boards of different timber species

The species of wood from which scaffold boards are made will not affect their working loads to a significant degree, but there may be cases where it is desirable to know what strength is available. Average failure stresses in bending and bending moment for clear specimens of wood are given in Tables 33.7, 33.8 and 33.9.

The working stresses have to take into account the water content of the boards in their exposed location and damage due to reuse. The use of boards on the flat as opposed to on edge increases the significance of any defect. There is the fact that boards have to act separately, i.e. adjacent boards cannot come to their rescue, as would be the case with rafters in a roof or joists beneath a floor.

Boards are for supporting persons working with heavy materials. Having regard to all these points, a safety factor of 5 is reasonable. This is applied to the failure stresses in the tables, and the working stresses in bending are shown in round figures in the third column.

The bending moments that should be applied to 225 mm

Table 33.3. Working stresses and resistance moments in timber

Nominal size (mm)	Minimum modulus of sections		Maximum working stress			Working moment of resistance		
	mm³	in³	N/mm²	kg/mm²	lb/in²	Nm	kg m	lb in
38 × 225	47520	2.897	9.85	1	1429	468	47.5	4123
50 × 225	80997	4.939	9.85	1	1429	798	81.0	7030
63 × 225	132000	8.049	9.85	1	1429	1300	132.0	11456
75 × 225	190080	11.599	9.85	1	1429	1872	190.1	16499

Table 33.4. Scaffold board characteristics, based on the nominal sizes

Property	Units		Thickness				
		(mm)	25	38	50	63	75
		(in)	1	1.5	2	2.5	3
Weight at a relative	kg/m		3.375	5.130	6.750	8.505	10.125
density of 0.6	kg/m²		14.999	22.798	29.997	37.796	44.996
	N/m		33.906	50.305	66.191	83.401	99.287
	N/m²		147.08	223.559	294.153	370.631	441.235
	lb/ft		2.268	3.447	4.536	5.715	6.804
	lb/ft²		3.073	4.671	6.146	7.744	9.219
Cross-section area	mm²		5625	8550	11250	14175	16875
(225 mm wide)	in²		8.72	13.25	17.44	21.97	26.16
Modulus of section	mm³		23438	54150	93750	148838	210938
	cm³		23.44	54.15	93.75	148.84	210.94
	in³		1.430	3.304	5.72	9.081	12.870
Moment of inertia	cm⁴		29.30	102.89	234.38	468.84	791.02
	in⁴		0.704	2.472	5.631	11.264	19.004
Length per tonne	m		296.30	194.93	148.15	117.58	98.77
Number of 3.96 m boards per tonne			75	49	37	30	25
Maximum bending	Nmm		230864	533378	923438	1466054	2077739
moment at	kgm		23.534	54.371	94.132	149.445	211.798
9.85 N/mm² stress	lb in		2042.5	4718.8	8169.7	12970.3	18381.9
Maximum bending	Nmm		356223	823002	1424865	2262121	3205951
moment at	kgm		36.313	83.894	145.246	230.594	326.804
15.2 N/mm²	lb in		3151.6	7281.1	19450.8	20013.2	28363.3

Table 33.5. Relative moduli of section and percentage strengths of nominal size boards and boards cut to the minimum tolerance

Nominal thickness (mm)	32	38	50	63	75
Nominal width (mm)	225	225	225	225	225
Section modulus (mm³)	38400	54150	93750	148838	210938
Minimum thickness (mm)	30	30	47	60	72
Minimum width (mm)	220	220	220	220	220
Section modulus (mm³)	33000	47520	80997	132000	190080
Percentage strength of the minimum board compared with the nominal board	86	88	86	89	90
Relative strengths of various thicknesses compared with the nominal 38 × 225 mm board	71	100	173	275	390

Table 33.6. Modulus of section of 225 mm wide scaffold boards on edge

Thickness of boards used on edge (mm)	25	35	38	50	63	75
Section modulus of boards on edge (mm³)	210 938	295 313	320 625	421 875	531 563	632 813

Table 33.7. Scaffold board strengths

Species of timber	Typical failure stress in bending of clear specimens (N/mm²)	Working stresses in bending (N/mm²)
Spruce	35	7
Pine and Fir	40	8
Redwood and Whitewood	45	9
Western Hemlock	50	10
Larch	55	11

Table 33.8. Allowable bending moments on boards on the flat, in kNmm

Timber	Nominal board thickness (mm)					
	25	35	38	50	63	75
Spruce	164	322	379	656	1042	1477
Pine and Fir	188	368	433	750	1191	1688
Redwood and Whitewood	211	413	487	844	1340	1898
Western Hemlock	234	459	542	938	1488	2109
Larch	258	505	596	1031	1637	2320

Table 33.9. Allowable bending moments on boards on edge, in kNmm

Timber	Nominal board thickness (mm)					
	25	35	38	50	63	75
Spruce	1477	2067	2244	2953	3721	4430
Pine and Fir	1688	2363	2565	3375	4253	5063
Redwood and Whitewood	1898	2658	2886	3797	4784	5695
Western Hemlock	2109	2953	3206	4219	5316	6328
Larch	2320	3248	3527	4641	5847	6961

Table 33.10. Regulation spacing of supports for boards

Thickness of board (mm)	31.75	38.1	50.8
Maximum span of supports (mm)	991	1524	2591

wide scaffold boards on the flat giving a safety factor of 5 are given in Table 33.8.

The bending moments that should be applied to 225 mm wide scaffold boards on edge giving a safety factor of 5 are given in Table 33.9.

33.8 The statutory requirement for the support of boards

Irrespective of the strengths given in the preceding tables, the Construction (Working Places) Regulations require the supports to be at not greater spacing than the value given in Table 33.10.

Because the spacings given are prescribed by law, they must not be exceeded, even if the timber is not overstressed by the particular load applied.

The applied bending moment is governed by the loading for which the scaffolding is designed. Table 33.11 gives the values for boards at a 1500 mm span.

33.9 Boards in slung scaffolds and birdcages

In some structures, the boards may have to make a contribution to the strength of the top lift of a slung scaffold or a birdcage, or to any bay of a scaffold wider than about 2.1 m. Where the strength of boards has to be utilised to transfer the load from the centre of the ledgers to the transom at the uprights, they must span the bay without any board joints. Figure 33.1 shows various locations of board joints in relation to the ledger span. In some cases, the board can contribute to the strength of the platform and, in other cases, it cannot. In some cases, the load will pass directly through to the ledgers.

In Figure 33.1, the board length is shown as 3.96 m (13 ft) in all cases.

Figure 33.1(a) shows a bay length of 3 m (10 ft), which may occur in a lightly loaded slung or birdcage scaffold. A point load such as a pile of fire sprinkler pipes may become a centre point load on the ledger if the board joints are in the middle of the bay. On the other hand, if the boards span the bay, they can share the load equally with the ledger.

Figure 33.1(b), (c) and (d) are similar cases indicating that the board joints will be at progressively changing locations in adjacent bays. In any of these cases, placing materials in the centre of a 7.92 m (26 ft) wide scaffold is a danger because there will be a joint in the 3.96 m (13 ft) long boards down the centre of the scaffold.

Table 33.11. Applied bending moments on scaffold boards

Scaffold duty	Distributed load (kN/m^2)	Total load on one board of 1500 mm span, 225 mm wide (kN)	Bending moment applied to a board from a uniformly distributed load (kNmm)
Inspection and very light duty	0.75	0.25	46.9
Light duty	1.5	0.51	95.6
General purpose	2.00	0.68	127.5
Heavy duty	2.50	0.84	157.5
Masonry or special duty	3.00	1.01	189.4

Figure 33.1(e) shows a bay length of 1.98 m (6 ft 6 in), i.e. a half board length. If the scaffold is carefully designed, the board joints can always be arranged adjacent to an upright or suspension point. If there is an option as to where the support or suspension points can be, this spacing is the best for normal loads.

Figure 33.1(f) and (g) are cases where the board joint comes in various places along adjacent bays.

In Figure 33.1(h), if the bay length can be arranged at 1.32 m (4 ft 4 in), the board joints can be arranged to coincide with every third standard. Such an arrangement, where the boards can take a share of the load, is useful for a vehicle ramp. If thicker boards are used with different lengths, a comparable layout can be devised.

In this case, if the deck is cross-boarded as well as longitudinally, all the longitudinal boards can take up a share of the load. A load may thus be supported by ten boards and two ledgers.

Table 33.8 gives the working bending moment of a 38 mm pine board at 433 kNmm. Table 31.19 gives the allowable bending moment on a BS 1139 4 mm steel tube as 792 kNmm.

In the case of the platform ten boards wide, all ten boards contribute 4330 kNmm and the two ledgers 1584 kNmm, giving a total of 5914 kNmm. The boards contribute about 73% of this total.

To achieve this, the placing of boards will require very close engineering supervision on site.

33.10 Hazards with utilising the strength of boards

- Some scaffold bays may contain board joints because they have not been built as the designer intended. This difficulty can be put right on the job by erecting an A frame below the board-end transoms.
- If a second level of boards is placed over a set of board joints, there will be a trip in the platform surface.
- A hazard sometimes occurs in trying to spread wheel loads by using cross-boarding. The wheel loads will pass over individual top-layer boards, and these will transmit the load through to the lower layer of boards as a point

load and not a uniformly distributed load, thereby halving the safe bending strength of the lower boards.

33.11 Modulus of section of timber beams

Table 33.12. Modulus of section of timber beams

Nominal timber section				Modulus of section	
Width		Depth		cm^3	in^3
mm	in	mm	in		
50	2	38	1.5	12.03	0.75
50	2	50	2	20.83	1.33
50	2	75	3	46.88	3.00
50	2	100	4	83.33	5.33
50	2	114	4.5	108.30	6.75
50	2	150	6	187.50	12.00
50	2	203	8	343.41	21.33
50	2	228	9	433.20	27.00
75	3	38	1.5	18.50	1.13
75	3	50	2	31.25	2.00
75	3	152	6	288.80	18.00
75	3	228	9	649.80	40.50
100	4	38	1.5	24.07	1.50
100	4	50	2	41.67	2.67
100	4	100	4	166.67	10.67
114	4.5	38	1.5	27.44	1.69
114	4.5	50	2	47.50	3.00
152	6	38	1.5	36.58	2.25
152	6	50	2	63.33	4.00
152	6	75	3	142.50	9.00
152	6	100	4	253.33	16.00
203	8	50	2	84.58	5.33
203	8	63	2.5	134.28	8.33
228	9	50	2	95.00	6.00
228	9	75	3	213.75	13.50

33.12 Strength of poles used in timber scaffolds and temporary suspended scaffold roof rigs

The timber poles customarily used for scaffolding, especially suspended scaffold roof rigs, have a butt diameter of 115 mm and top diameter of 65 mm.

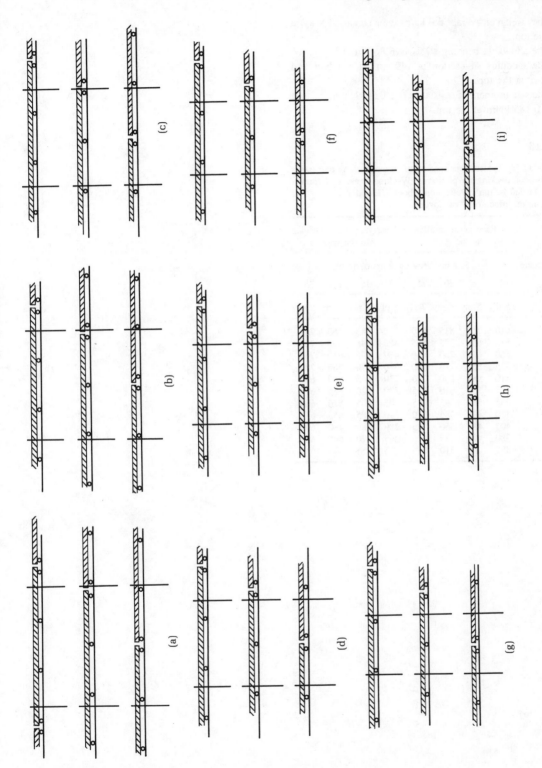

Figure 33.1 Various ways that scaffold boards can apply loads to transoms and ledgers.

They weigh on average 6.9 kg/m at the butt and 1.9 kg/m at the top.

The allowable bending stress is 6.86 N/mm^2.

The modulus of section is 149 cm^3 at the butt and 27 cm^3 at the top.

The safe moment of resistance is 1.023 kNm at the butt and 0.185 kNm at the top.

33.13

Table 33.13. Typical safe spacing of supports along two edges of plywood decking for platforms. Working stress 12 N/mm^2, $E = 11.000$ N/mm^2. Deflections about 1/12500 Y span (consult manufacturers' catalogues).

Distributed load (kN/m^2)	Face grain parallel to the span				Face grain perpendicular to the span			
	Spacing between supports (mm)							
	12	16	20	25	12	16	20	25
	5 ply	5 ply	5 ply	7 ply	5 ply	5 ply	5 ply	7 ply
1.00	610	710	815	915	510	585	660	735
1.50	585	660	785	865	480	545	635	725
2.00	560	625	760	810	430	495	610	715
2.50	535	585	710	760	380	470	585	700
3.00	510	560	660	750	330	445	560	685
3.50	485	535	610	735	320	420	535	675
4.00	455	510	595	725	305	405	510	660
4.50	430	495	585	710	290	385	455	650
5.00	405	485	560	685	280	380	480	635
5.50	380	470	535	660	265	370	465	620
6.00	355	455	510	635	255	355	455	610

34 Technical data on prefabricated beams

34.1 Types of beam

Section 9.4 lists the types of beam dealt with and gives diagrams of their shapes.

34.2 Method of evaluating safe working loads

The basic principle in evaluating the strength of a deep scaffold beam is to assess the strength of the top chord as a strut, having regard to the manner and frequency of the means to prevent lateral displacement. Assuming the lower chord is evenly stabilised, the safe bending resistance of the beam is the safe top chord load as a strut multiplied by the depth of the beam. This principle is adequate for beams of 600 mm and greater depth.

For beams shallower than 600 mm, the strength should be found by testing. They are usually stronger than the value calculated by the method described above. This is particularly true of ladder beams.

If a ladder beam is calculated by the method described above or by the more strict Vierendeel method and subsequently tested, it will be found to be stronger than expected. This is due to the large section areas of the elements at the welded joints, i.e. the joints are stiffer than expected, and the effective length of the elements between the joints is shorter.

Test results are acceptable proof of beam performance, but they must be done in a series with various degrees of top chord stabilisation. Manufacturers should produce these test results so that purchasers and users can rig the beams correctly for the anticipated load.

34.3 Methods of rigging beams

Figure 34.1 shows six methods of rigging unit beams, fish-bellied beams and ladder beams in which the frequency of the chord stiffening is increased from none to a close spacing. Where there is no stabilisation of the top chord, the beam will roll over on to its side when loaded and so will have little strength as a beam. The beam strength increases as the spacing of the chord stiffening is progressively reduced.

From this it will be seen that there is no single safe working load for a scaffold beam. The allowable load varies with the manner of its rigging, and tables of the various safe loads are given below.

The chord stiffeners are preferably attached to the beams with right-angle couplers, but the top chord stiffeners can sometimes double as purlins or transoms. In this case, it will be necessary to attach the stiffeners with putlog couplers so as not to interfere with the fixing of roof sheeting on the decking boards. If this is done, care should be taken to ensure proper tightening of the couplers. The band-and-plate type of right-angle coupler can be used so as not to interfere with the sheeting or decking.

In a family of beams, the beams should be braced from the top chord to the lower chord, and this bracing should also be attached by right-angle couplers.

34.3.1 Rigging type 1

This beam of any depth has only one coupler at its end, having been hung between two scaffolds and coupled only on the top chord. It will turn flat when loaded, with the lower chord swinging upwards. It will then have no beam strength and will act as a single scaffold tube between two supports. For this condition, load Table 31.18 applies, using the line detailed 4 mm, 116 N/mm^2, used steel simply supported. It will be seen that there is no advantage rigging a beam in this way.

34.3.2 Rigging type 2

This beam has both chords fixed at each end. Its top chord

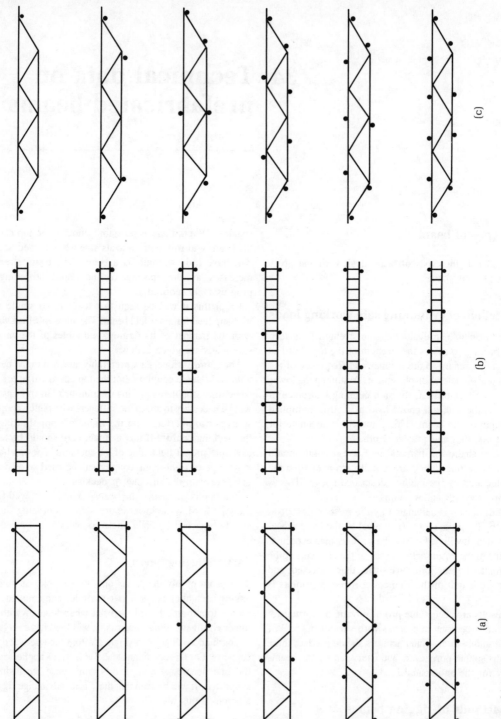

Figure 34.1 Types of prefabricated beam.

is a strut of the length of the span. The following tables give the safe load ratings for various beams with various spans.

This method of rigging is frequently seen in practice, but reference to the load tables shows how small is its strength compared with other types of rigging.

34.3.3 Rigging type 3

This beam is fixed on both its chords at its ends and in the centre. The shortening of the top chord strut produces a very large increase in the top strut strength, the safe moment and the safe load.

Fish-bellied beams are shown in Figure 34.1. In this case, the lower chord is shown stabilised in three places, but the bottom chord coupler cannot be at its ends.

If the depth of the fish-bellied beam is the same as that of the unit beam, the safe load will be the same. If the depth is less, the strength will be reduced in proportion.

34.3.4 Rigging type 4

This beam is coupled on both chords at its ends and at its third points.

34.3.5 Rigging type 5

This beam is coupled on both chords at its ends and at its quarter points.

34.3.6 Rigging type 6

This beam is coupled on both chords at its ends and at intervals of 2 m, 1.5 m, 1.2 m or 1 m.

34.4 Splicing of beams

Unit beams are joined end to end using the fish plates welded into the beams and the appropriate bolts specified by the manufacturer.

Fish-bellied beams can be lapped side by side using parallel couplers − preferably those that clamp the tubes closely together. The loads in the chords must be calculated and the correct number of parallel couplers fixed on both chords to sustain the calculated loads. An overlap of 2 m should be the minimum.

Ladder beams can be lapped side by side. An overlap of 2 m or six panels should be made. In the case of ladder beams, the chord loads must be calculated and the correct number of parallel couplers fixed on both chords. If the panels in a ladder beam are 10 in clear space, two bands and plates 5 in long each in each panel will provide metal-to-metal contact along each chord in each panel, with much enhanced strength. If the panels in the beam are 12 in clear space, beams should be overlapped by half a panel bonding, which will enable two bands and plates to be inserted to give metal-to-metal contact in each chord in each panel.

34.5 Load tables for prefabricated beams

The following tables give the characteristics of the three families of beams shown in Figure 34.1 for various spans, depths and rigging types. These apply to beams made of 4 mm used steel with a permissible stress of 116 N/mm².

34.6 The strength of angular joints in structures made from beams

When a portal frame is constructed from prefabricated beams, or in any structure where the elements are at fixed angles, the corners frequently have greater bending moments on them than the centres of the beams. The bending strength of a corner is governed by the number of couplers used to make the joint and the distance apart at which these are fixed.

If only single couplers are used to make a right-angle joint between two beams in a portal frame, the joint will have very little stiffness. With the most rigid right-angle coupler, the joint stiffness could be 500 Nm. With the least rigid, the joint stiffness may be zero.

When two couplers are used, the joint stiffness is the sum of that for each coupler, plus their slipping resistance multiplied by the lever arm between the couplers.

The tables in the preceding section show that in many cases the chord load is greater than 6.3 kN (635 kg), i.e. a single coupler on the chord will not safely transfer its thrust or tension around the corner of the portal. It is always good practice to use at least four couplers at every corner. Figure 34.2, which is a cross-section of the frontispiece, is an excellent example of the use of beams to make stiff-angle portals.

34.7 Hazard with portals in temporary structures

- The use of insufficient couplers on temporary buildings and roofs has led to many collapses. If the corners of a portal are not stiff enough, there is no relief of the moment in the centres of the beams, and the effective length of the side scaffold may be induced to rise to twice its real length.

34.8 Properties and loads for prefabricated beams used as columns

Both unit beams and ladder beams can be coupled together

Table 34.1. Unit beams. Rigging type 1. Any depth. Ends coupled on the top chord only.

		Span (m)					
		2	2.5	3	4	5	6
Safe UDL	kN	2.7	2.2	1.8	1.4	1.1	0.9
	kg	276	220	184	138	110	92
Self weight at 16 kg/m		32	40	48	64	80	96
Safe imposed UDL	kg	244	180	136	74	30	0
Safe imposed CPL	kg	122	90	68	37	15	0

Table 34.2. Unit beams. Rigging type 2. Coupled at the ends on both chords.

		Span (m)					
		3	4	5	6	7	8
Depth 500 mm							
Top chord safe strut strength	kN	11.5	6.7	4.3	3.0	2.0	1.0
Safe moment	kNm	5.8	3.4	2.2	1.5	1.0	0.5
Safe UDL	kN	15.5	6.8	3.5	2.0	1.1	0.5
	kg	1581	694	357	204	112	51
Self weight at 16 kg/m		48	64	80	96	112	120
Safe imposed UDL	kg	1533	630	277	108	0	0
Safe imposed CPL	kg	767	315	138	54	0	0
Depth 600 mm							
Top chord safe strut strength	kN	11.5	6.7	4.3	3.0	2.0	1.0
Safe moment	kNm	6.9	4.0	2.6	1.8	1.2	0.6
Safe UDL	kN	18.4	8.0	4.2	2.4	1.4	0.6
	kg	1877	816	428	245	143	61
Self weight at 16 kg/m		48	64	80	96	112	128
Safe imposed UDL	kg	1829	752	348	149	31	0
Safe imposed CPL	kg	915	376	174	75	15	0

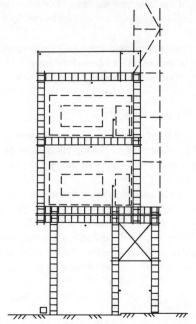

Figure 34.2 Use of prefabricated beams in portal-type structures.

into the shapes given in Figure 9.9 and erected as columns. In these configurations, each beam stiffens the other and, if they are coupled together frequently, a strong column results. The closer the spacing of the couplers up the column, the stronger is the resulting assembly.

For a composite column to work as an integral member, it must be able to resist internal longitudinal shears. Even with a coupler placed in every panel of a ladder beam, i.e. every 300 mm, it is unlikely that this condition will be met. Accordingly, it is not safe to rely on calculated values for the moment of inertia, the section modulus and the radius of gyration of the whole column assembly. Consequently, the length to radius of gyration ratio may be assessed wrongly, resulting in unsafe loads being allowed.

For these composite columns, it is safer to assess the l/r ratio for each of the vertical tubes separately, taking into account the spacing of the couplers and adding the values to give the supporting value of the column assembled from the beams.

Common beams are designed with panels modulated on 300 mm. It is convenient to assume spacings of the couplers between the beams using this modulation. For each interval, the l/r value and the strut value of the beam chord are assessed, and hence the supporting ability of the whole column.

Table 34.20 gives the allowable loads, based on a value of r of 1.57 cm, a sectional area of 5.57 cm^2 and steel with a yield strength of 210 N/mm^2.

34.9 Hazards using prefabricated beams as composite columns

- Table 34.20 shows that the range of loadings for each type of column is large, e.g. 41.25 to 12.55 tonnes for a triple-beam column. The former has nine couplers per storey in a building, the latter only two, one at the top and one at the bottom. The loss of strength if either of these couplers is left off is hazardous.
- The various legs forming a column are assumed to be evenly loaded. This requires careful detailing to achieve in practice, and errors on site are common.
- If the values of I, z and r are worked out for the whole of the composite column and used to assess the strength for any value of l, for long columns the corresponding loads may be estimated at twice the values given in Table

Table 34.3. Unit beams. Rigging type 3. Coupled at the ends and the centre on both chords.

		Span (m)							
		4	5	6	7	8	9	10	12
Depth 500 mm									
Top chord safe strut strength	kN	23.7	16.1	11.5	8.6	6.7	5.3	4.3	3.0
Safe moment	kNm	11.9	8.0	5.7	4.3	3.4	2.6	2.2	1.5
Safe UDL	kN	23.8	12.8	7.6	4.9	3.4	2.3	1.8	1.0
	kg	2428	1306	775	500	347	235	184	102
Self weight at 18 kg/m		72	90	108	126	144	162	180	216
Safe imposed UDL	kg	2356	1216	667	374	203	173	0	0
Safe imposed CPL	kg	1178	608	333	187	102	86	0	0
Depth 600 mm									
Top chord safe strut strength	kN	23.7	16.1	11.5	8.6	6.7	5.3	4.3	3.0
Safe moment	kNm	14.2	9.7	6.9	5.2	4.0	3.2	2.6	1.8
Safe UDL	kN	28.4	15.5	9.2	5.9	4.0	2.8	2.1	1.2
	kg	2897	1581	938	602	408	286	214	122
Self weight at 18 kg/m		72	90	108	126	144	162	180	216
Safe imposed UDL	kg	2825	1491	830	476	264	124	34	0
Safe imposed CPL	kg	1413	746	415	238	132	62	17	0

Table 34.4. Unit beams. Rigging type 4. Coupled at the ends and the third points on both chords.

		Span (m)								
		4	5	6	7	8	9	10	12	14
Depth 500 mm										
Top chord safe strut strength	kN	40.4	31.3	23.7	18.3	14.5	11.5	9.5	6.7	
Safe moment	kNm	20.2	15.7	11.8	9.2	7.3	5.8	4.8	3.4	
Safe UDL	kN	40.4	25.1	15.7	10.5	7.3	5.2	3.8	2.3	
	kg	4121	2560	1601	1071	745	530	388	235	
Self weight at 20 kg/m		80	100	120	140	160	180	200	240	
Safe imposed UDL	kg	4041	2460	1481	931	585	350	188	0	
Safe imposed CPL	kg	2020	1230	740	465	293	175	94	0	
Depth 600 mm										
Top chord safe strut strength	kN	40.4	31.3	23.7	18.3	14.5	11.5	9.5	6.7	5.0
Safe moment	kNm	24.2	18.8	14.2	11.0	8.7	6.9	5.7	4.0	3.0
Safe UDL	kN	48.4	30.1	18.9	12.6	8.7	6.1	4.6	2.7	1.7
	kg	4937	3068	1931	1282	887	626	469	275	173
Self weight at 20 kg/m		80	100	120	140	160	180	200	240	280
Safe imposed UDL	kg	4857	2968	1811	1142	727	446	269	35	0
Safe imposed CPL	kg	2429	1484	906	571	364	223	135	18	0

34.25. This is dangerous if any of the couplers joining the beams together show even a small degree of slipping.

34.10 Beams used as scaffolding elements

Where scaffolds have to carry loads in excess of those that can be supported by single tubes, the entire structure can be assembled using ladder beams as standards, ledgers and transoms. For each of these elements, care must be taken to incorporate adequate stiffness to prevent the beams buckling sideways.

When bracing cannot be incorporated in a structure, as when traffic access has to be provided through the legs, the ladder beams forming the portal frames should be fixed with four couplers at each joint to assist the lateral stability of the whole structure. If this is not sufficient, a place for bracing must be found.

The frontispiece is an excellent example of this use of ladder beams as scaffolding elements. The structure is a storage bay for precast concrete wall slabs, a starter frame for an ordinary access scaffold and an unloading bay for lorries.

Table 34.5. Unit beam. Rigging type 5. Coupled at the ends and the quarter points on both chords.

		\multicolumn{6}{c}{Span (m)}					
		4	6	8	10	12	14
Depth 500 mm							
Top chord safe strut strength	kN	49.1	35.7	23.7	16.1	11.5	8.6
Safe moment	kNm	24.5	17.9	11.9	8.1	5.8	4.3
Safe UDL	kN	49.0	23.9	11.9	6.5	3.9	2.5
	kg	4998	2438	1214	663	398	255
Self weight at 22 kg/m		88	132	176	220	264	308
Safe imposed UDL	kg	4910	2306	1038	443	134	0
Safe imposed CPL	kg	2455	1153	519	222	67	0
Depth 600 mm							
Top chord safe strut strength	kN	49.1	35.7	23.7	16.1	11.5	8.6
Safe moment	kNm	29.5	21.4	14.2	9.7	6.9	5.2
Safe UDL	kN	59.0	28.6	14.2	7.7	4.6	3.0
	kg	6018	2913	1450	788	469	301
Self weight at 22 kg/m		88	132	176	220	264	308
Safe imposed UDL	kg	5930	2781	1274	568	205	0
Safe imposed CPL	kg	2965	1391	637	284	103	0

Table 34.7. Ladder beam. Rigging type 1. Any depth. Ends coupled on the top chord only.

		\multicolumn{6}{c}{Span (m)}					
		2	2.5	3	4	5	6
Safe UDL	kN	2.7	2.2	1.8	1.4	1.1	0.9
	kg	276	220	184	138	110	92
Self weight at 13.5 kg/m		27	34	41	54	68	82
Safe imposed UDL	kg	249	186	143	84	42	10
Safe imposed CPL	kg	125	93	74	42	21	5

Table 34.8. Ladder beam. Rigging type 2. Various depths. Coupled at the ends on both chords.

Beam depth (mm)			\multicolumn{3}{c}{Span (m)}		
			4	5	6
300	UDL	kg	427	82	50
	CPL	kg	214	41	25
350	UDL	kg	509	214	74
	CPL	kg	255	107	37
400	UDL	kg	592	256	98
	CPL	kg	296	128	49
450	UDL	kg	1673	298	121
	CPL	kg	337	149	61

Table 34.6. Unit beams. Rigging type 6. Coupled at the ends and at intervals not greater than 2 m on both chords for a depth of 500 mm or 1.2 m on both chords for a depth of 600 mm.

		\multicolumn{7}{c}{Span (m)}						
		4	6	8	10	12	14	16
Depth 500 mm								
Top chord safe strut strength	kN	23.7	23.7	23.7	23.7	23.7	23.7	23.7
Safe moment	kNm	11.9	11.9	11.9	11.9	11.9	11.9	11.9
Safe UDL	kN	23.8	15.9	11.9	9.5	7.9	6.8	6.0
	kg	2428	1622	1214	971	806	694	612
Self weight at 20 kg/m		80	120	160	200	240	280	320
Safe imposed UDL	kg	2348	1502	1054	771	566	414	292
Safe imposed CPL	kg	1174	751	527	386	283	207	146
Depth 600 mm								
Top chord safe strut strength	kN	44.1	44.1	44.1	44.1	44.1	44.1	44.1
Safe moment	kNm	26.46	26.46	26.46	26.46	26.46	26.46	26.46
Safe UDL	kN	52.9	35.3	26.5	21.2	17.6	15.1	13.2
	kg	5396	3601	2703	2162	1996	1540	1346
Self weight at 22 kg/m		88	132	176	220	264	308	352
Safe imposed UDL	kg	5308	3469	2527	1942	1732	1232	994
Safe imposed CPL	kg	2654	1735	1263	971	866	616	497

Table 34.9. Ladder beam. Rigging type 3. Various depths. Coupled at the ends and centres on both chords.

Beam depth (mm)			Span (m)		
			4	5	6
300	UDL	kg	1675	864	466
	CPL	kg	838	432	233
350	UDL	kg	1966	1022	559
	CPL	kg	983	511	280
400	UDL	kg	2256	1180	654
	CPL	kg	1128	590	327
450	UDL	kg	2548	1337	748
	CPL	kg	1274	669	374

Table 34.10. Ladder beam. Rigging type 4. Various depths. Coupled at the ends and at the third point on both chords.

Beam depth (mm)			Span (m)		
			4	5	6
300	UDL	kg	2905	1757	1063
	CPL	kg	1453	878	532
350	UDL	kg	3400	2053	1256
	CPL	kg	1200	1027	628
400	UDL	kg	3895	2370	1450
	CPL	kg	1948	1185	725
450	UDL	kg	4388	2677	1643
	CPL	kg	2194	1339	822

Table 34.11. Ladder beam. Rigging type 5. Various depths. Coupled at the ends and at the quarter points on both chords.

Beam depth (mm)			Span (m)		
			4	5	6
300	UDL	kg	3545	2432	1650
	CPL	kg	1773	1216	825
350	UDL	kg	4145	2852	1942
	CPL	kg	2073	1426	971
400	UDL	kg	4747	3271	2680
	CPL	kg	2374	1636	1340
450	UDL	kg	5347	3690	2568
	CPL	kg	2674	1845	1284

Table 34.12. Ladder beam. Rigging type 6. Various depths. Coupled at the ends and stabilised every 2 m on both chords. (*Note:* These values are 1.2 × the calculated loads. Test values confirm this.)

Beam depth (mm)			Span (m)		
			4	5	6
300	UDL	kg	1675	1310	1063
	CPL	kg	838	655	532
350	UDL	kg	1966	1542	1256
	CPL	kg	983	771	628
400	UDL	kg	2256	1775	1450
	CPL	kg	1128	888	725
450	UDL	kg	2546	2000	1643
	CPL	kg	1273	1000	822

Table 34.13. Ladder beam. Rigging type 6. Various depths. Coupled at the ends and stabilised every 1.2 m on both chords. (*Note:* These values are 1.2 × the calculated loads. Test values confirm this.)

Beam depth (mm)			Span (m)		
			4	5	6
300	UDL	kg	3172	2508	2060
	CPL	kg	1586	1254	1030
350	UDL	kg	3710	2939	2419
	CPL	kg	1855	1470	1210
400	UDL	kg	4248	3370	2778
	CPL	kg	2124	1685	1389
450	UDL	kg	4788	3802	3138
	CPL	kg	2894	1901	1569

Table 34.14 Fish-bellied beam. Rigging type 1. Ends coupled on the top chord only.

		Span (m)			
		6	8	10	12
Depth 450 mm					
Safe imposed UDL	kg		Zero for any span		
Safe imposed CPL	kg		Zero for any span		
Depth 600 mm					
Safe imposed UDL	kg		Zero for any span		
Safe imposed CPL	kg		Zero for any span		

Table 34.15. Fish-bellied beam. Rigging type 2. Coupled at the ends on the top chord and twice on the bottom chord.

		Span (m)			
		6	8	10	12
Depth 450 mm					
Top chord safe strut strength	kN	3.0	0	0	0
Safe moment	kNm	1.35			
Safe UDL	kN	1.80			
	kg	184			
Self weight at 14 kg/m					
Safe imposed UDL	kg	100			
Safe imposed CDL	kg	50			
Depth 600 mm					
Top chord safe strut strength	kN	3.0	0	0	0
Safe moment	kNm	1.8			
Safe UDL	kN	2.4			
	kg	245			
Self weight at 15 kg/m		90			
Safe imposed UDL	kg	155			
Safe imposed CPL	kg	78			

Table 34.16. Fish-bellied beam. Rigging type 3. Coupled at the ends and the centre on the top chord and three times on the bottom chord.

		Span (m)			
		6	8	10	12
Depth 450 mm					
Top chord safe strut strength	kN	11.5	6.7	4.3	3.0
Safe moment	kNm	5.18	3.02	1.94	1.35
Safe UDL	kN	6.88	3.02	1.55	1.08
	kg	703	308	158	110
Self weight at 14 kg/m		84	112	140	168
Safe imposed UDL	kg	619	196	18	0
Safe imposed CPL	kg	310	98	9	0
Depth 600 mm					
Top chord safe strut strength	kN	11.5	6.7	4.3	3.0
Safe moment	kNm	6.5	4.02	2.58	1.8
Safe UDL	kN	9.20	4.02	2.06	1.20
	kg	938	410	210	122
Self weight at 15 kg/m		90	120	150	180
Safe imposed UDL	kg	848	290	110	0
Safe imposed CPL	kg	424	145	55	0

Table 34.17. Fish-bellied beam. Rigging type 4. Coupled at the ends and at the third points on the top chord and four times on the bottom chord.

		Span (m)			
		6	8	10	12
Depth 450 mm					
Top chord safe strut strength	kN	23.7	14.5	9.4	6.7
Safe moment	kNm	10.67	6.53	4.23	3.02
Safe UDL	kN	14.23	6.53	3.38	2.01
	kg	1451	666	345	205
Self weight at 14 kg/m		84	112	140	168
Safe imposed UDL	kg	1367	554	205	37
Safe imposed CPL	kg	684	277	103	18
Depth 600 mm					
Top chord safe strut strength	kN	23.7	14.5	9.4	6.7
Safe movement	kNm	14.22	8.70	5.64	4.02
Safe UDL	kN	18.96	8.70	4.51	2.68
	kg	1934	887	460	273
Self weight at 15 kg/m		90	120	150	180
Safe imposed UDL	kg	1844	767	310	93
Safe imposed CPL	kg	922	384	155	47

Table 34.18. Fish-bellied beam. Rigging type 5. Coupled at the ends and at the quarter points on the top chord and four times on the bottom chord.

		Span (m)			
		6	8	10	12
Depth 450 mm					
Top chord safe strut strength	kN	35.7	23.7	16.1	11.5
Safe moment	kNm	16.07	10.67	7.25	5.18
Safe UDL	kN	21.43	10.67	5.80	3.45
	kg	2186	1088	592	352
Self weight at 14 kg/m		84	112	140	168
Safe imposed UDL	kg	2102	976	452	184
Safe imposed CPL	kg	1051	488	226	92
Depth 600 mm					
Top chord safe strut strength	kN	35.7	23.7	16.1	11.5
Safe movement	kNm	21.42	14.22	9.66	6.90
Safe UDL	kN	28.56	14.22	7.73	4.60
	kg	2913	1450	788	469
Self weight at 15 kg/m		90	120	150	180
Safe imposed UDL	kg	2823	1330	638	289
Safe imposed CPL	kg	1412	665	319	145

Table 34.19. Fish-bellied beam. Rigging type 6. Coupled at the ends. Stabilised every 2 m (depth 450 mm) or 1.2 m (depth 600 mm) on the top chord and four times on the bottom chord.

		Span (m)			
		6	8	10	12
Depth 450 mm					
Top chord safe strut strength	kN	23.7	23.7	23.7	23.7
Safe moment	kNm	10.665	10.665	10.665	10.665
Safe UDL	kN	14.22	10.665	8.535	7.11
	kg	1450	1088	871	725
Self weight at 14 kg/m		84	112	140	168
Safe imposed UDL	kg	1366	976	731	557
Safe imposed CPL	kg	683	488	366	278
Depth 600 mm					
Top chord safe strut strength	kN	44.1	44.1	44.1	44.1
Safe moment	kNm	26.46	26.46	26.46	26.46
Safe UDL	kN	35.28	26.46	21.17	17.64
	kg	3601	2703	2162	1996
Self weight at 15 kg/m		90	120	150	180
Safe imposed UDL	kg	3511	2583	2012	1816
Safe imposed CPL	kg	1756	1292	1006	908

Table 34.20. Allowable loads in composite columns in BS 1139 scaffold tubes

Column type (see Figure 9.9)	Vertical spacing of the couplers along the beams (mm)							
	300	600	800	1200	1500	1800	2100	2400
(a) kN	134.8	124.2	119.8	97.6	84.00	66.8	56.26	41.00
tonnes	13.75	12.67	12.22	9.96	8.57	6.81	5.74	4.18
(b) kN	269.6	248.4	239.60	195.2	168.00	133.60	112.52	82.00
tonnes	27.50	25.34	24.44	19.19	17.14	13.63	11.44	8.36
(c) kN	404.4	372.6	359.4	292.9	252.00	200.4	168.8	123.00
tonnes	41.25	38.01	36.66	29.88	25.70	20.44	17.22	12.55
(d) kN	539.2	496.8	479.2	390.4	336.00	267.2	225.04	164.00
tonnes	55.00	50.67	48.88	39.82	34.27	27.25	22.95	16.73

Note: For any load above 20 tonnes and for any height above 6 m, the *l/r* value of the composite column must be assessed and the appropriate allowable compressive stress not exceeded for any eccentricity of load that may occur on site.

35 Prefabricated frames and scaffold systems

35.1 Frames

Many prefabricated frames have been designed to enable a scaffold to be built by unskilled labour. Most of these frames are intended to be placed at right angles to the building, i.e. they contain the transoms that carry the boards. Two types of frame have been popular, 1.25 m or 1.5 m high for bricklaying, and 1.8 m or 2 m for walk-through frames for painting and rendering.

The frames are joined longitudinally by cross braces and where there are platforms by prefabricated ledgers.

Some systems have required special-length boards. Others have used standard scaffold boards. Guard rails, guard rail posts and brick guards have been prefabricated to couple into the frames.

To deal with changes in ground level, adjustable screw bases are used at the bottom of the uprights.

Figure 35.1 shows various scaffold frames and systems:

(a) H frames with built-in ledger couplers.
(b) Walk-through H frames with ledger couplers.
(c) Brick wall frames and cross braces.
(d) Walk-through frames and cross braces.
(e) Mobile transoms with ledger couplers.
(f) Transoms without couplers, extended to provide for an inside board.
(g) Triangular frames for brickwork and for birdcages.
(h) Hop-up brackets.
(i) Ledgers and transoms without couplers.
(j) Adjustable braces.
(k), (l) and (m) Scaffold systems without couplers.

35.2 The strength of frames

The user should follow two lines of investigation. Firstly, he should obtain from the manufacturer or supplier the technical properties of all the materials and components in the system so that after assembly the characteristics of the scaffold can be compared with the standard tube-and-coupler scaffolds described in this book. For instance, it may be found that the tube in the frames is made of a high-tensile steel, or that it is made of tubes with a thin wall. Secondly, the manufacturer's or supplier's rules for assembling the elements must be obtained and followed.

Under the Health and Safety at Work Act, the supplier is required to have available the rules relating to the use of the system he supplies. These rules will be derived from previous experience with assembly and will be a better guide to producing a safe scaffold than technical calculations, which will be made unreliable by the lack of data on the multiplicity of joints of unknown stiffness.

35.3 Scaffold systems without frames

Systems based on straight tubes with half fittings on them that interlock with half fittings on other tubes are similar to tube-and-coupler scaffolds and should be constructed using the same rules.

The procedures described in Section 35.2 should be followed to obtain the necessary technical data to ensure that safe structures are built.

35.4 Stability

Many systems are available that use a lighter, thinner-walled tube than conventional scaffold tube. These systems will have greater flexibility than conventional tubes and couplers. Therefore, they will require a higher degree of lateral and transverse stability. The user must ensure that the means built into the system to achieve this stability are adequate. The tying of a prefabricated scaffold to a building should follow the same rules as those for conventional scaffolding.

Another feature to be borne in mind when using scaffold systems with lighter tubes is that the mass of the assembled scaffold will be considerably less with such systems than

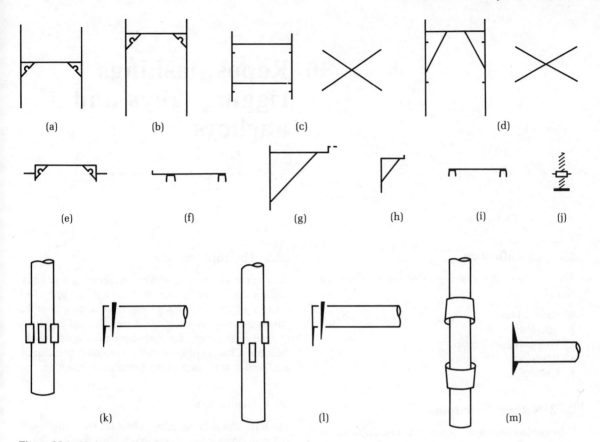

Figure 35.1 Various scaffold frames and systems.

with conventional scaffolds on account of the thinner-walled tube, and sometimes because of the lighter jointing method. The mass inertia of a lightweight scaffold is less, so if it is knocked accidentally, it will react more violently, perhaps to the point of collapse if it is fully loaded.

35.5 Hazards with system scaffolds

- Collapses of system scaffolds have resulted from an earnest belief that because they are prefabricated they are foolproof. The manufacturer's rules for bracing have

been disobeyed, and the scaffolds have not been adequately tied to the building.

- Sometimes, components that have been correctly incorporated into a finished scaffold have been removed in the belief that, because the system is made of prefabricated components, it can be taken to pieces during use.
- Collapses have been caused by having all the joints in the uprights at a similar height, or on two levels. Longitudinal continuity is also relatively low, and this requires special consideration in many cases.

36 Ropes, lashings, rigging, guys and anchors

36.1 Classification

The ropes used in scaffolding and allied industries can be classified into:

- Natural fibre.
- Man-made fibre.
- Wire-cored wire rope.
- Fibre-cored wire rope.
- Lashings.

36.2 Natural fibre ropes

Four subdivisions need recognition: manila, hemp, sisal and coir. (Cotton rope is ideal for manual use externally, wet or dry, but is not a commonly used type in the construction industry.) Manila is the highest quality, hemp and sisal follow, and coir is the least reliable.

There are three uses in the industry: for man riding on suspended scaffolds but not slung scaffolds, for materials handling using gin wheels or pulley blocks, and for lashing timber poles in pole scaffolds and pole roof rigs for suspended scaffolds.

For suspended scaffolds, 20 mm (0.75 in) manila rope is the most suitable in continuous lengths without splices. These systems are for raising and lowering and are a means of suspension, and therefore are subject to the Construction (Lifting Operations) Regulations 1961. The ropes are lifting gear and require a means of identification to be attached. This is best done by fixing a collar on to each end with the number imprinted on it. The collar can be of copper, aluminium, plastic or rubber about 25 mm long that is crimped or stretched on to the rope to act as an end whipping.

The rope should be inspected thoroughly when it is taken out of store to go on to a job and again when it is returned to store from a job. The inspector should record his findings against the date on the register.

36.3 The rope register

Because ropes deteriorate with age, a register of their actual and in-use life must be kept. Each rope should be booked out to a job and booked back into store at the end of the job. In this way, its actual age is recorded and the length of time it has been in use. Inspections should also be recorded. Changes of use should be recorded, particularly a relegation from man riding to materials handling.

36.4 Useful life

The limit should be set at two years for the active life of any man riding rope, except that it must be taken out of service earlier if inspections reveal defects. A limit of four years should be set on the actual life of the rope, i.e. its active life plus the time it is in storage.

After these limiting times have been reached, the rope should still be in good condition. It need not be discarded. It can be used for materials handling with gin wheels.

36.5 Inspection

The Construction (Lifting Operations) Regulations 1961 require that ropes, as an element of lifting gear, be inspected every six months and that a record of each inspection be completed and kept. This can be done from the data in the rope register. Current records for all ropes in use by any contractor must be maintained.

The inspection should be visual, and the whole length of the rope should be inspected carefully, including its inside surfaces. It is difficult to set down what degree of defects can be tolerated. Obviously, ropes begin to wear as soon as they are put into use. A figure of 5% worn-out strands is a useful guide for an inspector to regard as a limit, but his own judgement should be applied to any observation. Rotting is difficult to assess but, if the fibres show visible signs of rot or chemical deterioration, the rope must be discarded. The best way of judging is to open up the lay of the rope over a 150 mm length and pick at the fibres with a spike, first in a new rope and then in the rope under inspection. This comparison will guide the inspector.

When a rope has been left on a puddled pavement or in a flooded light well, or it has been run over by traffic or factory trucks, the inspector must be more careful. If the rope has been used in a chemical works or an area with a polluted atmosphere, or it has been on a job a very long time − during which time it should have been inspected every six weeks − it should receive special attention from the inspector.

36.6 Material characteristics

Cotton is the most flexible. Both cotton and hemp are soft ropes. Of the harder ropes, the top grade of manila is the strongest. Sisal has average characteristics, and coir (coconut fibre) is the least reliable, particularly when used over pulleys.

Ropes that have been treated with a water repellant will give longer service. This treatment should be done by the manufacturer and not by using chemicals on the building site.

The handiest diameter in any of the materials is 20 mm. Gin wheel ropes 16−25 mm are in use, but 16 mm is rather thin to grasp firmly, and 25 mm is bulky to store.

As a general guide, the breaking strength of a 25 mm rope will be 3.5−4.5 tonnes. For a 20 mm rope, it will be 2.5−3.5 tonnes, and for a 16 mm rope, it will be 1.5−2.5 tonnes.

The safety factors that should be used for building applications are given in Section 36.19.

36.7 Rope tension

The actual load lifted should be assumed to be greater than it really is by a factor of 1.05 to cover manually applied impacts and a further factor of 1.05 for each pulley over which the rope passes. This factor for the friction of each pulley is for well-maintained and oiled pulleys. Badly maintained pulleys may require a 15% increase in rope tension.

If the ropes have to be knotted or tied around goods, the strength should be assumed to be reduced by 20%.

36.8 Storage

Ropes must be stored under cover in a well-ventilated store that will ensure the drying out of ropes that have been lying around on wet ground. The ropes should be formed into hanks and kept on wooden pegs.

Ropes of a smaller diameter than those referred to above are used as traversing lines in suspended scaffolds and as rigging ropes generally. Any degree of damage destroys a greater percentage of a small rope than of a thicker one, so the care and storage of ropes below 16 mm in diameter must be carefully supervised.

36.9 Strength

These values are approximate only. The manufacturer must be consulted because there are several grades of fibre quality.

Differences in the strengths of various grades and materials should not necessarily rule out the use of the lower qualities. The safety factor adopted can be varied. For instance, if a duty requires a safety factor of 6 for manila, a hemp rope could be used at a factor of 7, a sisal rope at a factor of 8 or a coir rope at a factor of 10.

36.10 Man-made fibre ropes

Man-made fibre ropes divide roughly into three classes: those with a polished, shiny surface, those with a natural fibre type of finish, and those with a film or raffia type of surface. In the scaffolding industry, the middle group is preferred because it is safer to handle in all weather conditions. Staple polypropylene is best suited for manually operated suspensions.

It is not possible to give guidance on the strength of these ropes. The manufacturer's tables must be obtained and used.

The safety factors should follow the recommendations of Section 36.19.

Man-made fibre ropes are more durable than natural fibre ropes and so can be retained in use for man riding for four years, subject to a satisfactory periodic inspection that reveals no deterioration, wear or chafing. Their total age can be up to six years before relegation to materials handling.

Special applications such as use in a chemically polluted environment should be subject to the manufacturer's agreement. Failure due to heat, particularly from the flame of a blow lamp or torch, must be guarded against. The resis-

Table 36.1. Typical strengths of natural fibre ropes

Diameter (mm)	Breaking load (kg)
10	550
12	850
16	1600
20	2500
25	3600

Table 36.2. Typical strengths of wire ropes.

Diameter (mm)	Breaking load (kg)
6	2000
8	3500
10	6000
12	8500
16	14000

tance of some man-made fibre ropes to a blow-lamp flame may be only a few seconds.

The recommendations given for natural fibre ropes concerning registration, inspection, rope tension and storage also apply to man-made fibre ropes.

36.11 Wire ropes

Wire ropes fall generally into two classes: fibre-cored and wire-cored. Both are used in the scaffolding industry. Fibre-cored ropes are more flexible and generally more satisfactory.

36.12 Strength

The safety factor should be as indicated in Section 36.19.

The lifted load should be increased by a factor of 1.1 for impacts due to the manual operation of winches. Some climbing devices may require a factor of 1.3 to 1.4, and the machine manufacturer should be consulted about this. The increase in rope tension due to the wire rope passing around pulleys should be 1.05 in general, but for higher rope speeds this should be increased to 1.1 per pulley.

36.13 Inspection, certification and the rope register

The movements of wire ropes should be recorded in a register as for fibre ropes. Accordingly, they must be tagged with a serial number and a safe working load. They should be of a certified strength and inspected every six months, and on issue to and return from a contract site. Storage should be under cover in a well-ventilated dry store.

36.14 Rope terminations

These should be applied in the manner and using the equipment recommended by the rope manufacturer. The special components appropriate to the rope size must be used.

The terminations should have a strength of 80% of the breaking load of the rope.

36.15 Special rules for wire ropes for suspended scaffolding

In suspended scaffolding, precautions must be taken to prevent the platform falling free when it is lowered to the bottom limit. Where possible, the platform should be grounded before all the rope is fed out. When this is impossible, the rope must be prevented from coming away from the winch or lifting device. In the case of a drum winch, the rope must be correctly terminated on the drum, i.e. mechanically fixed to it and, in addition, there should be two turns of the rope left on the drum when the platform is used at its lowest prescribed level. Red paint on the last 2 m of the rope will also serve as a warning.

With a climbing device, the lower end of the rope has to be threaded through the device and so cannot have a permanent termination on it. A temporary loop can be formed with a bulldog grip. The end 2 m of rope can be painted red. In the case of a double rope suspension, the two ropes can be bulldogged together.

The ropes at the two ends of a suspended platform should be of the same length. The end stops should be fixed at the same height, and the warning paint should come into the view of the operator at the same time.

When safety ropes are used, these should preferably also have warning paint that comes into view at the same time. This is not a Code requirement, but it is an obvious precaution to take.

36.16 Lashings in a wire rope

The term 'lashing' is applied to both the piece of wire rope and the finished assembly. The rope is a 4–6 m length of a special wire 6 mm or 8 mm in diameter. The ends should be whipped, fused, or finished off with hard or soft nipples. Having one end terminated with an eye is a considerable advantage.

Because the rope is not of a certified strength, it cannot be used as a single straight element for lifting or lowering, or as a means of suspension, but when wrapped around two members to be joined together in a scaffold structure, it can be used as a means of suspension. Three complete loops are recommended, i.e. six parts of the rope, and the ends should be properly anchored or tied in and the six parts moused together.

Steel lashings should not be used to lash steel structural elements together when they are too small to be held tightly by the wire.

The strength of a single piece of lashing wire can be determined by testing. Breaking loads of the order of 1.5–2 tonnes are found for 6 mm diameter lashing wire and 2.5–3 tonnes for 8 mm diameter lashing wire. Applying a safety factor of 6 to the 6 mm wire for man riding applications and using a treble loop of six parts, the whole assembly gives a tensile strength of 1500 μg. For other elements, e.g. lashing poles together that are not directly supporting the platform, the same six-part lashing should be used, but a safety factor of 3 can be applied, giving a safe load of 3000 kg.

36.17 Lashing tubes

The friction between a steel wire rope and a steel element is dangerously low, and lateral forces will cause a lashing to slip along the element. Only direct tension loads should be allowed.

36.18 Lashings in fibre ropes

Fibre rope lashings can be of 10−12 mm diameter rope. If the lashings are used repeatedly, they should have the ends whipped or finished with rubber sleeves.

When wooden pole scaffolds were commonly used, the fibre rope lashing was started on the upright with a round turn and spiral as shown in Figure 32.1(a), which gives a seat for the ledger and a direct pull on the first two parts of the rope. This makes a better start than a clove hitch on the upright, from which the long end of the rope pulls out at a right angle, which may damage the rope.

The whole lashing should not contain any knots because these cannot be easily undone after the lashing has been loaded, especially if the rope is wet. The spiral avoids knots, and the short end is buried tightly in the rest of the lashing and cannot come free.

The clove hitch is not too difficult to undo when wet and is sometimes used as an alternative. It must be fixed on the upright in exactly the right position around the pole so that the long end pulls straight from the hitch and the knot does not roll.

Having started the lashing off with the round turn and spiral, it is continued as a square pattern as shown. It should have at least two circuits, giving eight parts, for 12 mm rope, and three circuits, giving twelve parts, for 10 mm rope. The circuits should be made with the loose end of the rope, which finally should be drawn under the loaded side of the lashing so that the applied weight traps it.

Diagonal bracing poles are usually fixed to the uprights and thus need a diagonal lashing. In scaffolding, lateral stability is achieved not only by the fixed length of the bracing elements, but also by the rigidity of the joints. The square lashing described above provides some degree of stability to the standard and ledger joint in a wood pole scaffold, but it is not as rigid as the forced right-angle coupler in a steel tube scaffold. The diagonal bracing pole may supply stability to the uprights by its fixed length between its lashings but, if these lashings are also of the type described above, they will contribute to the stability by their angular rigidity. Figure 32.1(d) shows the diagonal lashing formed in the same way as the square lashing, which achieves this objective. A cross lashing also does this.

36.19 Safety factors for ropes in scaffolding

- Traversing lines for cradles.
  Steering lines for lifted loads.
  Rigging ropes.   3
- Guy wires for stabilising towers.   4
- Guy wires for stabilising structural steelwork
  and masts during erection.   5
- Man riding ropes greater than 6 mm diameter.
  Crane ropes for simple work.   6
- Man riding ropes 6 mm or less in diameter.
  Fibre ropes for man riding.   8
- Crane ropes for rough work and scaffold jibs.   8−10
- Cableway ropes.   10−12
- Permanent lift ropes, man riding hoists.   12−15

36.20 Using ropes instead of tubes as bracing

There is little merit in using a rope set diagonally across a scaffold as a bracing element instead of a tube. Firstly, it is usually only possible to connect it to the top and bottom of the scaffold, so the intermediate lifts are not restrained laterally. Secondly, the rope has to stretch considerably before it takes up load and this may allow sufficient distortion in the scaffold to cause it to collapse before the rope can do its work.

36.21 Methods of attaching wire ropes to scaffolding

Indisputably, the most reliable way of attaching a rope to a scaffold is to use a rope with a hard eye at one end, which is passed over a tube.

In the centre of a length of rope, a short length of the same rope can be used looped around the scaffold and coupled to the main rope by bulldog grips.

Figure 36.1 gives various methods of attaching ropes to avoid damage by knots.

36.22 Component forces

Ropes can take only tension loads themselves, but they can apply compression loads in structures. The tension in a rope can create two component forces, as determined by the parallelogram of forces. These are frequently horizontal and vertical. Whenever ropes are joined to a structure, as in Figure 36.1, horizontal tension will be applied to the horizontal element of the scaffold framework. In the case

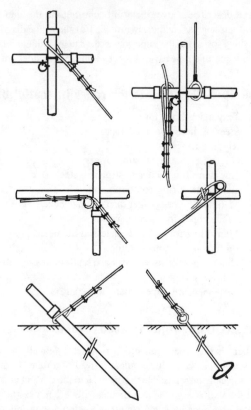

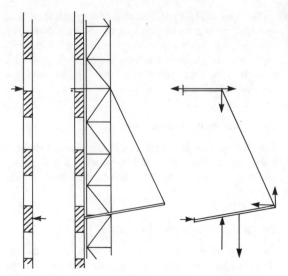

Figure 36.1 Attaching ropes to scaffolding and anchors.

shown, the transom across the top of the tower has to take the full horizontal component of the rope tension. In this case, the transom must be coupled to the framework with a right-angle coupler and not a putlog coupler.

The vertical component of the rope tension passes down the lower leg, and this load has to be added to the sum of the leg loads. Moreover, if the rope is attached around a ledger/standard joint, the whole joint tends to slip down the standard. Supplementary couplers may be needed below the ledger coupler to assist it.

36.23 Hazards in structures with rope elements

- A case that frequently escapes attention is in the support for protection fans. Figure 36.2 shows a protection fan supported by a steeply inclined rope from a higher part of the scaffold. The force diagram drawn at the ends of the rope shows that there is an inward force applied at the level of the fan, which must be transferred to the loading, as well as an uplift. At the top, there is an outward force on the scaffold as well as a vertical load. The scaffold must be tied to the building at the top to resist the force and butted on to the wall at lower levels.

Figure 36.2 The forces on and from a fan supported by wires.

Failure to estimate these inward and outward forces may lead to distortion in the scaffold, which may lead to its collapse.

- In a demolition job, the level of demolition progresses down the building and may reach a level where the top tie has been taken out or become inadequate. The fan may then collapse, and it may take the scaffold with it.

36.24 Guys

A study of the component forces resulting from guy tension shows that a guy at a flat angle has a greater horizontal component and a smaller vertical component than a steep guy. Some structures may need this at the top of the guy to minimise the loads in the standards. Some ground may permit only shallow anchors at the bottom of the guy, as in Figure 36.3(g). This situation needs the reduced uplift afforded by a low-angle guy.

In other circumstances, a steep guy may be needed for reasons of space. Special measures may then be needed at the top joint and special anchors at the bottom.

36.25 Anchors

Figure 36.3 shows several types of anchor suitable for use with either rope or tubular guys.

Type (a) is a scaffold tube 1500 mm long, which in typical grund will sustain about 200 kg. If this is inadequate, batches of three to six tubes can be driven and connected with lashings, as in (b), or preferably tubes, as in (c) and (d). Each tube driven contributes about 200 kg.

Type (e) is a cut from 75 by 75 mm structural angle

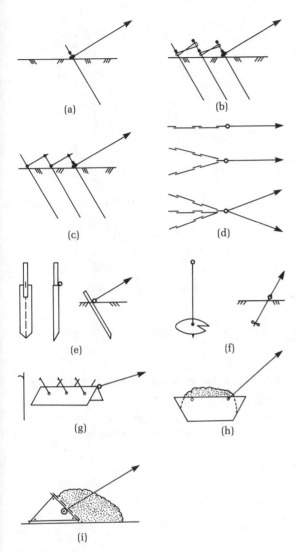

Figure 36.3 Ground anchors.

1.5 m long, presenting a 100 mm wide face against the pull. This will hold about 1500 kg in typical ground. This is suitable for driving into gravel, where tubes will not penetrate.

Type (f) is a spiral flight anchor with a diameter of 300–500 mm. It will hold 1000–1500 kg in typical ground. It is useful in clay but cannot be driven in stony ground.

Type (g) is for pinning down on to chalk and limestone, when this is near the surface. It needs a flat-angle guy because it has little resistance to uplift.

Types (h) and (i) are for cases where the ground cannot be penetrated by driven anchors, e.g. streets and floors. The weight of material in the skip or in front of the boarded scaffold frame is three times the pull required.

36.26 Internal guys

The normal concept of a guy is an inclined external rope. This has two disadvantages:

- The top of the structure is put into tension when the guy begins to resist wind forces.
- A lot of ground space is needed.

Both these disadvantages are overcome if the guys are attached to the downwind top corner of the structure and lie back at an angle through the structure.

36.27 Horizontal guys

Wire ropes are often used to span more or less horizontally across structures, as in the case of fixing safety nets. As the sag is taken out of the wire, the tension increases in inverse proportion. Thus a nearly tight wire requires a very high tension, which has to be accommodated in the supporting structure or otherwise taken through it by another guy to the ground.

The tension in a horizontal wire is given by the formula $T = wL/8H$, where L is the length, H is the sag, w is the total weight of the rope and anything it is supporting, such as a safety net.

As an example, a net across a motorway with a span of 30 m and weighing 2.5 kg/m^2, supported by 12 mm ropes at 4 m centres weighing 0.11 kg/m and rigged to sag 3 m, exerts a tension in the rope of 379 kg from the dead weight. Reducing the sag to 1 m triples the tension to 1137 kg.

The horizontal wind force on the net also has to be taken by the side guys, and 15% of the net area is sometimes taken as the effective area resisting the wind. In the case considered, this evaluates to $0.15 \times 30 \times 4 = 18$ m^2. At a typical wind pressure of 40 kg/m^2, the additional tension on the horizontal wire is 720 kg.

Total tension is 1099 kg at a sag of 3 m and 1857 kg at a sag of 1 m. The first of these forces needs two couplers and the second three couplers to transfer it to the structure.

36.28 Hazards with horizontal guys

- The force calculated above is very much greater than the slip resistance of a putlog coupler. If the supporting transom is coupled with putlog couplers, it will be pulled off.
- Slinging a safety net between scaffold towers exerts such large horizontal forces on the top of the towers that they are liable to be pulled over. Calculations must be made in relation to the stability and the number of couplers per guy.

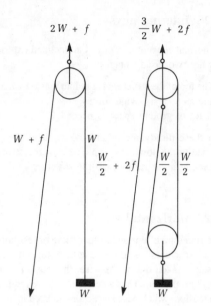
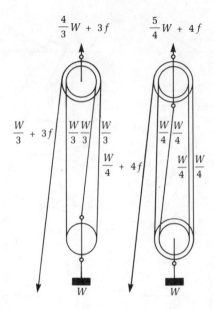

Figure 36.4   Rope tensions and supporting forces for rope block lifting gear. f is the pull required to overcome the friction of each arc of the pulley. It is usually taken as 2.5% of W per pulley.

36.29 Scaffold tube guys

Guys formed of scaffold tubes are frequently advantageous. In tension, they can sustain the load that their end couplers may carry. In thrust, they may be limited by the strut value of the length of tube installed.

36.30 Lifting rope tensions and pulley supports

Figure 36.4 gives some data on rope tensions when pulley blocks are used.

36.31 Hazard with pulley block mounting

- Failing to use a jib arm strong enough to resist the doubled or increased load exerted by the top pulley connection compared with the lifted load.

37 The weights on scaffolding

37.1 Imposed loads on normal scaffolds

Table 37.1.

Scaffold duty	Distributed load on the platform		
	kN/m^2	kg/m^2	lb/ft^2
Inspection and very light duty Painting, stone cleaning, access	0.75	76.5	15.7
Light duty Plastering, painting, stone cleaning, glazing, pointing	1.50	153	31.4
General purpose General building work, brickwork, window and mullion fixing, rendering and plastering	2.00	204	41.8
Heavy duty Blockwork, brickwork, heavy cladding	2.50	255	52.3
Special duty Masonry, concrete blockwork, very heavy cladding	3.00	306	62.7
Loading platforms	To be agreed with the building contractor		

The wind forces on scaffolding are dealt with separately in Chapter 38.

37.2 The self weight of scaffolding materials

Tubes

BS 1139 steel, 4 mm wall: 4.405 kg/m (2.98 lb/ft).
EN 39 steel, 3.2 mm wall: 3.58 kg/m (2.42 lb/ft).
BS 1139 aluminium, 4.5 mm wall: 1.674 kg/m (1.12 lb/ft).

Couplers

The weights in Table 37.2 are for use in assessing the self weight of a scaffold. Different types have slightly different weights.

Other fittings

Base plate:	1.00−1.81 (3.74 long stem)
Stair treads:	1.50−1.75
Parallel couplers:	0.80−0.90
Four-way couplers:	1.12−1.20
Finials:	0.73−0.82
Sheeting clips:	0.63−0.70
Sheeting bolts:	0.06−0.09
Timber rafter clips:	0.96−1.10
Girder couplers:	1.00−1.70 (3.20 for rollers)
Brick guards:	375−425
Castors:	8.00−10.00
Reveal pins:	1.00 (3.5 long stem)
Putlog ends:	0.54−0.70
Wire board clips and ladder clips:	0.44−0.65
Gin wheels:	4.88−6.35
Forkheads:	2.49−3.50

All weights in kilograms.

Boards

The values in Table 37.3 are sufficiently accurate to assess the self weight of a scaffold. They are based on a relative density of 0.6.

Table 37.2. Weights of various couplers, in kilograms

	Right angle	Swivel	Putlog	Spigot	Sleeve
Band and plate	1.40	1.47	0.96	1.16	1.06
Forged	1.25−1.35	1.16−1.50	0.62−0.70	1.10−1.20	1.10−1.30
Pressed	0.91−0.94	1.02−1.05	0.70−0.80	0.82−0.87	1.30−1.32
Mills type	1.00−1.10	1.10−1.20	0.70−0.75	1.10−1.15	1.10−1.20

Table 37.3. Weights of scaffolding boards

Thickness (mm)	Weight	
	kg/m^2	lb/ft^2
20	12.21	2.50
25	15.23	3.12
35	21.33	4.37
38	23.14	4.74
50	30.46	6.24
63	38.37	7.86
75	45.70	9.36

Quantities per tonne

Tube, steel:	228 m
Tube, aluminium:	607 m
Number of fittings, average weight 1.8 kg each:	560
Number of boards, 225 × 38 × 3.9 mm:	46

37.3 Weights and densities of materials stacked on platforms

Scaffold tube stacked
Steel: 2200 kg/m^3
Aluminium: 607 kg/m^3

Tarpaulins   1−1.2 kg/m^2

Ladders   8 kg/m

Pine poles
Butt end: 5.8 kg/m
Top end: 1.9 kg/m

Other timbers (softwood and pine)   500−650 kg/m^3

Mortar   2000 kg/m^3

Spotboard with mortar   30 kg

Wheelbarrow with mortar   150 kg

Concrete   2400 kg/m^3

Bricks
100 average bricks: 275 kg
100 engineering bricks: 400 kg
230 average bricks: scaffold coupler SWL
158 engineering bricks: scaffold coupler SWL

Water   1000 kg/m^3 (add 10% for containers)

Flooring tiles, slates, ceramics and roofing tiles
  1600 kg/m^3

37.4 Densities of various materials

Material	Density
Water:	1 tonne/m^3, 1 gm/cm^3, 1 kg/litre 62.4 lb/ft^3, 10 lb/gallon
Sea water:	1.05 tonne/m^3, 1.05 gm/cm^3, 1.05 kg/litre 65.5 lb/ft^3, 10.5 lb/gallon
Petrol and spirit:	0.8 kg/litre, 8 lb/gallon
Steel:	7.8 tonne/m^3
Iron:	7.5 tonne/m^3
Aluminium:	2.7 tonne/m^3
Zinc:	7.1 tonne/m^3
Lead:	11.4 tonne/m^3
Hardwood:	700 kg/m^3
Pine:	600 kg/m^3
Medium softwood:	500 kg/m^3
Very fast-growing wood:	450 kg/m^3
Slate:	2.6 tonne/m^3
Rock (solid):	2.5 tonne/m^3
Rock (crushed):	1.7 tonne/m^3
Concrete:	2.4 tonne/m^3
Masonry:	2.4 tonne/m^3
Sand:	2.0 tonne/m^3 (if dense and wet) 1.6 tonne/m^3 (if loose and dry)
Glass:	2.6 tonne/m^3
Ice:	0.92 tonne/m^3
Snow:	0.2 tonne/m^3 (if dense) 0.12 tonne/m^3 (if loose)
Bitumen and roofing compounds:	1.03 tonne/m^3 (add 10% for containers)

37.5 Weights of corrugated steel sheeting (without sheeting rails or purlins)

Table 37.4.

	Thickness (mm)	kg/m^2
26 g	0.5	4.82
24 g	0.6	5.77
22 g	0.7	6.70
20 g	1.0	9.62
18 g	1.2	11.56

37.6 Self weights of access scaffolding

Table 37.5. Weight of one lift of scaffolding one bay long without boards and supported on two standards, in kilograms

Bay length (m)	1.2	1.5	1.8	2.0	2.1	2.4	2.7
Lift height (m)							
1.20	61	63	66	68	69	71	74
1.37	62	65	67	69	70	73	75
1.5	63	66	69	70	71	74	77
1.8	66	69	71	73	74	77	79
2.0	68	70	73	75	76	78	81
2.1	69	71	74	76	77	79	82
2.4	71	74	77	78	79	82	85
2.7	74	77	79	81	82	85	87

Table 37.6. Extra weight to add for each boarded lift one bay long, including the imposed load on two standards, in kilograms. (Boards are assumed to be 38 mm thick.)

Width of scaffold	Loading (kN/m^2)	Length of bay (m)						
		1.2	1.5	1.8	2.0	2.1	2.4	2.7
		Extra mass of boarded lift (kg)						
3 boards	0.75	111	134	158	174	182	205	229
4 boards	0.75	139	168	202	220	230	261	291
4 boards	1.50	221	271	327	357	374	426	476
4 boards	2.00	276	339	410	450	470	536	600
5 boards	0.75	166	202	240	265	277	315	352
5 boards	1.50	269	331	393	437	456	521	583
5 boards	2.00	337	416	496	562	576	658	737
5 boards	2.50	405	501	599	667	695	796	891
5 boards	3.00	466	579	681	757	803	917	1031
6 boards	2.00	388	493	589	657	686	781	877
6 boards	2.50	480	595	711	791	830	946	1062
6 boards	3.00	562	696	834	927	974	1110	1247

37.7 Miscellaneous weights

- An average man weighs 80 kg (176 lb).
- A man carrying small tools weighs 90 kg.
- A man carrying a hod of bricks weighs 110 kg.
- A crowd standing closely weighs 480 kg/m^2.
- A wheelbarrow full of mortar weighs 150 kg. Allow 100 kg of horizontal load. Allow 50 kg extra on the man's weight for impact.
- A man carrying a wooden ladder — add 8 kg per metre of ladder.
- Two men together can load the same transoms or the same scaffold board.
- For wind load on a person, allow a vertical area of 1 m^2 per person.
- For a person drilling into a wall, allow 25 kg of horizontal force.
- For a person pulling on a rope, allow 25 kg.

38 Wind forces, and ice and snow loads

38.1 Importance

Scaffolding is usually outside. Lateral instability is its greatest enemy, so wind force, which is both horizontal and upwards, is a factor that induces many failures. Well-constructed, well-braced and properly tied scaffolds do not blow down. The wind finds out those that are substandard in one or other respect, even though they are carrying out their function satisfactorily.

The important points to keep in mind are that different parts of any country have different wind speeds and that when a building is erected, the wind speed around some of its surfaces may be increased because the wind is forced around its edges. Increases in ground altitude and in the height of a building expose it to increased wind speeds. There may be local areas of high exposure or shelter.

When the wind flow is obstructed, the surface obstructing the wind is subjected to pressure. This is proportional to the square of the wind speed. The wind force is the product of the surface area and the pressure.

38.2 Wind velocity

For the United Kingdom, the starting point for the assessment of the wind velocity for which one must design a scaffold is the basic wind speed map from the British Standard CP3, Chapter V, Part 2. This divides the United Kingdom into areas according to the maximum gust speeds that are exceeded on average only once in 50 years 10 m above the local ground level in open, level country. This map is Figure 38.1.

When the wind velocity has to be assessed for a particular site, the values given on this map must be corrected for three characteristics of the site:

(i) The general topography of the whole area.
(ii) The height of the building being served by the scaffold above the local ground level and the local topography.
(iii) The length of time that the scaffold will be standing.

Figure 38.1 Basic wind speeds in the United Kingdom in m/s.

Table 38.1. Values of the wind velocity coefficient S_2

Building height (m)	Open country with no obstructions	Some wind breaks	Many wind breaks, and small towns and suburbs	Large obstructions and city centres*
3	0.83	0.72	0.64	0.56
5	0.88	0.79	0.70	0.60
10	1.00	0.93	0.78	0.67
15	1.03	1.00	0.88	0.74
20	1.06	1.03	0.95	0.79
30	1.09	1.07	1.01	0.90
40	1.12	1.10	1.05	0.97
60	1.15	1.14	1.10	1.05
80	1.18	1.17	1.13	1.10
100	1.20	1.19	1.16	1.13
150	1.24	1.23	1.21	1.18
200	1.27	1.26	1.24	1.22

*Buildings around the perimeter of parks and open commons within city boundaries cannot be included under this heading.

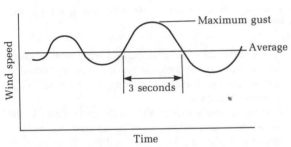

Figure 38.2 Wind gusting.

These variations are dealt with by velocity coefficients S_1, S_2 and S_3.

S_1 is taken as 1, unless the site is either exposed or sheltered. An exposed site is one on the top of a hill, on high moorland, in a marine estuary or in a location that may funnel the wind on to the site. In this case, S_1 is taken as 1.1.

A sheltered site is one in a depression or in a clearing in woodland. In this case, S_1 is taken as 0.9.

S_2 depends on the height of the building and on the local topography. The British Standard gives different factors for various parts of the building and different shapes of building. When the building walls are present, the scaffolding will have to resist forces similar to those affecting the building cladding, for which S_2 values are given in the Standard.

Table 38.1 gives the values of S_2 that are appropriate to scaffolding. The descriptions of local topography refer to distances of up to 250 m.

Figure 38.3 shows how S_2 is reduced by ground friction. Lower level wind speeds are also reduced by ground roughness and buildings or other obstructions.

The wind speeds shown in Figure 38.1 are for the maximum gust speed occurring once in 50 years. For short-term projects, i.e. of less than two years duration, a reduction factor S_3 is allowable. The value appropriate to access scaffolding up to this length of time is 0.77. Values less than this are not recommended.

S_1, S_2 and S_3 are applied to the gust speed obtained from the map to give the design speed V_s for the site:

$$V_s = V_{map} \times S_1 \times S_2 \times S_3.$$

This design speed may have to be modified for any

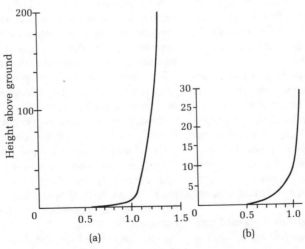

Figure 38.3 The S_2 coefficient at various heights above ground in open country. Diagram (b) is an enlargement of the lower part of diagram (a).

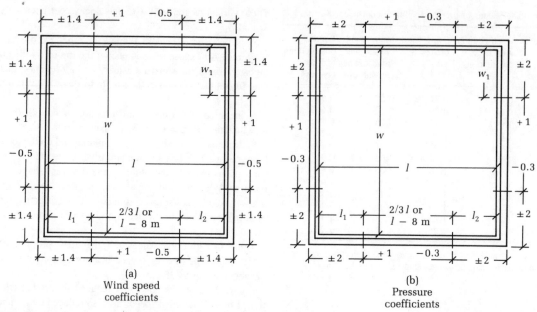

Figure 38.4 Velocity and corresponding pressure coefficients for various parts of a scaffold around a building. l_1 and l_2 should be $l/6$ up to a maximum of 4 m. W_1 should be $W/6$ up to a maximum of 4 m.

particular element of the scaffold after studying the wind stream diagrams. This modification can be applied by using further velocity coefficients S_x for various parts of the building, or it may be left until a later stage in the wind force calculations, when the equivalent force coefficients can be used instead. Figure 38.4 shows both the velocity coefficients and the force coefficients for various parts of the plan of a building. The designer must use one or the other coefficient and not both to find the effect of an isolated building on the wind force on the scaffold surrounding it.

When a wide building obstructs the wind, an upward velocity is created as the wind passes over the top. This must receive consideration by the scaffold designer, with special reference to the tendency of decking boards to be blown off, and of protective sheeting to be torn off or to drag the scaffold from the building.

Figures 38.5 and 38.6 give some guidance on this upward flow. Long buildings create a greater upward velocity than narrow ones because the air cannot escape around the sides of the building so easily. In long buildings, an upward velocity factor of 0.50 is probably representative of site conditions, corresponding to an upward pressure or force coefficient of 0.25. In narrow tower blocks, an upward velocity coefficient of 0.33 is a reasonable estimate, corresponding to an upward pressure or force coefficient of 0.1.

The upward velocity of the wind must be studied very carefully when high-level protection fans are included in the design.

The influence of holes in isolated walls must be taken into account because these increase the velocity of the wind stream through them, with consequent increase in the pressure and forces.

If the apertures exceed 40% of the surface area of the wall behind the scaffold, it can be assumed that the wall does not deviate the wind flow, and the whole length of the scaffold in front of or behind the wall will be subjected to the design velocity V_s and its consequent pressure. The extra velocity coefficient S_x is 1, and the force coefficient is 1.

If the apertures are less than 40% of the surface, the wind velocity will be increased through the holes and reduced opposite the solid areas. For those areas of the scaffold opposite the holes, the extra velocity coefficient S_x should be taken as 1.4, corresponding to a force coefficient of 2. For the areas of the scaffold opposite the solid wall, the extra velocity coefficient S_x should be 0.55, corresponding to a pressure coefficient of 0.3.

These values are shown diagrammatically in Figure 38.7, in which it will be seen that a hole near a corner prevents the acceleration of the wind just outside the corner.

These considerations are very important in demolition jobs, particularly when there is a single wall to be left in place. The scaffold may be on the upwind or the downwind face of the wall, depending on which way the wind is blowing.

At this stage in the estimation of wind effects, the designer should have decided on the pattern of wind flow for every

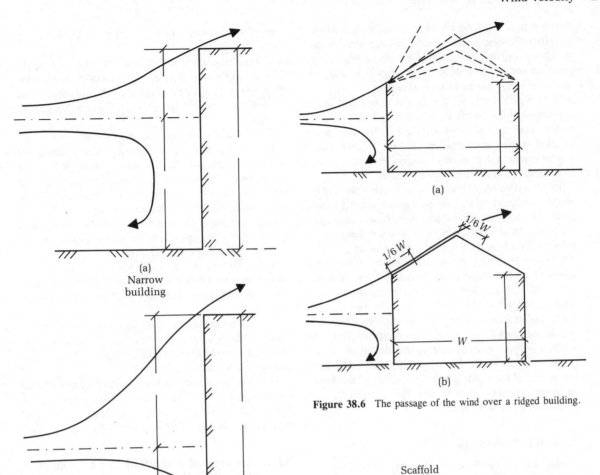

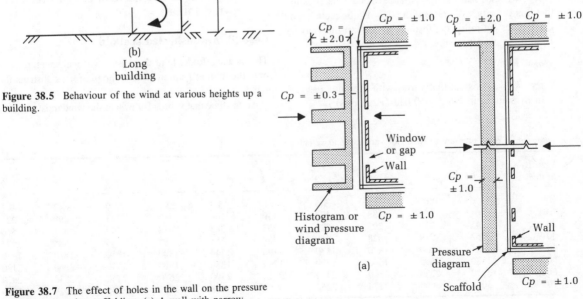

Figure 38.5 Behaviour of the wind at various heights up a building.

(a) Narrow building

(b) Long building

Figure 38.6 The passage of the wind over a ridged building.

Figure 38.7 The effect of holes in the wall on the pressure of the wind on the scaffolding. (a) A wall with narrow openings. (b) A wall with large openings.

direction from which the wind may come. He should have assessed the velocities directly at the building, deviated past the corners and diverted upwards. He should have studied the effect that adjacent buildings, openings in the walls and any other local circumstances will have.

If the wind velocity for the site has been found by the application of the factors S_1, S_2 and S_3 and further modified by the extra velocity coefficient S_x for any special location of the building or the site, the local pressures can be determined from this velocity. Pressure or force coefficients should not be applied as well.

This has been emphasised because in other parts of the world, there may not be a gust speed map to use as a starting point. The designer will have to use his local experience and apply factors S_2 and S_x to assess the design wind velocity for any particular part of the structure.

For new buildings, the structural engineer may have prepared a wind study of the building in its particular environment. If this includes a study of the wind speeds after the completion of the building, it can be used to assess the forces on the scaffolding. It is unlikely that the study will include data for the partly constructed building. The scaffold designer must be alert to this limitation.

The end point of the study of wind speeds adjacent to the surface of the building is the evaluation of the forces in the bracing and the ties for the tubular structure, with or without protection sheeting.

38.3 Wind pressure

Having assessed the velocity and direction of the wind at every part of the scaffold, it is possible to obtain the pressure that this velocity will create if the flow is obstructed. The Code gives the following formula to convert the velocity V_s into the wind pressure q:

$$q = K(V_s)^2,$$

where K is a constant that varies with the units used.

In SI units, q in $N/m^2 = 0.613 \ (V_s$ in $m/s)^2$.

In metric technical units, q in $kg/m^2 = 0.0625 \ (V_s$ in $m/s)^2$.

In Imperial units, q in $lb/ft^2 = 0.00256 \ (V_s$ in mph$)^2$.

To obtain the wind force on any element of the structure, the pressure q must be multiplied by the area obstructing the wind and by a shape pressure factor related to the element.

A scaffold tube has a shape pressure factor of 1.2.

The various parts of a scaffold around a completed building have pressure factors given in Figure 38.7, which can be used as an alternative to assessing the increase in wind speed on the various parts (p is the pressure coefficient and S_x is the increase in velocity factor).

Shape pressure factors or force coefficients for fences and signboards are given in Table 38.3 and Figure 38.10.

Plane surfaces have a shape factor of 1.

A scaffold tube with a shape factor of 1.2 can be considered to have a diameter of $1.2 \times 48.3 = 56$ mm. Its area is its length $\times$ 0.056 m^2. Couplers on the tube are ignored.

It is important to bear in mind that the pressure is proportional to the square of the velocity. If the velocity is increased around a corner of a building by 1.4 (40%), the pressure is doubled.

The Code gives a table to convert wind velocity to wind pressure with a shape factor of 1. This is reproduced as Table 38.2. To read the table, move down the left-hand column for intervals of 10 and across for units.

38.4 The area of a scaffold to be taken into account in the assessment of the wind force

Case (i) An unsheeted scaffold

This is a scaffold where there is no vertical sheeting, and only the tubular framework and the platforms obstruct the wind.

All the tube that is broadside on to the wind must be taken

Table 38.2. Wind pressure in N/m^2 for various wind velocities

Velocity in m/s	0	1	2	3	4	5	6	7	8	9
0	0	0.61	2.5	5.5	9.8	15.3	22.1	30.0	39.2	49.7
10	61	74	88	104	120	138	157	177	199	221
20	245	270	297	324	353	383	414	497	481	516
30	552	589	628	668	709	751	794	839	885	992
40	981	1030	1080	1130	1190	1240	1300	1350	1410	1470
50	1530	2590	1660	1720	1790	1850	1920	1990	2060	2130
60	2210	2280	2360	2430	2510	2590	2670	2750	2830	2920
70	3004	3090	3178	3267	3357	3448	3541	3635	3729	3826

Note: For intermediate velocities, interpolation can be used.

into account. The face area of the toe boards and the edges of the platforms should be included where they are broadside on to the wind. Braces should be evaluated as their full length if they lie in a plane at right angles to the wind, but taken at their projected length if they lie in the plane along the wind flow direction. The tube area should be increased by 1.2 to include the shape factor. The toe boards and platform edge areas should be increased by 1.5 for their shape factors.

Tubes that are end on to the direction of the wind and scaffold couplers can be ignored, except those in the band and plate system. For these, the length of the tube in the wind calculations should be increased by 2% to compensate for the couplers.

Case (ii) A sheeted scaffold

This is where there is vertical protection sheeting fixed around a scaffold, or the scaffold contains buildings such as contractor's cabins or other solid screens.

The whole sheeted or solid area should be taken as obstructing the wind. None of the tubular area of the scaffold framework that is sheltered by the solid area need be considered.

After the areas have been corrected by the various shape factors, the wind force is given by the product of the pressure and the modified areas.

The shape factors are greater than 1 to take into account that there is suction on the downwind side of the element as well as pressure on the upwind face. Tubes are slightly streamlined and require only a 20% increase, whereas toe boards have considerable suction on their downwind face and require a 50% increase.

38.5 Shadowing

The question as to whether any part of the scaffold structure is protected from the wind by another part or by an external screen has to be considered. As a very general rule, it can be assumed that the wind 'bends' around each side of an

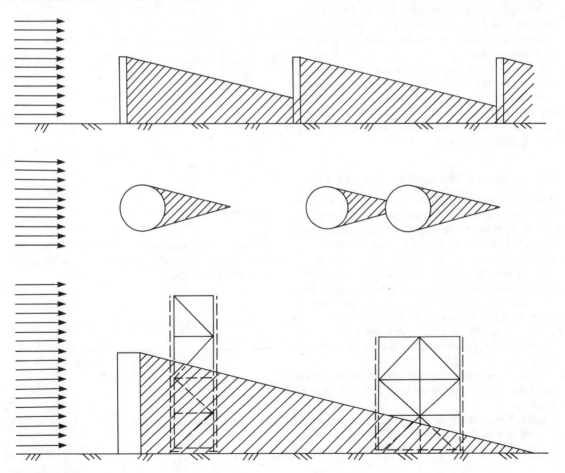

Figure 38.8 Wind shadowing.

obstruction at a slope of 1 in 4 if there is free passage for the air around both sides. If there is not a free flow path around one side of the obstruction, as in the case of a fence on the ground, the wind bends around only one side of the obstruction at the angle of 1 in 4.

Figure 38.8 shows examples of this characteristic. The areas shaded are in the wind shadow and need not be taken into account in wind force calculations. All other areas must be included.

This rule of 1 in 4 is empirical. Wind tunnel tests produce all manner of special cases, but the rule is adequate in practice, and evidence for it can be seen by observing the behaviour of sand in dunes or snow in windy conditions.

Figure 38.8 shows that if the wind bends around a tubular section at an angle of 1 in 4, it reunites on the lee side two diameters downwind. In practice, the consequence of this is that a scaffold tube can never be considered to shelter an adjacent tube because tubes are not usually fixed so closely together.

Thus in calculating the area of an unsheeted scaffold, the total area of all the tubes that are broadside on to the wind has to be evaluated and not only that in the front framework on the upwind side.

In the case of an extensive external birdcage, all the frames contribute to the wind force. This raises the question that if there are enough frames in the birdcage, the effective area may be greater than the face area of the scaffold if it is sheeted over on the upwind face. In access scaffolding, this is unlikely in practice.

38.6 Location of the centre of the wind force

After calculating the wind force, taking account of all the factors described above, it is necessary to find where its resultant is applied to the structure, what its overturning moment will be and what internal forces within the scaffold result from it.

The centre of the wind force on the scaffold is unlikely to be at the centre of the building surface. Two factors have to be taken into account. One is that the wind speed is greater at higher levels in the scaffold. For this, it is usually satisfactory to assume that the centre of the wind force is 0.625 of the height of the scaffold above the ground.

If there is any sheeting on the scaffold, the final location of the centre of the wind force must be found by calculation.

Figure 38.9(a) shows an unclad free-standing scaffold subjected to a wind force centred at 0.625 of the height. The consequential uneven leg loads and lateral forces are shown.

Figure 38.9(b) shows the effect of adding sheeting, which

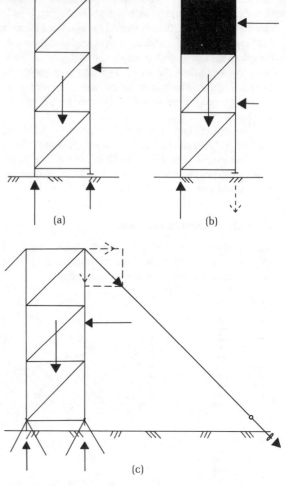

Figure 38.9 Wind forces on towers.

collects a further wind force at a higher level. This may be sufficient to overturn the tower, as shown by the tension in the upwind standards.

Table 38.3. Wind force coefficients or shape factors for fences, hoardings and signboards

| Case | Length/height ratio | | | |
	2	4	8	20
a	1.20	1.24	1.30	1.38
b	1.24	1.30	1.36	1.62
c	2.0	2.0	2.0	2.0
d	1.24	1.30	1.36	1.62
e	1.30	1.36	1.54	1.74
f	2.0	2.0	2.0	2.0

Table 38.4. The front elevation area (m²) of one bay and one lift of scaffold tubes, including the shape factor

Bay length (m)	Lift height (m)							
	1.2	1.5	1.8	2.0	2.1	2.4	2.7	3.0
1.26	0.323	0.360	0.397	0.422	0.434	0.472	0.510	0.546
1.37	0.347	0.384	0.422	0.446	0.460	0.508	0.533	0.570
1.50	0.366	0.403	0.439	0.464	0.478	0.515	0.552	0.589
1.80	0.410	0.448	0.484	0.509	0.521	0.558	0.596	0.613
2.00	0.439	0.476	0.512	0.538	0.550	0.588	0.625	0.662
2.10	0.454	0.490	0.527	0.552	0.559	0.602	0.640	0.676
2.40	0.496	0.534	0.571	0.596	0.608	0.646	0.682	0.720
2.70	0.540	0.477	0.614	0.640	0.682	0.689	0.726	0.769
3.00	0.583	0.620	0.671	0.683	0.695	0.732	0.770	0.808

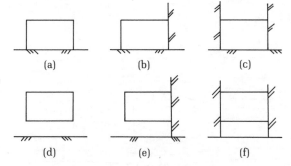

Figure 38.10 Wind forces on hoardings and signboards.

Figure 38.9(c) shows a tower stabilised by a guy and anchor. This applies a stabilising horizontal force at the top of the tower. It also introduces a vertical leg load in the upwind standards, which has to be added to the self weight and imposed leg loads.

Structures are frequently fully sheeted and situated at or above ground level, and they are sometimes in contact with other buildings. These cases are dealt with by Newberry and Eaton, who give the coefficients listed in Table 38.3 for the six cases shown in Figure 38.10. They are important

Table 38.5. The front elevation area (m²) of a scaffold that has a working platform fixed in it, including the shape factor. The areas given must be added to those given in Table 38.4 whenever there is a working platform.

Bay length (m)	Face area of the platform (m²)	
1.26	0.593	
1.37	0.645	
1.50	0.707	*Note:* Each metre length of
1.80	0.848	platform adds 0.471 m² of
2.00	0.942	effective area. These values
2.10	0.989	must be added for every
2.40	1.130	boarded bay.
2.70	1.1272	
3.00	1.413	

in scaffolding because they are for hoardings and sheeting attached to scaffolds for protection or advertisement.

38.7 The effective area of the tubes and platforms resisting the wind in an access scaffold

For the case of an unsheeted access scaffold with the wind blowing on to its broad face, Table 38.4 gives the effective tube area obstructing the wind for one bay length and one bay height.

This has been evaluated from the length of two uprights, two ledgers, one-tenth of a face brace and half of a ledger brace. The one-tenth of a face brace allows for a longitudinal brace every ten bays. The half ledger brace allows for one ledger brace in alternate standards. The total has then been multiplied by a shape factor of 1.2.

Table 38.5 gives the effective area of the platform edge, toe boards and guard rails for one bay of working platform of various lengths. The total of these areas has then been multiplied by a shape factor of 1.5.

Table 38.6 gives the effective tube area for each bay of scaffolding with the wind blowing on to its end, including a 1.2 shape factor.

Table 38.7 gives the effective area of the platform end,

Table 38.6. The end elevation area (m²) of one bay and one lift of scaffold tubes, including the force coefficient

Lift height (m)	Area (m²)	
1.20	0.368	
1.37	0.396	
1.50	0.418	*Note:* The bay length does
1.80	0.466	not affect this area, but
2.00	0.498	every bay has this area
2.10	0.515	resisting the wind end on.
2.40	0.563	
2.70	0.612	
3.00	0.661	

Table 38.7. The end elevation area (m²) of one end of each working platform fixed in the scaffold, including the shape factor. The areas given must be added to those given in Table 38.6 whenever there is a working platform.

Width of the platform (m)	End area of the platform (m²)	
1.0	0.471	
1.2	0.565	*Note:* Each metre width of
1.5	0.707	platform adds 0.471 m² of
1.8	0.848	effective area. This is only
2.0	0.942	added once to the values in
3.0	1.413	Table 38.6.

	Values of Cp for wind direction	
	→	↑
O	1.2	1.2
□	2.0	2.0
▯	2.0	0.75
I	2.0	0.9
]	2.0	0.6
T	2.0	2.0
L	2.0	2.0
—	0.1	2.0

Figure 38.11 Coefficients for various structural sections.

the end toe board and quarter of an end guard rail, including a 1.5 shape factor for each working platform.

Various structural sections used in the construction of platforms, gantries, bridges and hoardings are shown in Figure 38.11. The pressure coefficients for these are given in the figure. It is particularly important to note that tubes have a coefficient of 1.2. This can be taken into account either in the pressure assessment or by assuming the diameter of the tube to be 48.6 × 1.2 = 58.32 mm.

38.8 Hazard in wind force assessment

- The hazard that causes the greatest number of over-turnings is when the user does not comply with the assumptions of the designer, or when the designer does not know the requirements of the user to begin with. Figure 38.12 is an example.

  The request to the scaffolding company is for a tower four lifts high. Figure 38.12(a) is the design, and the self weight is adequate for the location.

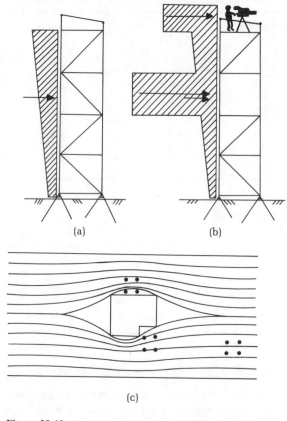

Figure 38.12 A camera tower in various locations.

Subsequently, a camera and cameraman are added on the top lift, with the necessary guard rails and toe boards. This adds 3 m² of area with a lever arm of 4.5 lifts. The tower is then found by calculation to be unstable. Crossed tube anchors are added.

At a third stage, sheeting is added around the third lift to accommodate a commentator.

At a fourth stage, the location is altered from being well away from a building to beside the building, where the wind flow is about twice the normal speed, giving about four times the pressure.

38.9 Height to base ratios for free-standing scaffold towers without cladding

It is useful to bear in mind the wind speeds that will blow over free-standing scaffolds of various height to base ratios. Such scaffolds may be work towers or camera towers. For a tower to blow over, the resultant of the horizontal wind force and the vertical weight must pass outside the base.

It has been pointed out that the wind force is centred at about five-eighths of the height of the tower, so for a tower with a height to base ratio of 1, the resultant of the two forces must be included less steeply than 0.625 in 0.5*b*, which for a scaffold with a height to base ratio of 1 ($h = b$) gives a slope less steep than 1 in 0.8 (1.25 in 1), i.e. the wind force must be greater than 80% of the weight of the tower.

Taking the weight of the tubes as 4.38 kg and the tube content of a 3 m cube-shaped tower as 65 m, the weight of the tower will be 284.9 kg for the tubes plus 57.1 kg for 28 couplers = 342 kg. The wind force must be greater than 273.6 kg.

The area of the tube broadside on to the wind, on which the wind will have an effect, is 53 m. Each 1 m will have an effective area of $1.0 \times 0.0483 \times 1.2 = 0.058$ m². The effective area of the tower obstructing the wind will be $53 \times 0.058 = 3.074$ m². The wind pressure to overturn the tower must be greater than $\dfrac{2684 \text{ N}}{3.074 \text{ m}^2} = 873$ N/m². This is derived from a wind speed of 37.74 m/s (84 mph, 73 knots).

The conclusion to draw from this is that any structure whose height to base ratio is only 1 will be blown over by a wind speed of 37.74 m/s, or by a lower speed if there are toe boards and other items obstructing the wind.

A structure with a height to base ratio of 2 can sustain a wind force of only about half this value, which results from a wind speed of 26.7 m/s (59.7 mph, 52 knots).

A tower with a height to base ratio of 3 can sustain a wind speed of about 21.8 m/s (48.7 mph, 42 knots).

A tower with a height to base ratio of 4 can sustain a wind speed of about 18.9 m/s (42.2 mph, 37 knots).

A tower with a height to base ratio of 5 can sustain a wind speed of about 17 m/s (38 mph, 33 knots).

The above figures are for steel towers. Aluminium weighs only one-third as much as steel, so the above speeds must be divided by the square root of 3 (1.73) to give the wind speeds that will overturn the corresponding towers built in aluminium.

The wind speeds calculated above overturn the various towers. There is no safety factor. Applying a safety factor of 1.5 means that the allowable wind speed must be reduced by 1.22.

These calculations have been included here to alert the designer to the instability of free-standing towers, especially aluminium work towers. Chapter 13 and Figure 13.1 give examples for various towers with platforms and an operative.

38.10 Canopies

Six types of canopy are shown in Figure 38.13. Three of these have skirts. The bottom of the skirt is sometimes higher than the solid building that the canopy protects. The

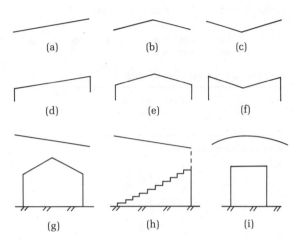

Figure 38.13 Types of canopy.

skirts sometimes come below the top of the building to form a completely clad building and can be treated as part of an enclosed protection building.

There will be wind forces on the canopy, with vertical and horizontal components arising from the wind pressures. Horizontal wind forces will arise from the friction of the wind over the skin and from the vertical surfaces of the skirts.

The uplift forces will depend not only on the slope and shape of the canopy, but also on the shape and degree of obstruction under the canopy. The designer must ascertain the nature of this obstruction.

The wind pressure corresponding to the design wind speed is found, and coefficients are applied to give the pressures applied in various directions and to various areas of the canopy.

Table 38.8 gives suitable pressure coefficients for empty and completely blocked areas below the canopy. The designer should be aware of hazards such as those represented in the bottom three diagrams of Figure 38.13.

38.11 Hazards with canopies

The bottom three diagrams of Figure 38.13 are commonly occurring circumstances.

- Case (g) represents a mono pitch canopy placed over a ridge roof. Before the roof is stripped and often as it is reclad, the roof may deflect the wind on to the canopy, whichever way the wind is entering, increasing the uplift. There may be venturi suction at the apex of the roof.
- Case (h) is that of a grandstand sloping upwards below a canopy sloping backwards. If there is a skirt along the back of the stand that has imperfectly sealed and

Table 38.8. Vertical wind pressure and lateral pressure coefficients for the projected vertical area and skirt area of canopies. + is downward, − is uplift.

(a) *Monopitch canopies without skirts*

Roof slope	Degree of obstruction of the downwind exit	Vertical pressure coefficient	Edge areas (1/6th of the plan area)	Lateral pressure coefficient
Wind entering the high side				
0	0	−1.0	−1.50	1.2
	Nearly full	−1.5	−2.00	1.5
10°	0	−1.1	−1.60	1.3
	Nearly full	−1.6	−2.50	1.6
20°	0	−1.2	−1.70	1.4
	Nearly full	−1.7	−2.75	1.7
30°	0	−1.3	−1.80	1.5
	Nearly full	−1.8	−3.00	1.8
Wind entering the low side				
0	0	+1.0	−1.00	1.2
	Nearly full	−0.5	−1.25	1.5
10°	0	+1.1	−1.05	1.3
	Nearly full	−0.6	−1.50	1.0
20°	0	+1.2	−1.50	1.6
	Nearly full	−0.7	−1.70	1.7
30°	0	+1.3	−1.45	1.5
	Nearly full	−0.8	−2.00	1.8

(b) *Double-pitch canopies without skirts*

Roof slope	Degree of obstruction of downwind exit	Vertical pressure coefficient Upwind slope	Downwind slope	Edge areas (1/6th of the plan area)	Lateral pressure coefficient
With a ridge					
10°	0	$\begin{cases} +1.4 \\ -2.0 \end{cases}$	−1.0	−1.5	1.3
	Nearly full	−0.6	−1.0	−1.5	1.6
20°	0	$\begin{cases} +1.6 \\ -1.8 \end{cases}$	−1.0	−1.5	1.4
	Nearly full	−0.7	−1.0	−1.5	1.7
30°	0	+1.8 −1.4	−1.0	−1.5	1.5
	Nearly full	−0.8	−1.0	−1.5	1.8
With a valley					
10°	0	−1.6	+1.0	−1.6	1.3
	Nearly full	−1.4	+1.0	−2.0	1.0
20°	0	−1.8	+1.0	−1.8	1.4
	Nearly full	−1.6	$\begin{cases} +1.0 \\ -1.0 \end{cases}$	−2.0	1.7
30°	0	−2.0	+1.0	−2.0	1.5
	Nearly full	−1.8	$\begin{cases} +1.0 \\ -1.0 \end{cases}$	−2.0	1.8

Notes: Where both positive and negative coefficients are given, design the structure against both downward and upward forces. When skirts are installed, use the same coefficients. Add the skirt area to the vertical projected area of the canopy slope.

tied panels of sheeting, there may be dangerous wind forces along the edges of the panels.

- Case (i) is typical of a temporary storage bay. Wind acceleration may damage the contents of the store as well as the canopy.

38.12 Enclosed protection buildings

If buildings are completely surrounded by scaffolding that is also completely covered in protection sheeting, the whole assembly is comparable to a completed building, and the wind forces on it should be calculated as for a building using the Code of Practice CP3, Chapter V, Part 2. Figure 38.2 gives the velocity or alternative pressure coefficients for this purpose.

If there are holes in one or other side of the covering, the forces on the outside of the covering should be designed as for a completely covered building and, in addition, an inside pressure or suction should be added.

An empirical rule is to evaluate the area of the hole and express it as a fraction of the area of the wall. This fraction is the multiplier of the basic pressure to give the internal pressure or suction.

Figure 38.14 shows the case of a hole in the upwind wall. If this hole has an area of one-third of the upwind wall area, the internal pressure coefficient is 0.33.

A hole in the downwind face will cause a reduction in the internal pressure. If the hole area to wall area ratio is 0.25, there is a suction of 0.25 times the basic pressure.

The external roof pressure is uplift in both the cases shown in Figure 38.14. For the purpose of temporary scaffold buildings, the overall external uplift pressure coefficient can be taken as −0.8 and the edge area of the one-sixth of the plan area of the roof as −2.0.

38.13 Wind flow patterns

The problem of assessing the wind force on a scaffold is more complicated than that on the building it serves. The building itself already affects the wind flow before the scaffold is put in place. Adjacent buildings also have an effect. During construction, buildings grow in height and sometimes will have walls and sometimes and open frame. In demolition jobs, a wall with window openings may be left without a building frame behind it.

The effects can be assessed by sketching flow patterns of the wind. An example is given in Figure 38.12 showing how a tower placed near a building is influenced by twice the wind speed, and hence four times the pressure, than when sited well away from the building.

When an obstruction is placed in a wind flow, there are two effects. One is that the obstruction is subjected to pressure, suction and shear forces, or any or all of these.

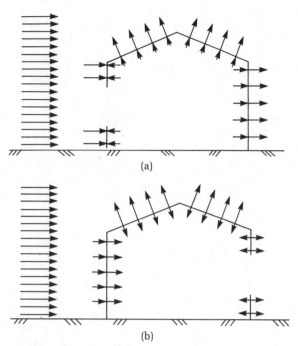

Figure 38.14 Internal wind pressures.

The other is that the wind is diverted around the obstruction, and its direction and velocity are changed. Any object in this diverted wind flow is subjected to different pressure, suction and shear from those on the obstruction.

Figure 38.15 shows a wind flow diverted by a building. There is pressure on the upwind face, represented by the plus sign, suction on the downwind face, represented by the minus sign and shear along the sides.

The flow lines are closer together where the wind is diverted around the building. This means that the quantity of air entering any space between two flow lines at the edge of the diagram has to pass through a smaller space when the lines converge and is therefore moving faster where the lines are closer. In consequence, a scaffold on the building is subjected to a greater wind speed than one built in the normal wind flow where it is not affected by the building.

Figure 38.16 shows the flow lines at the corner of a building. Where the flow lines are converging on the building, there is pressure. Where they are diverging from it, there is suction. These conditions need to be borne in mind, especially when cradles are to be rigged near the corners.

Figure 38.17 shows these effects exaggerated by the location of two lower blocks. Here the site wind speed is increased first by the plan arrangement of the buildings and second, by the corner effects.

Figure 38.18 is a flow pattern for a storage tank layout. Because the tanks are slightly streamlined, a large surface

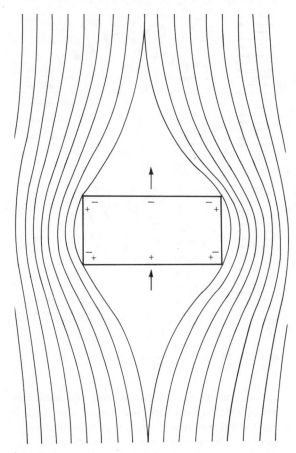

Figure 38.15 Acceleration of the wind speed by a building.

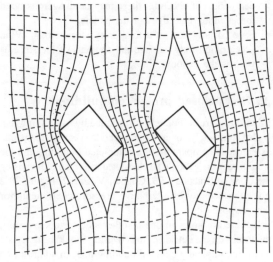

Figure 38.17 Flow lines around and through the gap between two buildings.

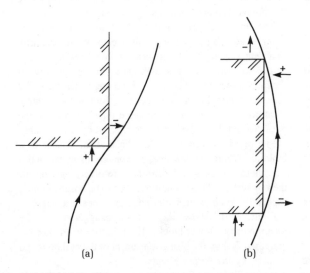

Figure 38.16 Pressure and suction near the corners of buildings.

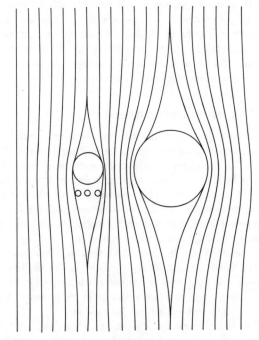

Figure 38.18 Flow lines around and between circular tanks.

area is subjected to a shear force. A scaffold along this face will need very much more than the normal amount of longitudinal bracing.

Exact flow diagrams of the type given can be prepared by aerodynamic specialists, or from wind tunnel tests if the job warrants it, but it is usually sufficient for the scaffolding engineer to prepare sketches himself so that he can recognise the potential hazards.

38.14 Ice loads on tubes

It will be seen in Figure 38.8 that there is a wedge-shaped wind shadow extending about two diameters downwind. There is a similarly shaped area upwind of reduced wind velocity. It is in these two areas that ice forms.

When scaffolds are to be built in exposed areas, usually inland and on high ground, an allowance for the accumulation of ice must be made.

The weight of ice that normally forms on scaffold tubes in severe conditions may be up to 2.5 kg/m of tube. The self weight of the tube-and-coupler assembly is about 5 kg/m. The ice adds 50% to the weight of the tubes that are broadside on to the wind, which is about 75% of the tubes in the whole assembly.

The increase in weight of the whole scaffold framework in severe ice conditions will be about 37.5%.

38.15 Snow loads

There is very little weight in the snow that accumulates on the tubular framework, but the platforms can collect a small amount. The top platform may accumulate up to 200 mm and the lower platform hardly any. This 200 mm of snow weighs 15.3 kg/m^2 (3.2 lb/ft^2). This is only 20% of the imposed load of a very light-duty scaffold. The imposed load will not be present when the snow is at this depth. Accordingly, in access scaffolding snow loads are not significant.

On temporary roofs, however, a considerable depth of snow may accumulate because it will not be cleared away each day. BS 6399 gives a map of the United Kingdom divided into areas with different basic snow loads. These range from 0.4 kN/m^2 in the south and in coastal areas to 0.6 kN/m^2 in the Pennines, and up to a maximum of 1 kN/m^2 in the Scottish Highlands. These figures are subject to correction for local conditions and altitude.

If the roofs are designed to support their self weight plus an imposed load of 0.75 kN/m^2 and are expected to be present through the winter, they will safely carry the snow load, except for the extreme conditions of mountain regions, where a special snow allowance must be added to the 0.75 kN/m^2 imposed load.

39 Units, symbols and conversion tables

39.1 Units and symbols

The units used in this book are SI, metric, technical and Imperial.

Length:	m: metres, mm: millimetres.
	ft: feet, in: inches.
Area:	m^2: square metres, mm^2: square millimetres.
	ft^2: square feet, in^2: square inches.
Volume:	m^3: cubic metres.
	ft^3: cubic feet.
Velocity:	in/s, ft/s, mph, knots.
Mass or weight:	N: newtons, kN: kilonewtons.
	kg: kilograms, T: tonnes.
	lb: pounds, T: tons.
Density:	kg/m^3, $tonnes/m^3$, lb/ft^3.
Pressure and stress:	kg/m^2, $tonnes/m^2$, lb/ft^2.
	N/mm^2, kN/m^2.
Mass/unit length:	kg/m, lb/ft.
Moments and torque:	kNm, Nm, lb ft.

39.2 Conversion tables

Tables are included in this section for:

- Weight, mass and force.
- Length.
- Area.
- Volume.
- Velocity.
- Weight, mass or force per unit length.
- Pressure, stress or weight per unit area.
- Density.
- Moment of force and torque.

To convert a value expressed in one system of units into a second system, find the column headed with the first system and look down it as far as the value 1. Now look along the horizontal line either way as far as the column headed with the second system of units. The value found here for the second system corresponds to 1 unit in the first system.

Example: To convert a distributed load of 36.25 lb/ft^2 into kg/m^2. Find column 2, line 2 in the pressure table and move to the right to the column headed kg/m^2 to find that 1 lb/ft^2 = 4.882 kg/m^2. 36.25 × 4.882 = 177 kg/m^2.

Basic conversion factors

Weight:	1 kg	= 2.2046 lb
	1 lb	= 0.45359 kg
	1 tonne	= 1000 kg
		= 0.9842 UK tons
	1 UK ton	= 2240 lb
Length:	1 m	= 3.2808 ft
		= 39.3696 in
	1 km	= 0.6214 miles
		= 3281 ft
	1 ft	= 0.3048 m
	1 in	= 0.0254 m
		= 2.54 cm
		= 25.4 mm
	1 mile	= 5280 ft
		= 1.609 km

Table 39.1. Mass, weight

ounces (oz)	lb	tons	grams (g)	kg	tonnes	newtons	kilonewtons
1	0.06250	2.7902×10^{-5}	28.349	0.02835	2.8349×10^{-5}	0.27801	2.7801×10^{-4}
16	1	4.4643×10^{-4}	453.59	0.45359	4.5359×10^{-4}	4.44822	0.00445
35840	2240	1	1.0161×10^{6}	1016.1	1.0161	9964.6	9.9646
0.03527	0.0022	9.8420×10^{-7}	1	0.00100	10^{-6}	0.00981	9.8067×10^{-6}
35.274	2.2046	9.8420×10^{-4}	1000	1	0.001	9.8067	0.00981
35279	2204.6	0.9842	10^{6}	1000	1	9806.7	9.8067
3.5968	0.22480	0.00010	101.97	0.10197	1.0197×10^{-4}	1	0.001
3596.8	224.80	0.10036	101970	101.97	0.10197	1000	1

Table 39.2. Length

in	mm
1/32	
1/16	
1/10	
1/8	
3/16	
1/4	
3/8	
1/2	
9/16	
5/8	
11/16	
3/4	
13/16	
7/8	
15/16	
1	25.4

in	ft	yd	miles (mi)	mm	cm	m	km
1	0.08333	0.02775	1.57824×10^{-5}	25.4	2.54	0.0254	2.54×10^{-5}
12	1	0.3333	1.894×10^{-4}	304.8	30.480	0.3048	3.048×10^{-4}
36	3	1	5.682×10^{-4}	914.4	91.440	0.9144	9.144×10^{-4}
63360	5280	1760	1	1.609×10^{-6}	160934	1609.34	1.6093
0.03937	0.00328	0.001094	0.6214×10^{-7}	1	0.1	0.001	10^{-6}
0.3937	0.0328	0.01094	6.214×10^{-6}	10	1	0.01	10^{-5}
39.4	3.28	1.094	6.214×10^{-4}	1000	100	1	0.001
39400	3280	1094	0.6214×10^{-4}	10^{6}	10^{5}	1000	1

One nautical mile = 1852 m
= 6080 ft

Table 39.3. Area

inches²	feet²	yards²	acres	miles²	hectares	mm²	cm²	m²	km²
1	0.006944	7.716×10^{-4}	0.1594×10^{-7}	2.49×10^{-10}	6.451×10^{-8}	645.16	6.4516	6.4516×10^{-4}	6.4516×10^{-10}
144	1	0.11111	2.296×10^{-5}	3.586×10^{-8}	9.26×10^{-6}	92900	929	0.0929	9.29×10^{-8}
1296	9	1	2.066×10^{-4}	3.228×10^{-7}	8.361×10^{-5}	8.361×10^{5}	8361	0.83613	8.3613×10^{-7}
6.2726×10^{6}	43560	4840	1	0.001563	0.404686	4.046×10^{9}	4.0469×10^{7}	4046.86	4.04686×10^{-3}
4.0145×10^{9}	2.788×10^{7}	3.0976×10^{6}	640	1	259	2.59×10^{12}	2.59×10^{10}	2.59×10^{6}	2.59
1.55×10^{7}	107640	11960	2.471	3.861×10^{-3}	1	10^{10}	10^{8}	10^{4}	0.01
1.55×10^{-3}	1.076×10^{-5}	1.196×10^{-6}	2.471×10^{-10}	3.861×10^{-13}	10^{-10}	1	0.01	10^{-6}	10^{-12}
0.155	1.076×10^{-3}	1.196×10^{-4}	2.471×10^{-8}	3.861×10^{-11}	10^{-8}	100	1	10^{-4}	10^{-10}
1550	10.764	1.19601	2.471×10^{-4}	3.861×10^{-7}	10^{-4}	10^{6}	10^{4}	1	10^{-6}
1.55×10^{9}	1.0764×10^{7}	1.196×10^{6}	2.471×10^{2}	0.3861	10^{2}	10^{12}	10^{10}	10^{6}	1

Table 39.4. Volume

in³	ft³	yd³	mm³	cm³	m³	Imperial gallons	litres	US gallons
1	5.787×10^{-4}	2.143×10^{-5}	16387	16.387	1.639×10^{-5}	3.605×10^{-5}	0.01639	4.329×10^{-3}
1728	1	0.03703	2.8317×10^{7}	2.8317×10^{4}	0.02832	6.229	28.32	7.4805
46656	27	1	7.6456×10^{8}	7.6456×10^{5}	0.7646	168.18	76464	201.97
6.102×10^{-5}	3.531×10^{-8}	1.3079×10^{-9}	1	0.001	10^{-9}	2.2×10^{-7}	10^{-6}	2.6422×10^{-7}
0.06102	3.5315×10^{-5}	1.308×10^{-6}	1000	1	10^{-6}	2.2×10^{-4}	10^{-3}	2.6422×10^{-4}
61023.7	35.315	1.308	10^{9}	10^{6}	1	220	1000	264.22
277.419	0.16054	0.0059416	4.545×10^{6}	4.545×10^{3}	4.545×10^{-3}	1	4.546	1.201
61.026	0.03532	1.3079×10^{-3}	10^{6}	1000	0.001	0.22	1	0.2642
231	0.1337	0.004952	3.786×10^{6}	3786	0.003786	0.8327	3.785	1

Table 39.5. Velocity

knots	mph	ft/s	km/hr	m/s
1	1.1515	1.6889	1.8532	0.5148
0.8684	1	1.4667	1.6093	0.4470
0.5921	0.6818	1	1.0973	0.3048
0.5396	0.6214	0.9113	1	0.2778
1.394	2.237	3.281	3.6	1

Table 39.6. Weight per unit length

lb/in	lb/ft	kg/cm	kg/m	N/mm	N/cm	N/m	kN/m (N/mm)
1	12.00	0.1788	17.874	0.1752	17508	175.098	0.1752
0.0833	1	0.0149	1.4895	0.0146	0.1459	14.5915	0.0146
5.5997	67.196	1	100	0.9806	9.8061	980.61	0.9806
0.05591	0.6719	0.01000	1	0.0098	0.0981	9.8061	0.0098
5.7117	68.540	1.0200	102.00	1	10.000	1000.0	1.0000
0.5712	6.8540	0.102	10.2	0.10	1	100	0.1
0.00571	0.06854	0.001	0.102	0.0010	0.01	1	0.00
5.7117	68.54	1.02	102	1	10	1000	1

Table 39.8. Density

Specific gravity g/cm^3 (tonnes/m^3)	lb/in^3	lb/ft^3	UK tons/yd^3	lb/Imperial gallons	kg/mm^3	tonnes/m^3	N/cm^3	kN/m^3
1	0.0361	62.428	0.7525	10.022	1000	1	0.009806	9.806
27.68	1	1728	20.829	277.419	27680	27.68	0.27143	271.43
0.01602	5.79×10^{-4}	1	0.01205	0.1605	16.078	0.01602	1.57×10^{-4}	0.1571
1.3289	0.04801	82.963	1	13.319	1328.9	1.3289	0.01303	13.031
0.0998	0.003605	6.2288	0.0751	1	99.8	0.0998	9.786×10^{-4}	0.9786
0.001	3.6122×10^{-5}	0.06243	7.525×10^{-4}	0.01002	1	0.001	9.806×10^{-6}	0.009806
1	0.0361	62.428	0.7525	10.022	1000	1	0.00981	9.806
101.978	3.6814	6366.28	76.738	1021.82	101978	101.978	1	1000
0.102	0.00368	6.3663	0.07674	1.022	102	0.102	0.001	1

Table 39.9. Moments of force and torque

in lb	ft lb	ft tons	kg cm	kg m	tonnes m	N cm	N m	kNm
1	0.08333	3.72×10^{-5}	1.152	0.01152	1.15×10^{-5}	11.298	0.11298	1.13×10^{-4}
12	1	4.464×10^{-4}	13.824	0.1382	1.38×10^{-4}	135.576	1.35576	0.001356
26880	2240	1	30965.76	309.658	0.30912	303690	3036.9	3.0369
0.86806	0.07233	3.23×10^{-5}	1	0.01	9.9834×10^{-6}	9.8073	0.09807	9.807×10^{-5}
86.806	7.233	0.00323	100	1	9.983×10^{-4}	980.73	9.8073	9.807×10^{-3}
86806	7233	3.23	10^5	1000	1	980730	9807	9.807
0.08851	0.00738	3.293×10^{-6}	0.102	0.00102	1.02×10^{-6}	1	0.01	10^{-5}
8.85075	0.737562	3.293×10^{-4}	10.2	0.102	1.02×10^{-5}	100	1	0.001
8850.75	737.562	0.3293	10200	102	0.102	10^5	1000	1

Table 39.7. Pressure, stress and weight or mass per unit area

lb/in²	lb/ft²	tons/in²	tons/ft²	kg/mm²	kg/cm²	kg/m²	tonnes/m²	N/mm²	kN/mm²	kN/m²
1	144	4.46×10^{-4}	0.06432	7.031×10^{-4}	0.07031	703.1	0.703	6.895×10^{-3}	6.895×10^{-6}	6.895
0.00694	1	3.1×10^{-6}	0.0004464	4.882×10^{-6}	4.882×10^{-4}	4.882	0.004882	4.788×10^{-5}	4.788×10^{-8}	0.04788
2240	3.2256×10^{5}	1	144	1.5749	157.49	1.575×10^{6}	1575	15.444	0.01544	1544.3
15.556	2240	0.00694	1	0.01094	1.0936	10936	10.936	0.1072	1.072×10^{-4}	107.2
1422.4	20483	0.635	91.44	1	100	10^{6}	1000	9.806	0.00981	9806
4.224	2048.3	0.00635	0.9144	10^{-2}	1	10^{4}	10	9.806×10^{-2}	9.806×10^{-5}	9806
0.001422	0.2048	6.35×10^{-7}	9.144×10^{-5}	10^{-6}	10^{-4}	1	0.001	9.806×10^{-6}	9.806×10^{-9}	9.806
0.422	204.77	6.35×10^{-4}	0.09144	0.001	0.1	1000	1	9.806×10^{-3}	9.806×10^{-6}	0.009806
145.087	2.0892×10^{4}	0.06477	9.327	0.102	10.2	10.2×10^{4}	102	1	0.001	1000
145087	2.0892×10^{7}	64.77	9324	102	10200	10.2×10^{7}	102000	1000	1	10^{6}
0.14509	20.892	6.477×10^{-5}	0.009324	1.02×10^{-4}	0.0102	102	0.102	0.001	10^{-6}	1

Table 39.10. Standard wire gauges and thicknesses

British SWG number	Diameter	
	mm	in
0	8.230	0.324
1	7.620	0.300
2	7.010	0.276
3	6.401	0.252
4	5.893	0.232
5	5.385	0.212
6	4.877	0.192
7	4.470	0.176
8	4.064	0.160
9	3.658	0.144
10	3.251	0.128
12	2.642	0.104
14	2.032	0.080
16	1.626	0.064
18	1.219	0.048
20	0.914	0.036
22	0.711	0.028
24	0.559	0.022

References

Statutory (UK)

The Factory Act	1961
The Construction (Working Places) Regulations	1966
The Construction (Lifting Provisions) Regulations	1961
The Construction (General Provisions) Regulations	1961
The Construction (Health and Welfare) Regulations	1966
Health and Safety at Work etc. Act	1974
Occupiers Liability Act	1957
Offices, Shops and Railway Premises Act	1963

British Standards Codes and other standards

BS 1139: Metal scaffolding, Parts 1 to 4.

BS 5973: Code of practice for access and working scaffolds in steel.

BS 5974: Temporarily installed suspended scaffolds and access equipment.

ISO 4054: Scaffold couplers.

EN 39 and EN 244: Steel tubes.

BS 2483: Scaffold boards.

BS CP3: Chapter 5, Part 2; wind loads.

BS 6399: Part 3, CP for imposed snow loads.

BS 1397: Industrial safety belts and harnesses.

BS 4074: Metal props and struts.

BS 2830: Cradles and safety chairs.

BS 3913: Industrial safety nets.

BS CP93: Safety nets.

BS 2095 and BS 2826: Light and heavy-duty safety helmets.

BS 449: Parts 1, 2 and 4 amendments: Structural steel.

BS 5507: Part 3, Falsework equipment.

BS 5975: Falsework.

BS 302: Wire ropes for cranes and general engineering purposes.

BS 4928: Specification for man-made fibre ropes.

BS 2052: Ropes made from coir, hemp, manila and sisal.

BS 4360: Weldable structural steels.

BS 1474 and BS 1476: Wrought aluminium and aluminium alloys.

BS 1839: London pattern pulley blocks for fibre rope.

BS 4978: Timber grades for structural use.

BS CP112: Structural use of timber.

BS 825 and BS 3032: Steel shackles and high-tensile steel shackles.

BS 1775 and BS 938: Structural steel tubes and general requirements for metal arc welding of structural steel tubes.

BS 1129: Timber ladder steps, trestles and lightweight stagings.

BS 449: (and its successor) Structural steel in building.

Other relevant codes (UK)

Mining Regulations.
Ports, Harbour Regulations.
Offshore Oil Regulations.
British Railways Code.
BBC Code.
Local Bye-laws.
Police Regulations.

HSE guidance notes

GS 10: Roofwork: Prevention of falls.

GS 15: General access scaffolds.

GS 25: Prevention of falls by window cleaners.

GS 28/3: Safe erection, Part 3: Working places and access.

GS 31: Safe use of ladders, step-ladders and trestles.

GS 42: Tower scaffolds.

PM 9: Access to tower cranes.

Textbooks

Steel Designer's Manual : CSRDO : Crosby Lockwood.

Echafaudages Tubulaires : Coppel and Coulon : Published by Dunod, Paris (in French).

Das Stalhronhrgerust : Paul Roloff : Published by Ullstein Fachberlagh, Berlin (in German).

Wind Loading Handbook : Newberry and Eaton : BRE : HMSO.

The Assessment of Wind Loads : BRE Digest 346, Parts 1 to 7 : HMSO.

Wind Loads on Canopy Roofs : BRE Digest 284 : HMSO.

Wind Environment around Tall Buildings : BRE Digest 141 : HMSO.

Wind Forces on Unclad Tubular Structures : Constrado Publication 1/75.

Scaffolding : R. Doughty : Site Practice Series : Longman.

Falsework and Access Scaffolds in Tubular Steel : R. E. Brand : McGraw-Hill.

Redgrave's Factories Acts : Fife and Machin : Butterworth.

Index